177
Advances in Polymer Science

Advances in Polymer Science

Recently Published and Forthcoming Volumes

Phase Behavior of Polymer Blends
Volume Editor: Freed, K.
Vol. 183, 2005

Polymer Analysis/Polymer Theory
Vol. 182, 2005

Interphases and Mesophases in Polymer Crystallization II
Volume Editor: Allegra, G.
Vol. 181, 2005

Interphases and Mesophases in Polymer Crystallization I
Volume Editor: Allegra, G.
Vol. 180, 2005

Inorganic Polymeric Nanocomposites and Membranes
Vol. 179, 2005

Polymeric and Inorganic Fibres
Vol. 178, 2005

Poly(arylene ethynylene)s
From Synthesis to Application
Volume Editor: Weder, C.
Vol. 177, 2005

Metathesis Polymerization
Volume Editor: Buchmeiser, M.
Vol. 176, 2005

Polymer Particles
Volume Editor: Okubo, M.
Vol. 175, 2005

Neutron Spin Echo in Polymer Systems
Authors: Richter, D., Monkenbusch, M., Arbe, A., Colmenero, J.
Vol. 174, 2005

Advanced Computer Simulation Approaches for Soft Matter Sciences I
Volume Editors: Holm, C., Kremer, K.
Vol. 173, 2005

Microlithography · Molecular Imprinting
Vol. 172, 2005

Polymer Synthesis
Vol. 171, 2004

NMR · Coordination Polymerization · Photopolymerization
Vol. 170, 2004

Long-Term Properties of Polyolefins
Volume Editor: Albertsson, A.-C.
Vol. 169, 2004

Polymers and Light
Volume Editor: Lippert, T. K.
Vol. 168, 2004

New Synthetic Methods
Vol. 167, 2004

Polyelectrolytes with Defined Molecular Architecture II
Volume Editor: Schmidt, M.
Vol. 166, 2004

Polyelectrolytes with Defined Molecular Architecture I
Volume Editor: Schmidt, M.
Vol. 165, 2004

Filler-Reinforeced Elastomers · Scanning Force Microscopy
Vol. 164, 2003

Liquid Chromatography · FTIR Microspectroscopy · Microwave Assisted Synthesis
Vol. 163, 2003

Radiation Effects on Polymers for Biological Use
Volume Editor: Kausch, H.
Vol. 162, 2003

Poly(arylene ethynylene)s

From Synthesis to Application

Volume Editor: Christoph Weder

With contributions by
L. Blankenburg · U. H. F. Bunz · E. Klemm · J. S. Moore · T. Pautzsch
C. R. Ray · T. M. Swager · G. Voskerician · C. Weder · I. Yamaguchi
T. Yamamoto · T. Yasuda · J. Zheng

Springer

The series presents critical reviews of the present and future trends in polymer and biopolymer science including chemistry, physical chemistry, physics and material science. It is addressed to all scientists at universities and in industry who wish to keep abreast of advances in the topics covered.

As a rule, contributions are specially commissioned. The editors and publishers will, however, always be pleased to receive suggestions and supplementary information. Papers are accepted for "Advances in Polymer Science" in English.

In references Advances in Polymer Science is abbreviated Adv Polym Sci and is cited as a journal.

The electronic content of APS may be found springerlink.com

ISSN 0065-3195
ISBN-10 3-642-43020-1 **Springer Berlin Heidelberg New York**
ISBN-13 978-3-642-43020-6
DOI 10.1007/b101353

Springer is a part of Springer Science+Business Media
springeronline.com

Softcover re-print of the Hardcover 1st edition 2005

Cover design: KünkelLopka GmbH, Heidelberg/design & production GmbH, Heidelberg
Typesetting: Fotosatz-Service Köhler GmbH, Würzburg

Printed on acid-free paper 02/3141 xv – 5 4 3 2 1 0

Volume Editor

Dr. Christoph Weder

Department of Macromolecular Science and Engineering
Case Western Reserve University
2100 Adelbert Road, Kent Hale Smith Building, Rm 416
Cleveland, OH 44106-7202, USA
christoph.weder@case.edu

Editorial Board

Advances in Polymer Science
Also Available Electronically

For all customers who have a standing order to Advances in Polymer Science, we offer the electronic version via SpringerLink free of charge. Please contact your librarian who can receive a password for free access to the full articles by registering at:

springerlink.com

If you do not have a subscription, you can still view the tables of contents of the volumes and the abstract of each article by going to the SpringerLink Homepage, clicking on "Browse by Online Libraries", then "Chemical Sciences", and finally choose Advances in Polymer Science.

You will find information about the

- Editorial Board
- Aims and Scope
- Instructions for Authors
- Sample Contribution

at springeronline.com using the search function.

Preface

Over the past three decades, π-conjugated semi-conducting polymers have attracted significant interest since these materials combine the processability and outstanding mechanical characteristics of polymers with the readily-tailored electrical, optical, and magnetic properties of functional organic molecules. In particular, the potential use of these materials in light-emitting diodes, field-effect transistors, photovoltaic cells, and other opto-electronic devices has motivated the development of synthesis and processing methods of conjugated polymer materials with unique properties. Among a variety of materials, poly(arylene ethynylene) (PAE) derivatives have attracted the attention of an ever growing number of research groups around the globe. Hundreds of different PAEs have been reported to date and, during the last ten years, this family of conjugated polymers has established itself as an important class of materials with interesting optical and electronic properties. The spectacular progress made on many frontiers has propelled PAEs into the scientific mainstream, and many technologically relevant applications that utilize these polymers have been spurred. In six chapters, which root in the authors' own research experience, this special volume of the series Advances in Polymer Science attempts to capture the most recent phase of this exciting evolution. The book does not claim to be a complete compilation of the extensive literature in this field, but rather attempts to document recent progress on the basis of selected, illustrative examples. On behalf of all the contributors to this volume, I ask for the understanding of those researchers whose work has not been included.

The first chapter, written by *Uwe Bunz*, summarizes the most recent progress in the synthesis of PAEs and covers the literature from 1999 through 2003. In the second chapter, *Elisabeth Klemm, Thomas Pautzsch,* and *Lars Blankenburg* describe current work in the field of organometallic PAEs. The next chapter by *Christian R. Ray* and *Jeffrey S. Moore* focuses on the supramolecular organization of foldable phenylene ethynylene oligomers and polymers. The application of PAEs in bio- and chemosensors is described in the chapter written by *Juan Zheng* and *Timothy M. Swager*. The penultimate chapter, written by *Takakazu Yamamoto, Isao Yamaguchi,* and *Takuma Yasuda* , reviews the synthesis and chemical properties of PAEs based on sulfur-, nitrogen- and silicon-containing heteroaromatic moieties. Finally, a summary of the electronic properties of PAEs and their potential applications in 'plastic electronic' devices is given in chapter written by *Gabriela Voskerician* and myself.

We hope that this volume will not only become a key reference for those in the field, but also serve its purpose as a source of inspiration for the design of future generations of advanced materials with unique and/or unusual optoelectronic properties.

Shaker Heights, March 2005 Christoph Weder

Contents

Adv Polym Sci (2005) 177: 1–52
DOI 10.1007/b101374

Synthesis and Structure of PAEs

Uwe H. F. Bunz (✉)

School of Chemistry and Biochemistry, Georgia Institute of Technology, Atlanta, GA 30332 USA
E-mail: uwe.bunz@chemistry.gatech.edu

Abstract This review discusses the progress in the synthesis of poly(aryleneethynylene)s (PAEs). It covers the literature from 1999 through 2003. The last comprehensive review of PAE synthesis appeared in 2000. The present review comprehensively updates the developments in the synthesis of PAEs. Synthetic methods are discussed first, followed by the description of PAEs with novel structures and topologies. Progress has been made in the field of water-soluble and bioavailable poly(*para*-phenyleneethynylene)s (PPEs) and in the field of heterocyclic PAEs. In the last section the polymer-analogous reactions of PAEs are treated. Reduction and metal complexation are discussed. Most of the novel PAEs are proposed for applications in "plastic electronics" and/or in sensors.

Keywords Pd catalysis · Alkyne metathesis · Conjugated polymers · Acetylene chemistry · Alkynes · Arene chemistry

1 Introduction

Poly(aryleneethynylene)s (PAEs) [1] are a class of polymers in which arene groups are separated by alkyne linkers. Figure 1 shows the general structure of a PAE and the structure of the most important representative of this class, the poly(*para*-phenyleneethynylene)s (PPEs), in addition to two classes of structurally related conjugated polymers, the poly(*para*-phenylenevinylene)s (PPVs) [2] and the poly(diacetylene)s (PDAs) [3]. While the structural relationship between the PPEs and both the PPVs and the PDAs is close, their properties are different. PPEs, here as an example for all PAEs, are fluorescent in solution and in the solid state, are often stable up to 300 °C in air, and generally show enhanced photostability when compared to the PPVs. The PAEs show a distinct chromogenic behavior that is expressed in their solvatochromicity, thermochromicity, ionochromicity, surfactochromicity, and biochromicity [4–8] (see also other chapters in this volume) not found in the PPVs, but somewhat reminiscent of the optical properties of the PDAs and of the polythiophenes [9].

This review deals with the synthesis of PAEs. Described are the developments this field has undergone in the last 4 years. A comprehensive review on PAEs appeared in 2000 (Table 1, entry 1) [1]. Several other noteworthy PAE reviews are those of Yamamoto (Table 1, entries 7, 8) that cover mostly heterocyclic representatives and Pinto and Schanze's review (Table 1, entry 6) about water-soluble conjugated polymers. The specific area of dialkyl PPEs and their synthesis by alkyne metathesis has been reviewed recently (Table 1, entries 2–4)

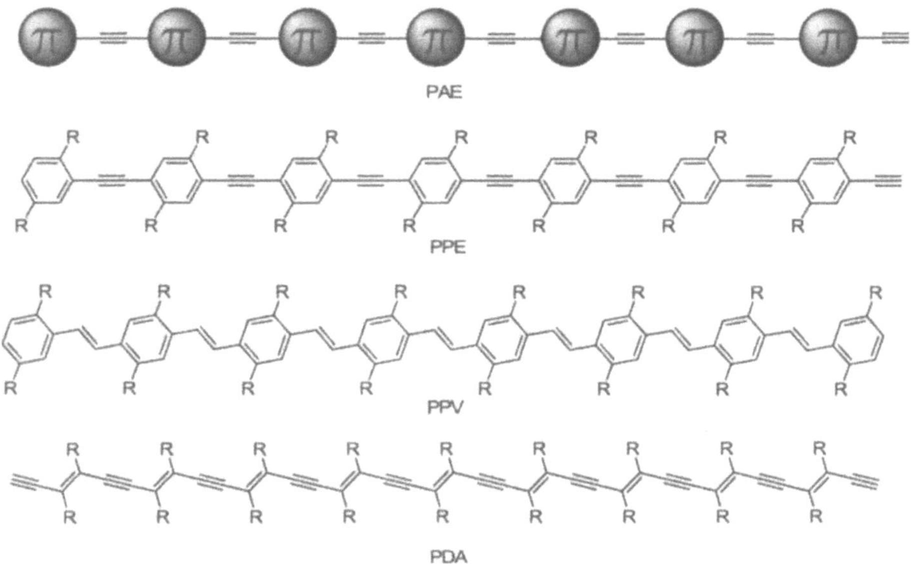

Fig. 1 PAEs, PPEs, and structurally related conjugated polymers PPVs and PDAs

Table 1 Reviews in the area of PAEs

Entry	Author(s)	Citation	Title
1	Bunz UHF	(2000) Chem. Rev. 100:1605–1644	Poly(aryleneethynylene)s: syntheses, properties, structures, and applications
2	Bunz UHF	(2001) Acc. Chem. Res. 34:998–1010	Poly(*p*-phenyleneethynylene)s by alkyne metathesis
3	Bunz UHF	(2002) In: Astruc D (ed) Modern arene chemistry. Wiley-VCH, pp 217–249	The ADIMET reaction: synthesis and properties of poly(dialkyl *para*-phenyleneethynylene)s
4	Bunz UHF	(2003) In: Grubbs RH (ed) Handbook of metathesis. Wiley-VCH, 3:354–370	Acyclic diyne metathesis utilizing in situ transition metal catalysts: an efficient access to alkyne-bridged polymers
5	Rusanov AL, Khotina IA, Begretov MM	(1997) Russ Chem Rev 66:1053–1068	The use of Pd-catalyzed cross coupling for the synthesis of polymers incorporating vinylene and ethynylene groups
6	Pinto MR, Schanze KS	(2002) Synthesis 1293–1309	Conjugated polyelectrolytes: synthesis and applications
7	Yamamoto T	(2003) Synlett 425–450	Synthesis of π-conjugated polymers bearing electronic and optical functionalities by organometallic polycondensations. Chemical properties and applications of the π-conjugated polymers
8	Yamamoto T	(1999) Bull Chem Soc Jpn 72:621–638	π-Conjugated polymers bearing electronic and optical functionalities. Preparation by organometallic polycondensations, properties, and their applications
9	Giesa R	(1996) Rev Macromol Chem Phys C36: 631–670	Synthesis and properties of conjugated poly(aryleneethynylene)s
10	Wosnick JH, Swager TM	(2000) Curr Opinion Chem Biol 4:715–720	Molecular photonic and electronic circuitry for ultrasensitive chemical sensors

and comprehensively. It will therefore be touched upon only cursorily here. If the reader is interested in the early attempts to make PPEs, Giesa's review (Table 1, entry 9) gives a good overview. Table 2 lists reviews that are tangentially important to the field of PAEs and/or deal with the synthesis, properties, and applications of oligomeric aryleneethynylenes (AE). The present review covers the field of *polymeric* AEs and the literature from mid-1999 through 2003. Oligomeric AEs – while interesting in themselves – are not included,

Table 2 Relevant reviews that are related to the field of PAEs

Author	Citation	Title
Negishi EI, Anastasia L	(2003) Chem Rev 103: 1979–2018	Palladium-catalyzed alkynylation
Moore JS	(1997) Acc Chem Res 30: 402–413	Shape-persistent molecular architectures of nanoscale dimension
Hill DJ, Mio MJ, Prince RB, Hughes TS, Moore JS	(2001) Chem Rev 101: 3893–4012	A field guide to foldamers
Swager TM	(2002) Chem Res Toxicol 15:125	Ultrasensitive sensors from self-amplifying electronic polymers
McQuade DT, Pullen AE, Swager TM	(2000) Chem Rev 100: 2537–2574	Conjugated polymer-based chemical sensors
Swager TM	(1998) Acc Chem Res 31: 201–207	The molecular wire approach to sensory signal amplification
Kingsborough RP, Swager TM	(1999) Prog Inorg Chem 48:123–231	Transition metals in polymeric π-conjugated organic frameworks
Marsden JA, Palmer GJ, Haley MM	(2003) Eur J Org Chem 2355–2369	Synthetic strategies for dehydrobenzo[*n*]annulenes
Haley MM, Pak JJ, Brand SC	(1999) Top Curr Chem 101:81–130	Macrocyclic oligo(phenylacetylenes) and oligo(phenyldiacetylenes)
Watson MD, Fechenkötter A, Müllen K	(2001) Chem Rev 101: 1267–1300	Big is beautiful – "aromaticity" revisited from the viewpoint of macromolecular and supramolecular benzene chemistry
Tour JM	(1996) Chem Rev 96: 537–553	Conjugated macromolecules of precise length and constitution. Organic synthesis for the construction of nanoarchitectures
Tour JM	(2000) Acc Chem Res 33:791–804	Molecular electronics. Synthesis and testing of components
Youngs WJ, Tessier CA, Bradshaw JD	(1999) Chem Rev 99: 3153–3180	*ortho*-Arene cyclynes, related heterocyclynes, and their metal chemistry

unless their discussion is necessary or useful for the general understanding of the topic, or where they serve as models for PAEs. For more information, Table 3 shows seminal contributions to the field of oligo AEs. The areas of porphyrin-containing ethynylated polymers and AE polymers containing metals *in the main chain* are not covered (see chapter in this volume by E. Klemm).

Table 3 Relevant primary publications in the field of oligomeric AEs

Author	Citation	Title
Zhang JS, Moore JS, Xu ZF, Aguirre RA	(1992) J Am Chem Soc 114:2273–2274	Nanoarchitectures 1. Controlled synthesis of phenylacetylene sequences
Schumm JS, Pearson DL, Tour JM	(1994) Angew Chem 33:1360–1363	Iterative divergent/convergent approach to linear conjugated oligomers by successive doubling of the molecular length
Huang SL, Tour JM	(1999) J Am Chem Soc 121:4908–4909	Rapid solid-phase synthesis of oligo-(1,4-phenyleneethynylene)s by a divergent–convergent tripling strategy
Hwang JJ, Tour JM	(2002) Tetrahedron 58: 10387–10405	Combinatorial synthesis of oligo-(phenyleneethynylene)s
Anderson S	(2001) Chem Eur J 7: 4706–4714	Phenyleneethynylene pentamers for organic electroluminescence
Melinger JS, Pan YC, Kleiman VD, Peng ZH, Davis BL, McMorrow D, Lu M	(2002) J Am Chem Soc 124:12002–12012	Optical and photophysical properties of light-harvesting phenylacetylene monodendrons based on unsymmetrical branching
Peng ZH, Pan YC, Xu BB, Zhang J H	(2000) J Am Chem Soc 122:6619–6623	Synthesis and optical properties of novel unsymmetrical conjugated dendrimers
Kukula H, Veit S, Godt A	(1999) Eur J Org Chem 277	Synthesis of monodisperse oligo-PEs using orthogonal protecting groups with different polarity for terminal acetylene units
Wagner RW, Johnson TE, Lindsey JS	(1996) J Am Chem Soc 118:11166–11180	Soluble synthetic multiporphyrin arrays. 1. Modular design and synthesis

2 Synthetic Methods

There are currently two methods available to make PAEs with significant molecular weights. The classical method, first used by Giesa and Schulz [10], employs the Heck–Cassar–Sonogashira–Hagihara (Table 2, entry 1) coupling to react an aromatic diyne with an aromatic dihalide in an amine solvent (Scheme 1). This method is general and compatible with most functional groups, save the heavier halides and unprotected alkynes. Good results are

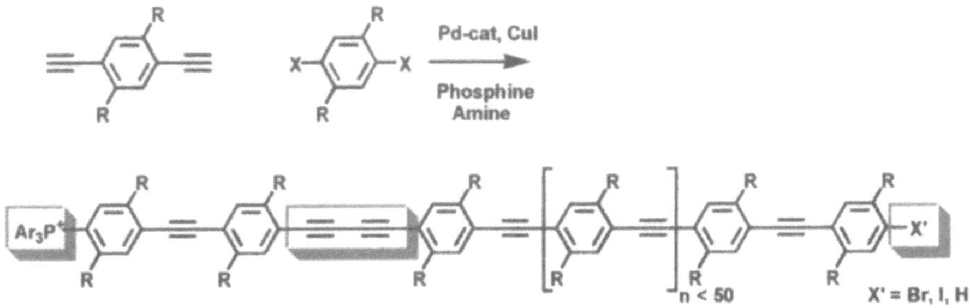

Scheme 1 The Pd-catalyzed synthetic method to make PAEs

obtained when diiodoarenes are utilized in piperidine, diisopropylamine (DIPA), or triethylamine as solvent and either $(PPh_3)_2PdCl_2$ or $(PPh_3)_4Pd$ as catalyst in combination with CuI. For an in-depth discussion of the parameters of this coupling and its mechanism see Table 1, entry 1 and Table 2, entry 1. The addition of THF to the reaction mixture, independently described by Thorand and Krause [11] and by LeMoigne et al. [12], increases the molecular weight and yield of the obtained PAEs.

The disadvantage of the Pd-catalyzed couplings is the often (but not always) low molecular weight or degree of polymerization (P_n) of the isolated PAEs. Typical P_n values in these reactions are approx. 20–50, even though independently Swager's and Weder's groups have reported high molecular weight materials when utilizing $(Ph_3P)_4Pd$ as catalyst [13].

A problem in these couplings is the identity of the end groups of the formed polymers. Pd-catalyzed dehalogenation and/or phosphonium salt formation are side reactions that are difficult to avoid. A concern for the structural integrity of the backbone is the formation of butadiyne defects. While there is no direct measure to determine the amount of butadiyne defects in PAEs, the numbers are estimated to range from 1 to 10% of all repeat units.

The formation of butadiyne defect structures in PAEs is always observed, even if Pd(0) catalysts are used. It is not clear where the necessary oxidant comes from. Often a small excess of diyne is utilized, and such preparations seem to give higher molecular weight materials [14]. In the process of the formation of the final PAE, very large oligomers or small polymers with alkyne end groups could be intermediates. These intermediates would oligomerize via a Pd-catalyzed variant of a Hay-type coupling [15] to give the observed high molecular weight PAEs. This is a hypothesis that could explain Swager's results in the formation of high molecular weight PPEs. The butadiyne defect structures would only have to occur every other 50–100 repeat units and would not be detected by ^{13}C NMR spectroscopy. Fortunately, the butadiyne defects do not seem to have a large influence on the optical and electronic properties of the PAEs, yet it would be desirable to have a more controlled way of making defect-free PAEs.

Table 4 PPEs containing only *para*-phenylene units in the main chain

	Monomer 1	Monomer 2		Polymer	Yield (%) P_n PDI
1[a]	H-C≡C-H	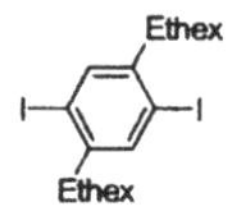	0.1 mol% $(Ph_3P)_2PdCl_2$ CuI, toluene, piperidine	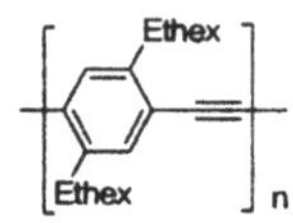	72, 181, 3.6
2[a]	H-C≡C-H	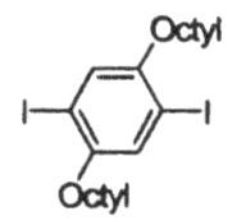	0.1 mol% $(Ph_3P)_2PdCl_2$ CuI, toluene, piperidine	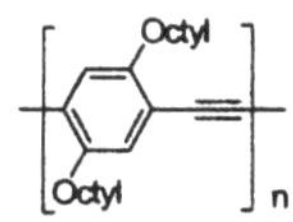	85, 127, 4.1
3[a]	H-C≡C-H	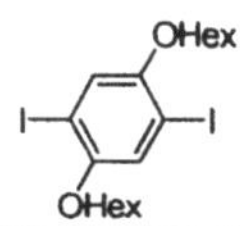	0.1 mol% $(Ph_3P)_2PdCl_2$ CuI, toluene, piperidine	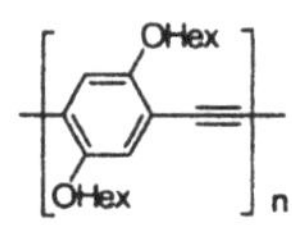	63, 33, 3.2
4[a]	H-C≡C-H	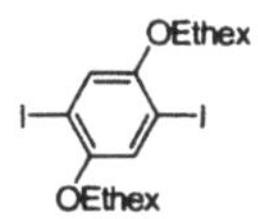	0.1 mol% $(Ph_3P)_2PdCl_2$ CuI, toluene, piperidine	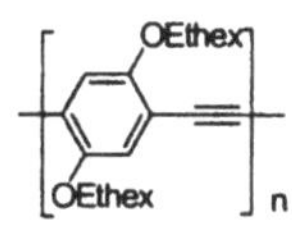	88, 104, 3.1
5[b]	H-C≡C-H	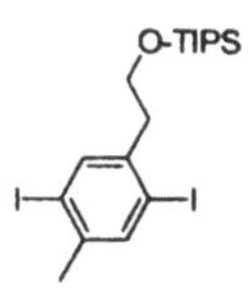	0.2 mol% $(Ph_3P)_2PdCl_2$ CuI, toluene, piperidine	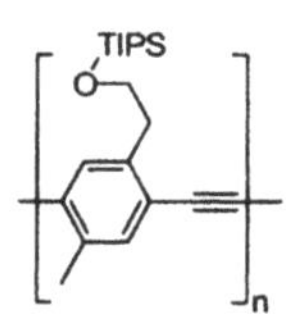	98, 118, 3.9
6[b]	H-C≡C-H	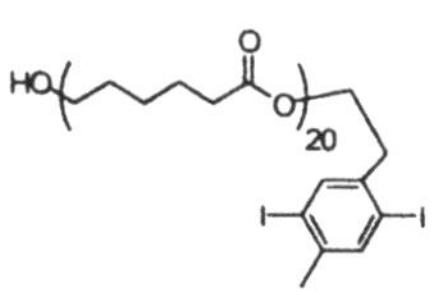	0.2 mol% $(Ph_3P)_2PdCl_2$ CuI, toluene, piperidine	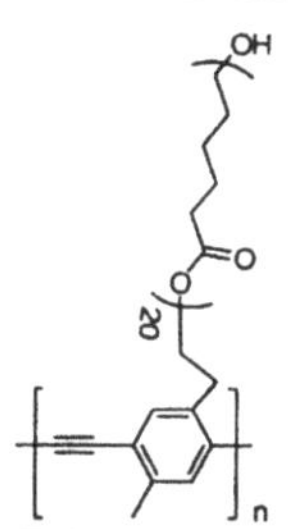	77, 140, 5.3
7[c]	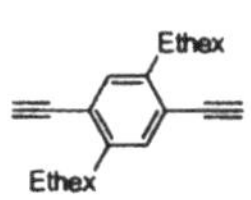	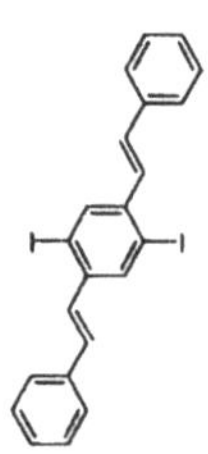	$(Ph_3P)_2PdCl_2$ CuI, toluene, piperidine	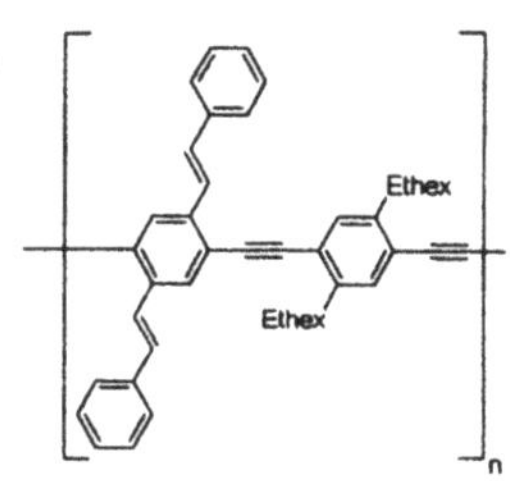	80, 55, 2.8

Table 4 (continued)

	Monomer 1	Monomer 2		Polymer	Yield (%) P_n PDI
8[c]	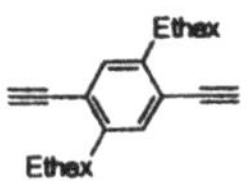	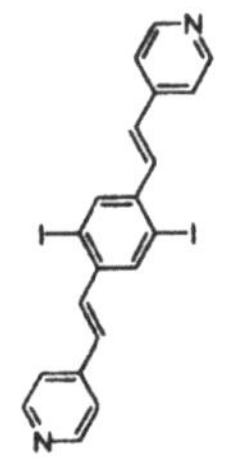	$(Ph_3P)_2PdCl_2$ CuI, toluene, piperidine	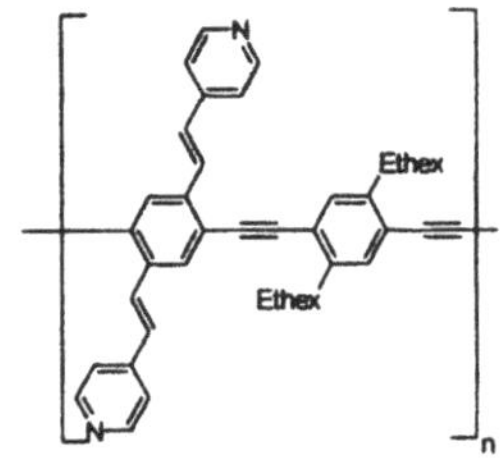	25, nd, nd
9[c]	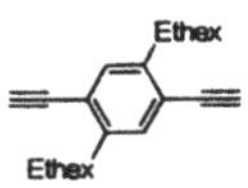	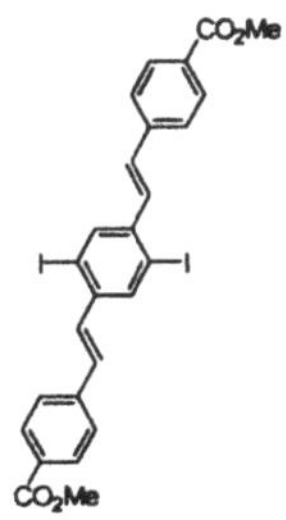	$(Ph_3P)_2PdCl_2$ CuI, toluene, piperidine	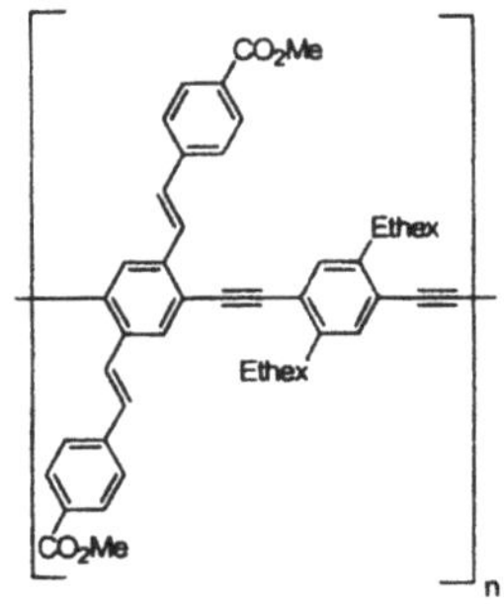	81, 38, 3.2
10[c]	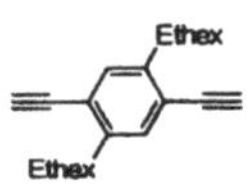	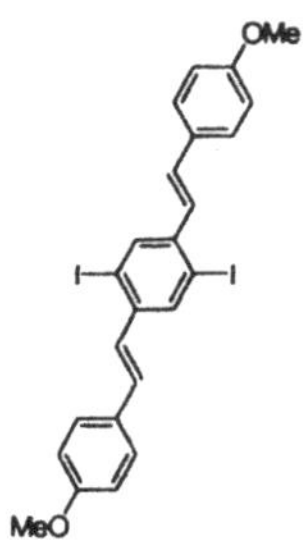	$(Ph_3P)_2PdCl_2$ CuI, toluene, piperidine	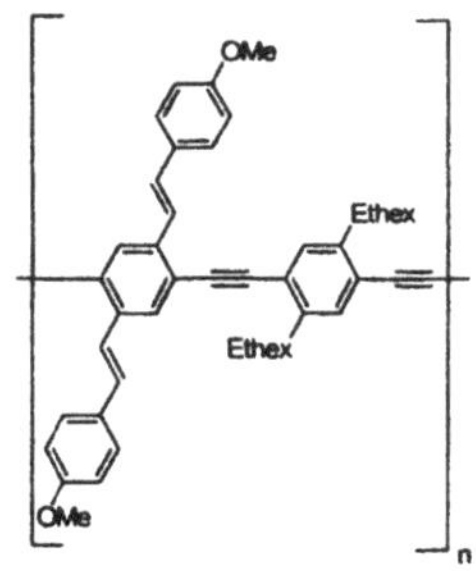	38, 50, 2.0
11[c]	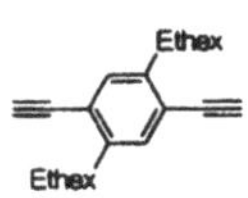	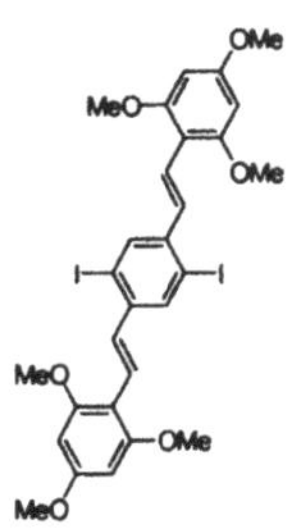	$(Ph_3P)_2PdCl_2$ CuI, toluene, piperidine	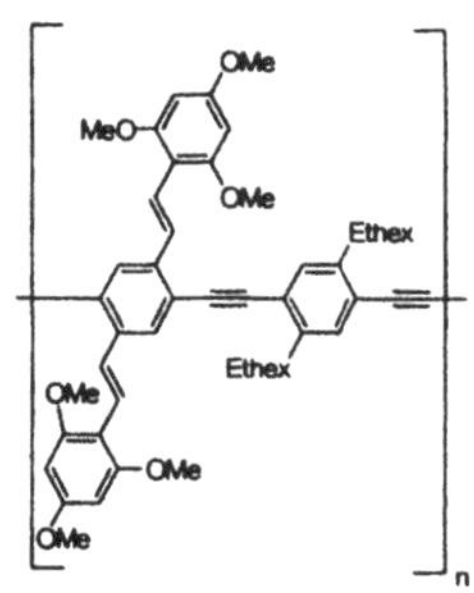	55, 52, 3.2

Table 4 (continued)

	Monomer 1	Monomer 2		Polymer	Yield (%) P_n PDI
12[c]	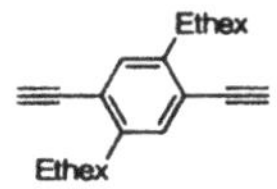	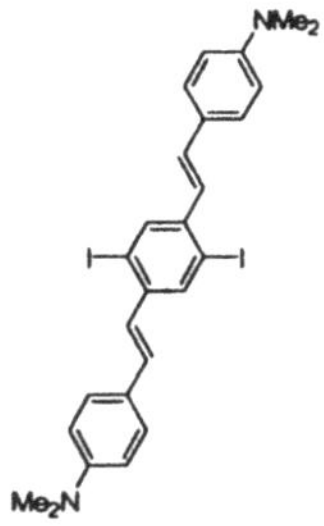	$(Ph_3P)_2PdCl_2$ CuI, toluene, piperidine	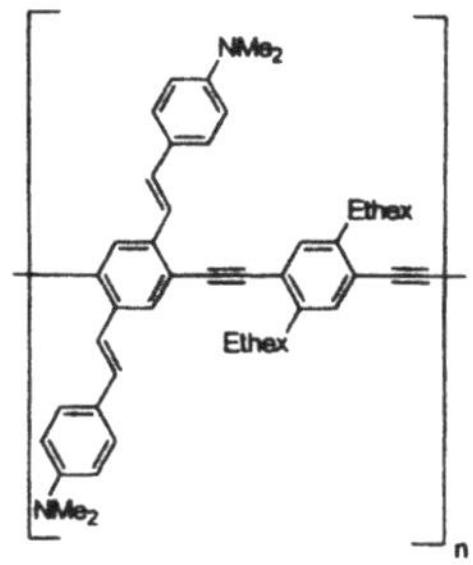	56, nd,nd
13[d]	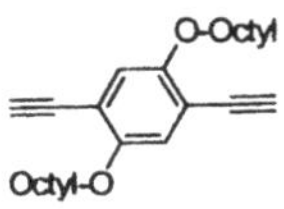	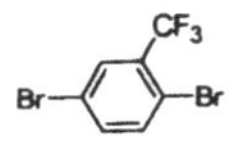	$(Ph_3P)_4Pd$ CuI, toluene, DIPA	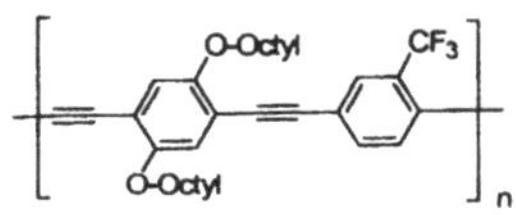	76, 15, ≈ 2
14[d]	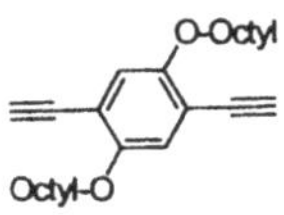	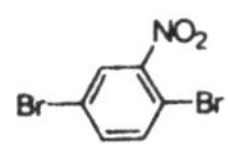	$(Ph_3P)_4Pd$ CuI, toluene, DIPA	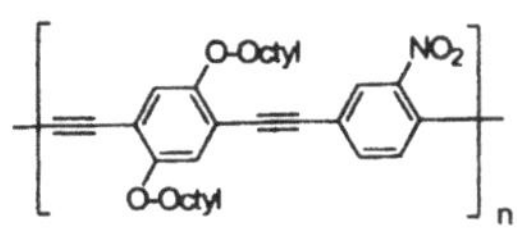	92, 13, ≈2
15[d]	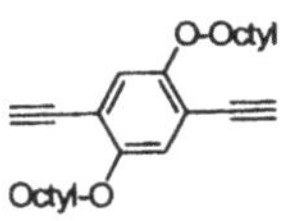	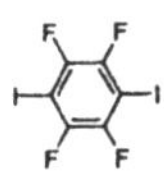	$(Ph_3P)_4Pd$ CuI, toluene, DIPA	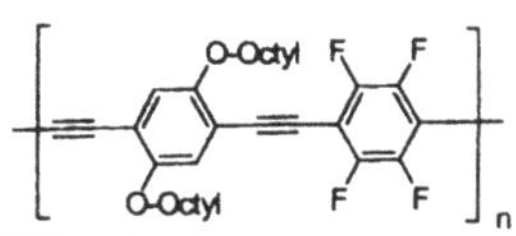	95 17 ≈2
16[d]	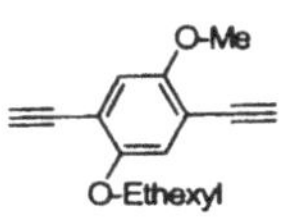	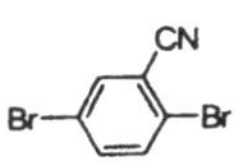	$(Ph_3P)_4Pd$ CuI, toluene, DIPA	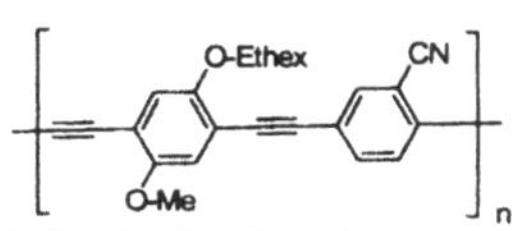	77 15 ≈2
17[d]	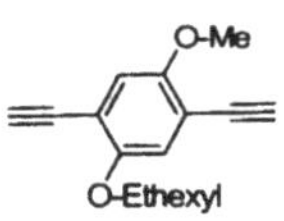	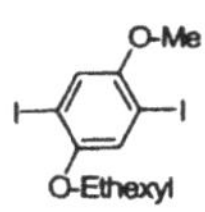	$(Ph_3P)_4Pd$ CuI, toluene, DIPA	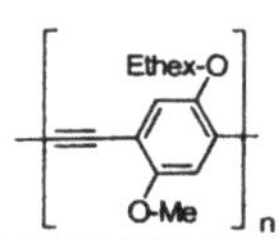	86 13 ≈2
18[e]		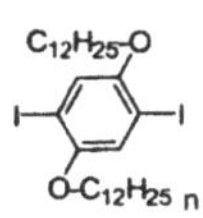	$(Ph_3P)_4Pd$ CuI, toluene, DIPA	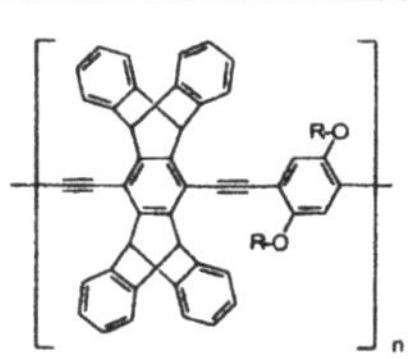	65 na na
19[e]	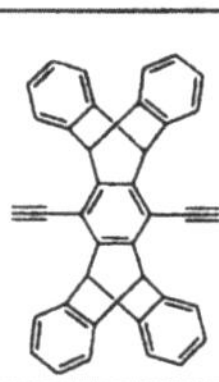	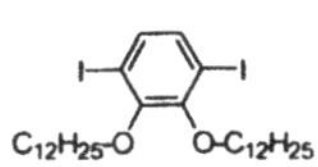	$(Ph_3P)_4Pd$ CuI, toluene, DIPA	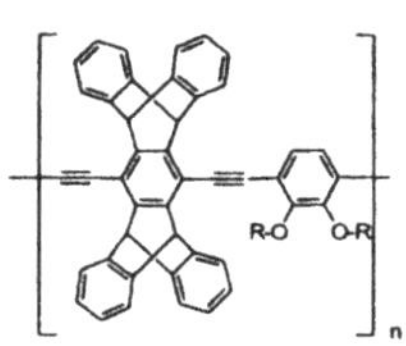	75 78 na

Table 4 (continued)

	Monomer 1	Monomer 2		Polymer	Yield (%) P_n PDI
20[f]	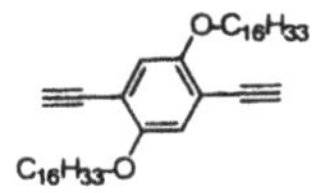	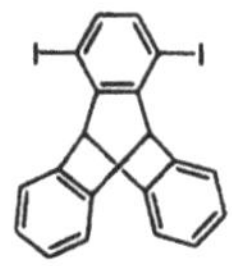	$(Ph_3P)_4Pd$ CuI, toluene, DIPA	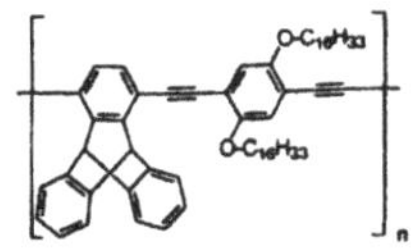	53 60 AE units 3.3
21[f]	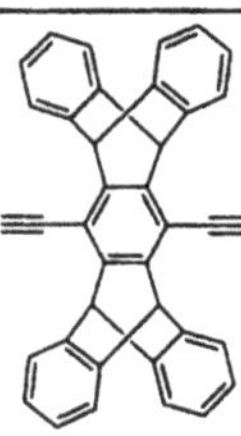	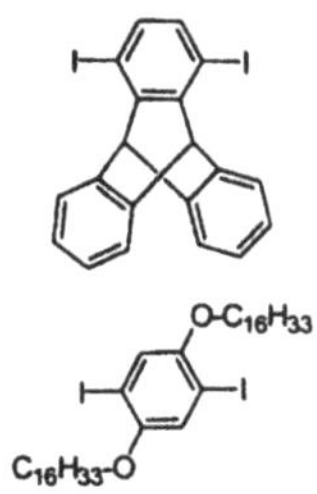	$(Ph_3P)_4Pd$ CuI, toluene, DIPA	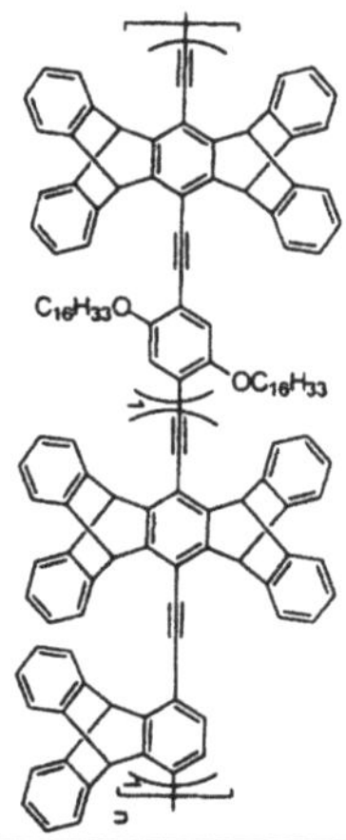	78 225 AE units 3.4
22[g]	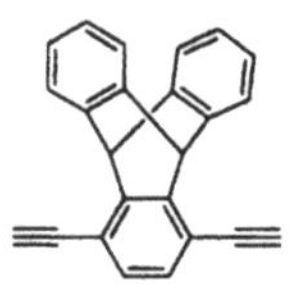	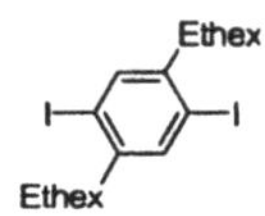	$(Ph_3P)_4Pd$ CuI, toluene, DIPA	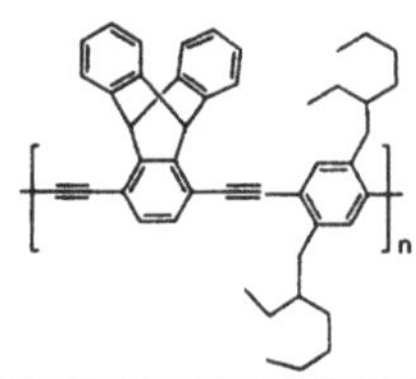	86 37 2.9
23[g]	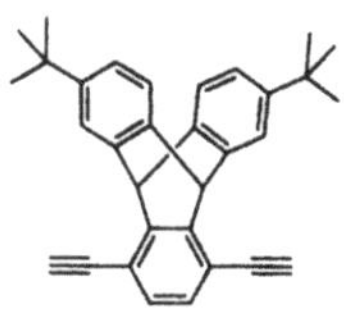	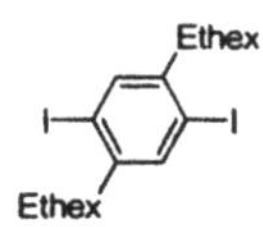	$(Ph_3P)_4Pd$ CuI, toluene, DIPA	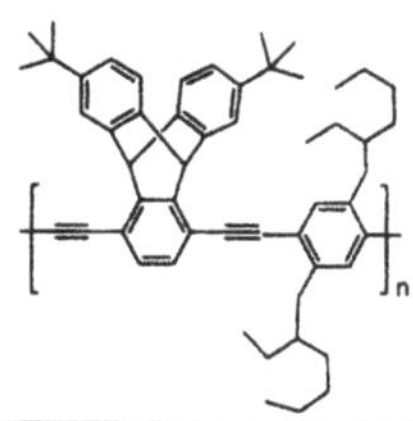	88 36 2.3
24[h]	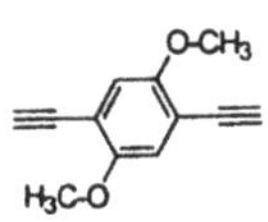	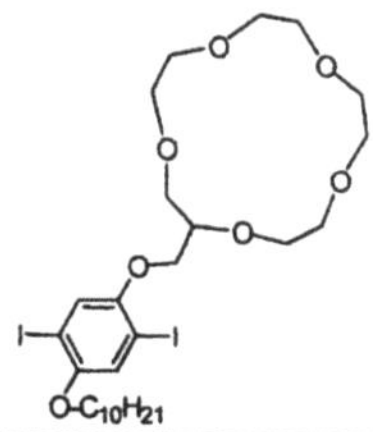	$(Ph_3P)_4Pd$ CuI, toluene, DIPA	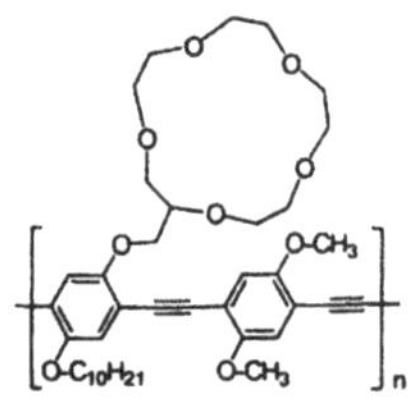	Nd 30 2.5

Table 4 (continued)

	Monomer 1	Monomer 2		Polymer	Yield (%) P_n PDI
25[h]	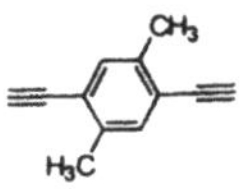	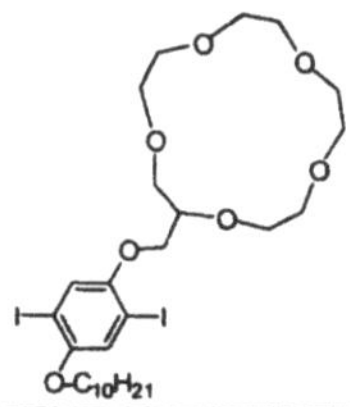	$(Ph_3P)_4Pd$ CuI, toluene, DIPA	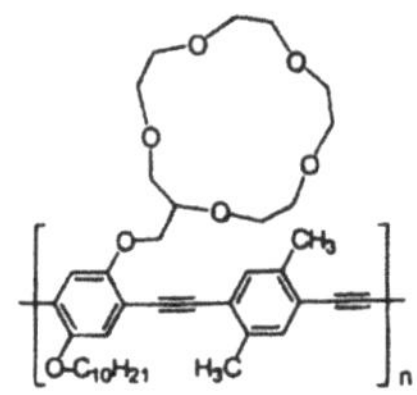	80% 57 3.2
26[h]	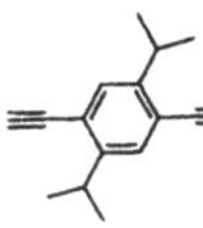	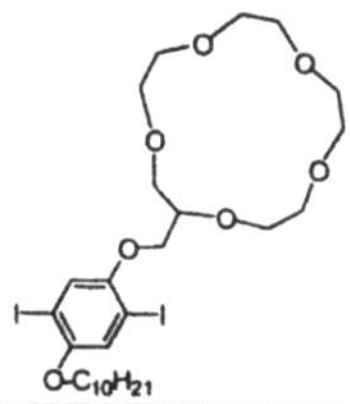	$(Ph_3P)_4Pd$ CuI, toluene, DIPA	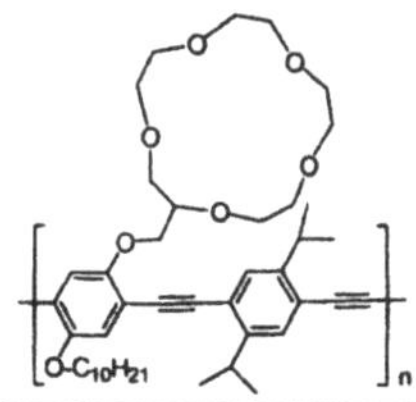	90% 96 1.9
27[h]	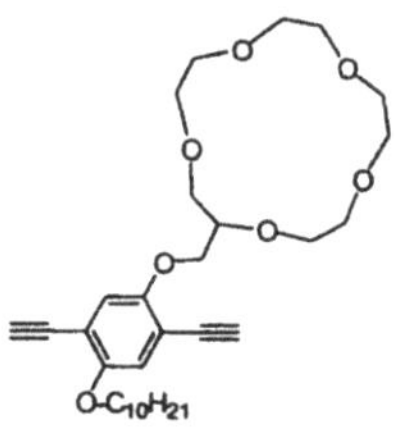	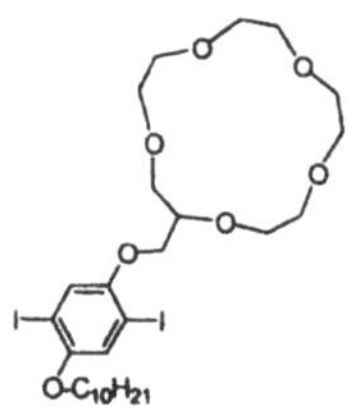	$(Ph_3P)_4Pd$ CuI, toluene, DIPA	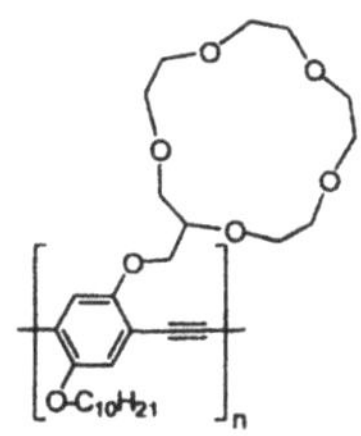	47% 16 2.0
28[i]	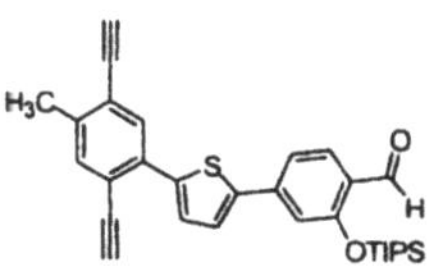	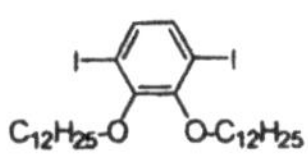	$(Ph_3P)_4Pd$ CuI, toluene, DIPA	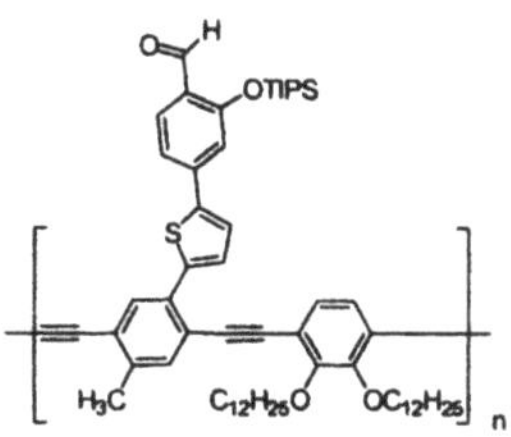	43% 21 PE units 2.0
29[j]	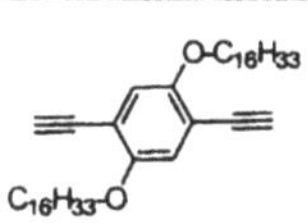	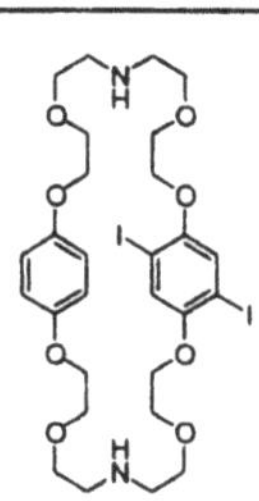	$(Ph_3P)_4Pd$ CuI, morpholine, DIPA	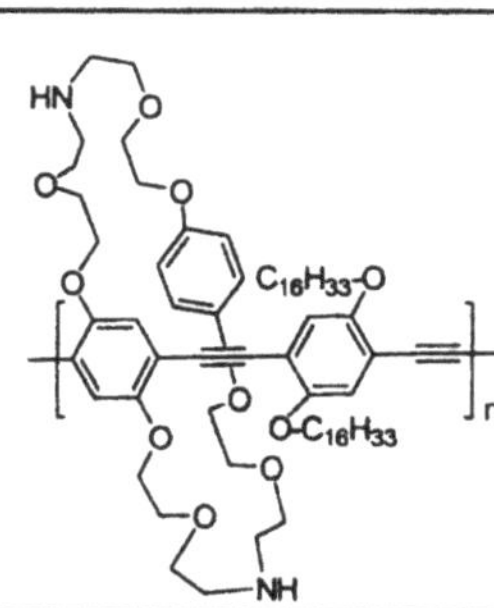	81 18 1.5

Table 4 (continued)

	Monomer 1	Monomer 2		Polymer	Yield (%) P_n PDI
30[k]	Not given, only structure of polymer.			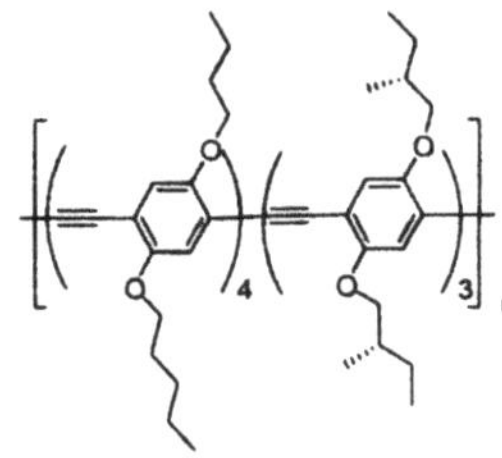	Nd Nd Nd
31[k]	Not given, only structure of polymer.			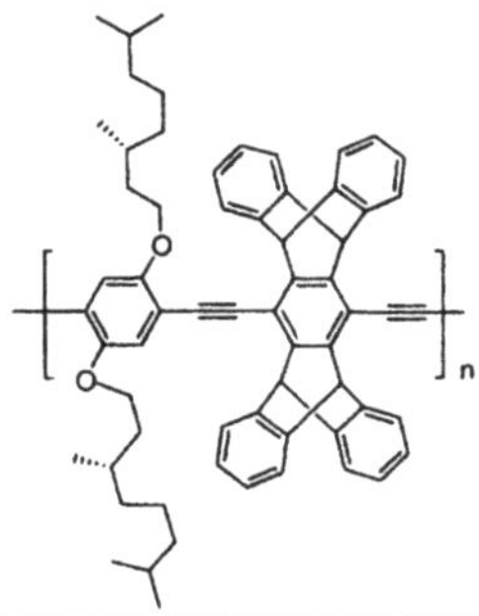	Nd Nd Nd
32[l]	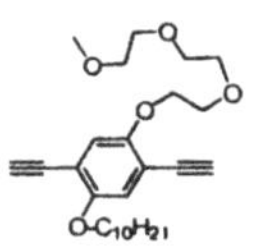	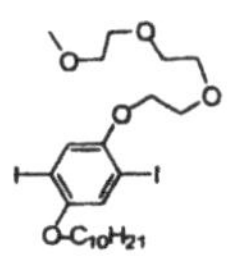	$(Ph_3P)_4Pd$ CuI, toluene, DIPA	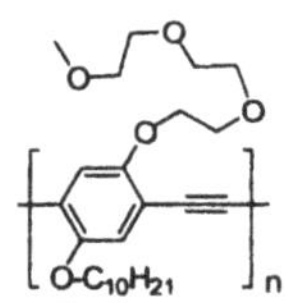	Nd 133 2.9
33[l]	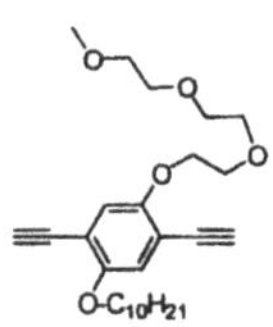	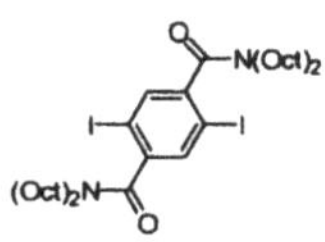	$(Ph_3P)_4Pd$ CuI, toluene, DIPA	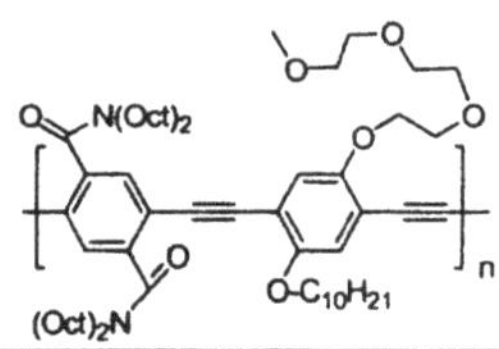	Nd 52 3.8
34[l]	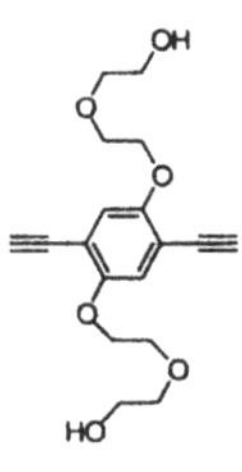	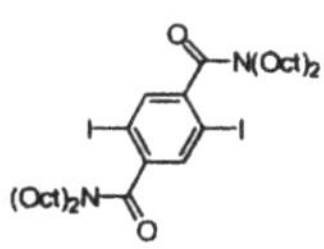	$(Ph_3P)_4Pd$ CuI, toluene, DIPA	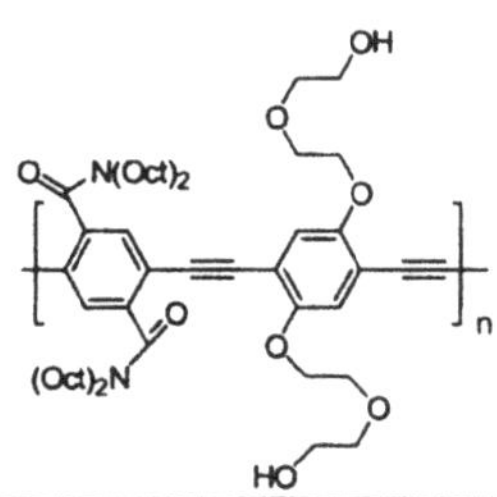	Nd 35 2.4
35[m]	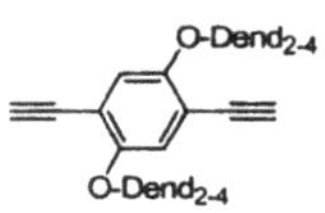	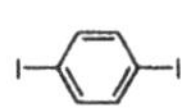	$(Ph_3P)_4Pd$ CuI, toluene, DIPA, THF	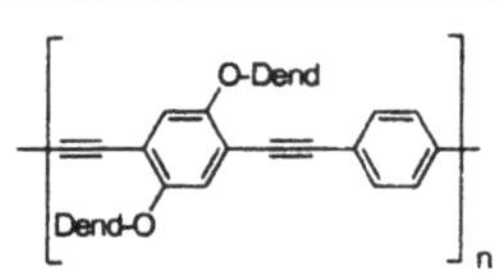	$Dend_2$ 30% 14 1.5

Table 4 (continued)

	Monomer 1	Monomer 2		Polymer	Yield (%) P_n PDI
36[m]		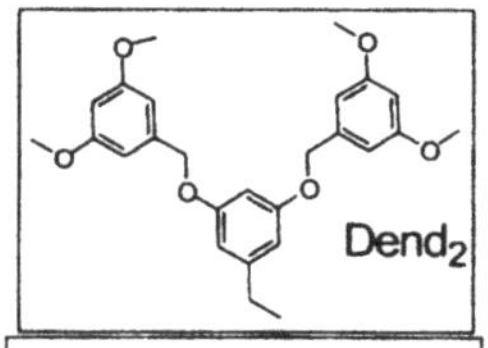		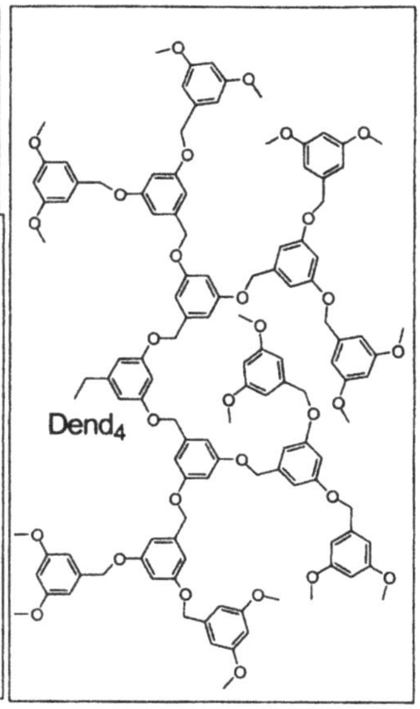	$Dend_3$ 85 41 6.5
37[m]		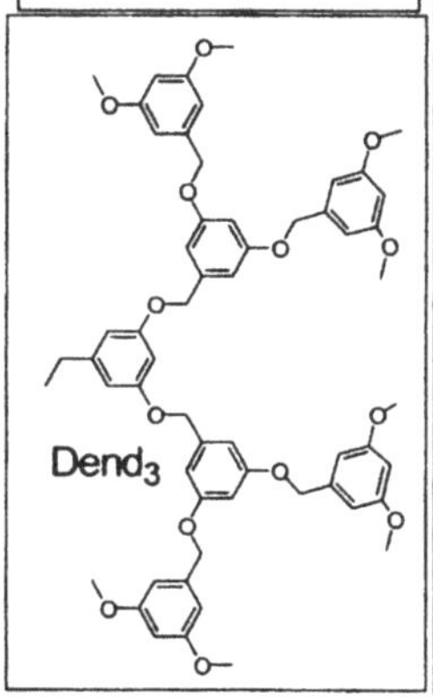			$Dend_4$ 90 11 1.5
38[n]	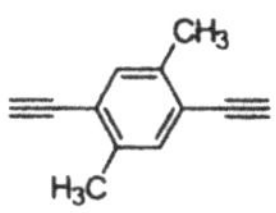	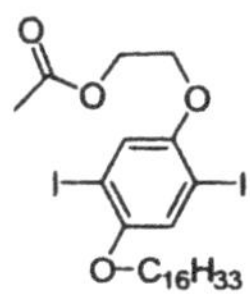	$(Ph_3P)_4Pd$ CuI, toluene, DIPA	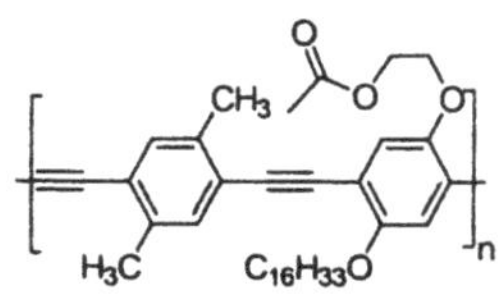	80% 83 AE units 2.2
39[n]	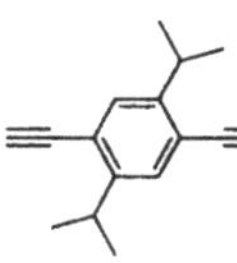	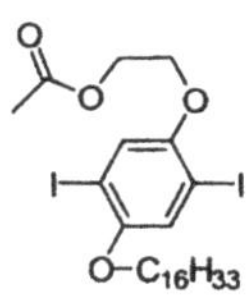	$(Ph_3P)_4Pd$ CuI, toluene, DIPA	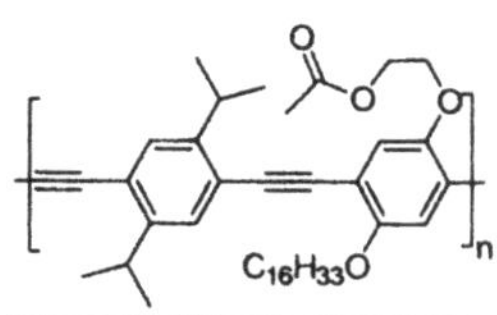	90% 718 AE units 2.4
40[n]	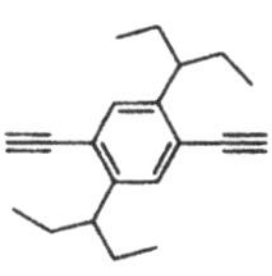	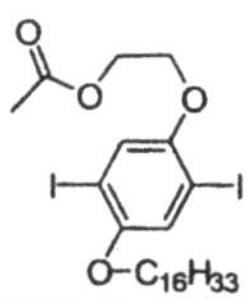	$(Ph_3P)_4Pd$ CuI, toluene, DIPA	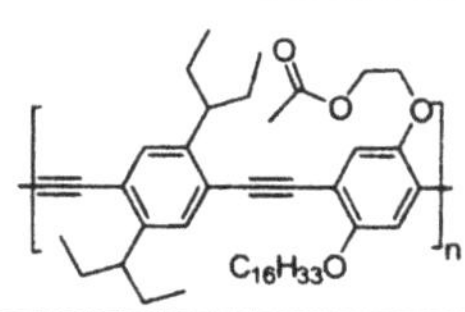	95% 404 AE units 3.1
41[o]	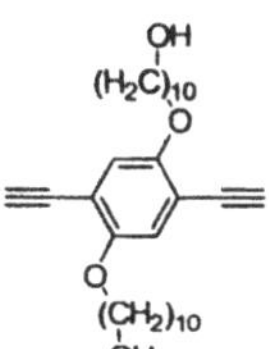	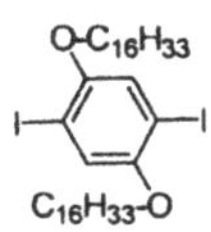	$(Ph_3P)_4Pd$ CuI, toluene, DIPA	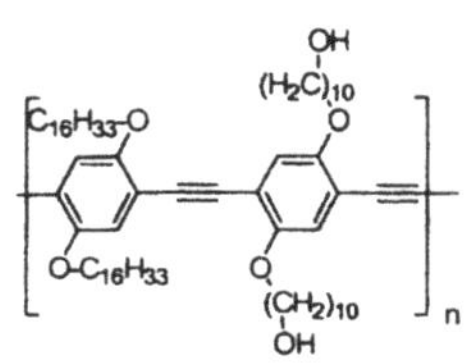	Nd 55 2.6

Table 4 (continued)

	Monomer 1	Monomer 2		Polymer	Yield (%) P_n PDI
42[p]	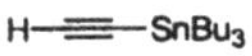	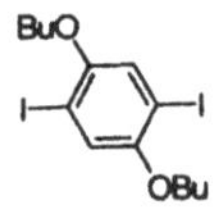	$(Ph_3P)_4Pd$ dioxane reflux then LDA then reflux 2-8h	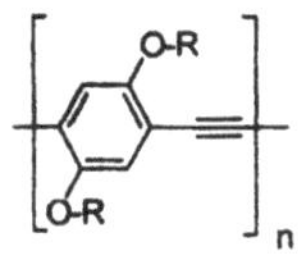	86, 25, 3.1
43[p]	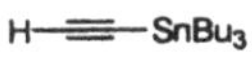	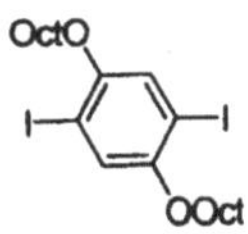	$(Ph_3P)_4Pd$ dioxane reflux then LDA then reflux 2-8h	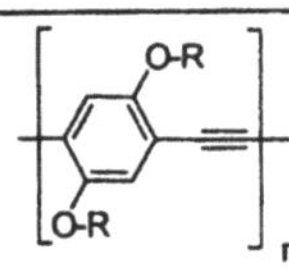	94, 35, 2.5
44[p]	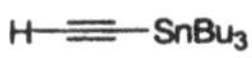	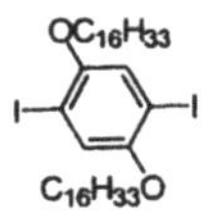	$(Ph_3P)_4Pd$ dioxane reflux then LDA then reflux 2-8h	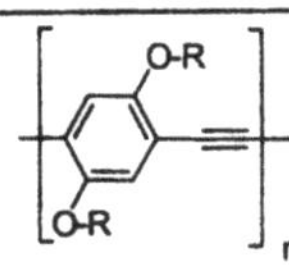	92, 22, 2.0
45[q]	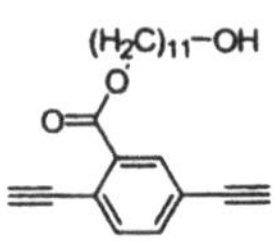	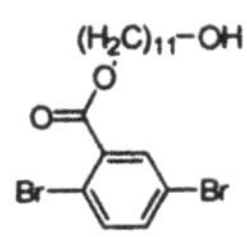	$PdCl_2$, CuI, NEt_3 PPh_3	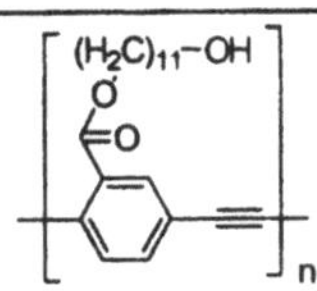	80-90% 8 1.71
46[r]	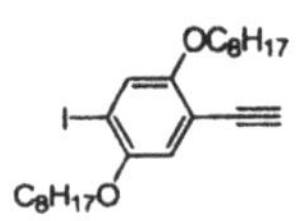	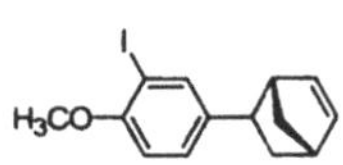	$(Ph_3P)_4Pd$ CuI, toluene, DIPA	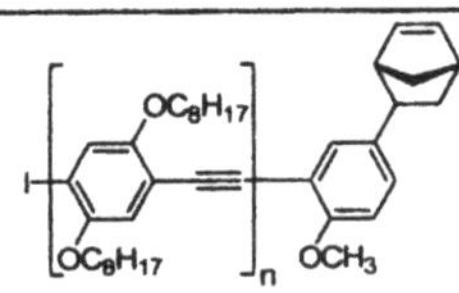	Yield nd 12×10^3 1.8

[a] Wilson JN, Waybright SM, McAlpine K, Bunz UHF (2002) Macromolecules 35:3799–3800.
[b] Wang Y, Erdogan B, Wilson JN, Bunz UHF (2003) Chem Commun 1624–1625.
[c] Wilson JN, Windscheif PM, Evans U, Myrick ML, Bunz UHF (2002) Macromolecules 35:8681–8683.
[d] Dellsperger S, Dötz F, Smith P, Weder C (2000) Macromol Chem Phys 201:192–198.
[e] Zhu Z, Swager TM (2001) Org Lett 3:3471–3474.
[f] Williams VE, Swager TM (2000) Macromolecules 33:4069–4073.
[g] Zhu Z, Swager TM (2002) J Am Chem Soc 124:9670–9671.
[h] Kim J, McQuade DT, McHugh SK, Swager TM (2000) Angew Chem 39:3868–3872.
[i] Kim TH, Swager TM (2003) Angew Chem 42:4803–4806.
[j] Deans R, Kim J, Machacek MR, Swager TM (2000) J Am Chem Soc 122:8565–8566.
[k] Zahn S, Swager TM (2002) Angew Chem 41:4226–4230.
[l] Kim J, Levitsky IA, McQuade DT, Swager TM (2002) J Am Chem Soc 124:7710–7718.
[m] Sato T, Jian DL, Aida T (1999) J Am Chem Soc 121:10658–10659.
[n] McQuade TD, Kim J, Swager TM (2000) J Am Chem Soc 122:5885–5886.
[o] Breen CA, Deng T, Breiner T, Thomas EL, Swager TM (2003) J Am Chem Soc 125:9942–9943.
[p] Pizzoferrato R, Berliocchi M, DiCarlo A, Lugli P, Venanzi M, Micozzi A, Ricci A, LoSterzo C (2003) Macromolecules 36:2215–2223.
[q] Arias Marin E, LeMoigne J, Maillou T, Guillon D, Moggio I, Geffroy B (2003) Macromolecules 36:3570–3579.
[r] Moon JH, Swager TM (2002) Macromolecules 35:6086–6089.

In most of these polymerization reactions 1–5 mol% of Pd catalyst is utilized. It was recently discovered that such an amount of Pd catalyst can be excessive. It is possible to make PPEs of excellent quality utilizing 0.1–0.2 mol% $(Ph_3P)_2PdCl_2$. This approach works well (see Table 4, entries 1–6) if acetylene gas is the difunctional alkyne and a diiodoarene is the coupling partner. This method is useful if the exact molecular weight of a monomer under consideration cannot be determined. Such a case arises if one utilizes a macromonomer that is already polydisperse (Table 4, entry 6). This polyester monomer can be successfully polymerized with acetylene gas to give a grafted PPE. In the case of the acetylene, PPEs of method, the precise amount of acetylene is not critical. As long as there is not a large excess of acetylene, PAEs of significant molecular weights are obtained. The available analytical data (^{13}C NMR, IR, UV-vis, fluorescence) show that the number of diyne defects is below the detection limit. Polymerization experiments in which larger amounts of Pd catalyst were utilized in combination with acetylene gas gave inferior materials, possibly due to the reaction of the Pd(0) species with the formed polymer to give multinuclear cluster compounds. Overall this is a valuable, effective, and attractive extension for the synthesis of PAEs.

Conventional Pd-based catalysts such as $(Ph_3P)_2PdCl_2$, $(Ph_3P)_4Pd$, and mixtures of $Pd(OAc)_2$ or $PdCl_2$ with triphenylphosphine are utilized in the synthesis of PAEs. In recent years the groups of Buchwald and of Hartwig have developed new catalysts for the Heck–Cassar–Sonogashira reaction [16, 17]. These catalysts have not been investigated in the synthesis of PAEs but might show fewer side reactions. This strategy would lead to defect-free PAEs of increased P_n. In addition, these catalysts might be able to make PAEs from inexpensive but less reactive dichloroarenes.

An alternative method to make PAEs is the acyclic diyne metathesis (ADIMET) shown in Scheme 2. It is the reaction of a dipropynylarene with $Mo(CO)_6$ and 4-chlorophenol or a similarly acidic phenol. The reaction is performed at elevated temperatures (130–150 °C) and works well for almost any hydrocarbon monomer. The reaction mixture probably forms a Schrock-type molybdenum carbyne intermediate as the active catalyst. Table 5 shows PAEs that have been prepared utilizing ADIMET with these "in situ catalysts". Functional groups (with the exception of double bonds) are not well tolerated, but dialkyl PPEs are obtained with a high degree of polymerization. The progress in this field has been documented in several reviews (Table 1, entries 2–4). Recently, a second generation of ADIMET catalyst has been developed that allows

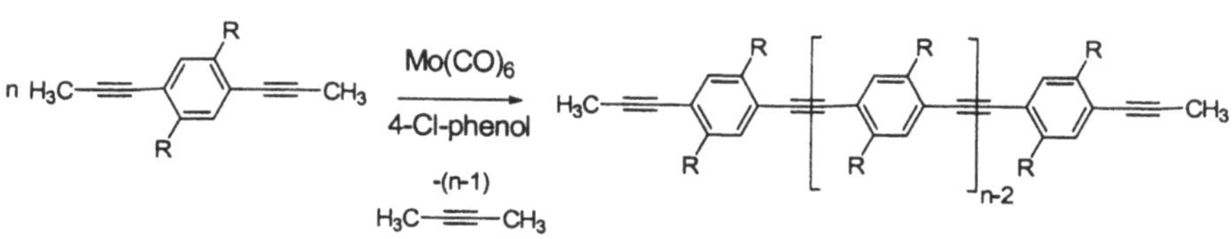

Scheme 2 Synthesis of PAEs by acyclic diyne metathesis (ADIMET).

Table 5 Synthesis of PPEs by alkyne metathesis

	Monomer 1	Monomer 2	Catalyst	Polymer	Yield (%) P_n PDI
1[a]	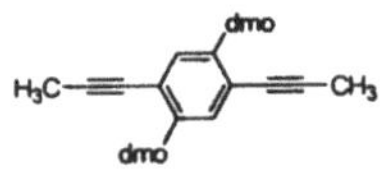		$Mo(CO)_6$ 4-Chloro- phenol		83 % yield 128 3.2
2[a]	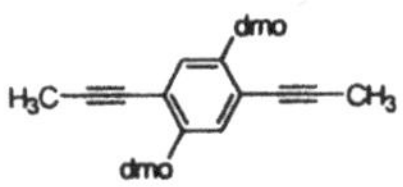	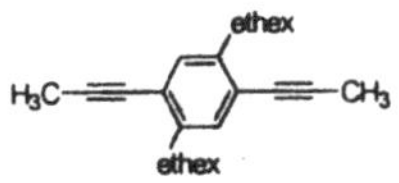	$Mo(CO)_6$ 4-Chloro- phenol	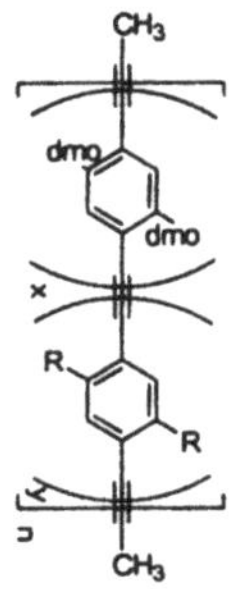	50% ethex 81% yield 304, 4.4
3[a]	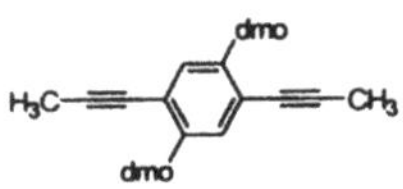	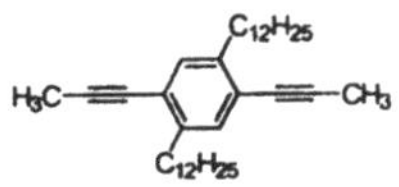	$Mo(CO)_6$ 4-Chloro- phenol		75% $C_{12}H_{25}$ 84%, 176 2.2
4[a]	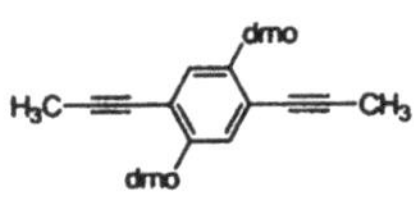	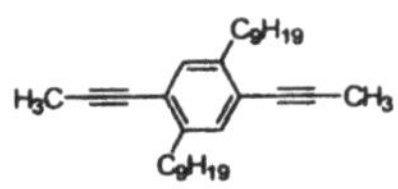	$Mo(CO)_6$ 4-Chloro- phenol		50% C_9H_{19} 70%,195 4.1
5[b]	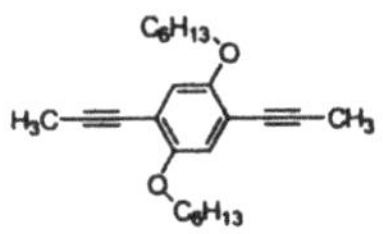	10 mol% $Mo(CO)_6$ 4-chlorophenol excess of 3-hexyne heat for 1 h then add monomer and react for 24 h at 130 °C		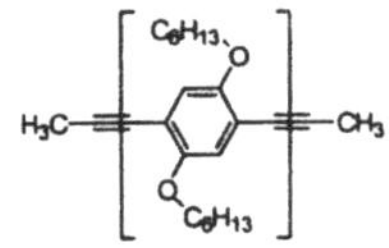	95% yield 140 nd
6[c]	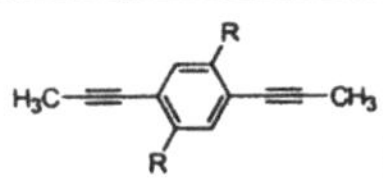	R = n-butyl	5 mol% $Mo(CO)_6$ 4-chlorophenol		95%, 201 2.84
7[c]	R = n-octyl		5 mol% $Mo(CO)_6$ 4-chlorophenol	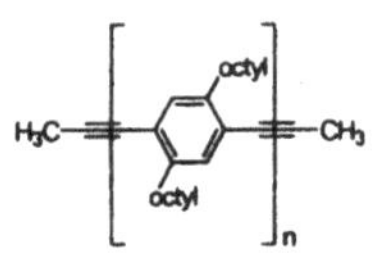	99%, 84 1.81
8[c]	R = n-tetradecyl		5 mol% $Mo(CO)_6$ 4-chlorophenol	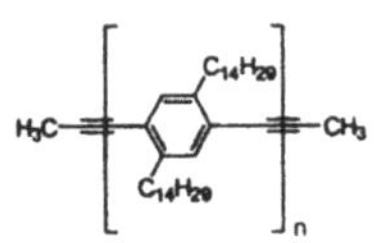	97% 37 1.44

Table 5 (continued)

	Monomer 1	Monomer 2	Catalyst	Polymer	Yield (%) P_n PDI
9[c]	R = 2-Ethex		5 mol% $Mo(CO)_6$ 4-chlorophenol	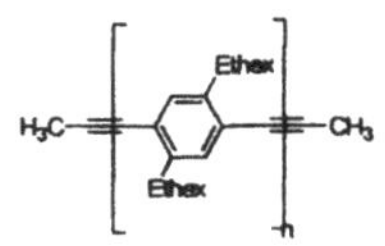	97% 48 1.92
10[c]	R = 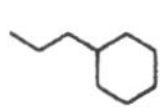		5 mol% $Mo(CO)_6$ 4-chlorophenol	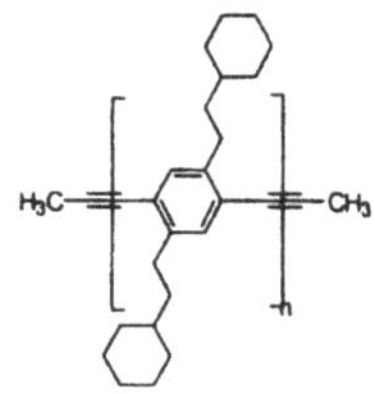	95, 51 (NMR), nd

[a] Wilson JN, Steffen W, McKenzie TG, Lieser G, Masao O, Neher D, Bunz UHF (2002) J Am Chem Soc 124:6830–6831.
[b] Brizius G, Bunz UHF (2002) Org Lett 4:2819–2831.
[c] Huang YW, Gao W, Kwei TK, Okamoto Y (2001) Macromolecules 34:1570–1578.

metathesis of dipropynyldi*alkoxy*benzenes to make dialkoxy PPEs. The trick is to utilize either 2-fluorophenol as a cocatalyst or to activate the catalyst mixture by heating 4-chlorophenol, $Mo(CO)_6$, and hexyne before adding the dipropynylated monomer [18, 19]. Both protocols work, but Grela's fluorophenol catalyst system is easier to use [19]. The dialkoxy PPEs are formed in high yield and with a high degree of polymerization (see Table 5, entry 5).

The ADIMET reaction often gives PAEs of high molecular weight in high yields. ADIMET's disadvantage is its lack of compatibility with sensitive functional groups and the high reaction temperatures necessary to prod the monomers to react. A solution to this problem would be the use of preformed alkylidyne complexes of the type used by Schrock, Fürstner, or Cummins and coworkers [20–22]. None of these carbyne complexes is commercially available, and their synthesis requires a significant investment of time and effort in addition to a well-working glove box or Schlenk line. These carbyne complexes are air and water sensitive. Moore et al. [23] have recently developed an alkyne metathesis strategy, where a preformed intermediate containing a Mo≡Mo triple bond is reacted with a suitable phenol. These "pseudo in situ" catalysts are an attractive alternative to either the in situ catalysts or the preformed Mo- or W-carbyne complexes. While these catalyst systems have not been utilized for PAE formation they show great promise in alkyne metathesis.

3 PAEs with Novel Structures and Topologies

The most vibrant development in the field of PAEs is the "explosion" of their structural features. There are conceptually different ways of how to vary PAE structures. The first approach deals only with PPEs, and the structural modification is introduced by the variation of the side chains. This concept is powerful for introducing macromolecular, amphiphilic, water-soluble, self-assembling, or protective side chains. In most cases, however, the *electronic* structure of the PPE is untouched. There are exceptions. If the substituent under consideration is conjugated, it will change the electronics of the backbone. Examples of this kind have been prepared by the Bunz and the Swager groups [24, 25].

The second approach involves backbone modification. The introduction of different arenes and/or vinyl groups into the main chain of PAEs and their different (*ortho*/*meta*/*para*) connectivity via the alkyne groups are subsumed in this section. This approach leads to PAEs featuring different electronic and optical properties as compared to the parent PPEs. Red, orange, and blue emitters have been made, as opposed to the PPEs that emit only greenish to yellowish in the solid state and blueish in solution. Depending on the utilized aromatic building block, the electronic properties can be effectively engineered.

3.1 Side-Chain Manipulation

Side-chain manipulation, while leading to a small diversity space with respect to structure, is valuable because PPEs can be made water-soluble, ionic, as Langmuir–Blodgett films, biogenic, and with polymeric or dendritic side chains. End-group modification, while not particularly popular, has also been reported and can change a PPE's properties behavior [26, 27]. Tables 4–6 show side-chain-modified PPEs that have been synthesized during the last 4 years. Alkyl- and alkoxy-substituted PPEs have been re-made, and in some cases the end groups have been changed to give ABA block copolymers [28]. More interesting, however, are developments in side-chain-functionalized PPEs that have been spearheaded by the groups of Aida, Swager, and Bunz [29–37]. Aida and coworkers have prepared a series of dialkoxy PPEs (Table 4, entries 33–35) that carry Frechet-type dendrons of different generations [29–33]. The synthesis of the monomers is straightforward, and the coupling of the dendron-substituted diyne to diiodobenzene gives "jacketed" PPEs. Swager et al. have made insulated PPEs, but instead of dendrimers they introduced ipticene-type monomers (Table 4, entries 18–23, 29) [38]. In this approach one of the monomers is an ipticene unit that is sterically demanding. The basic ipticene monomers and polymers were reported by Yang and Swager in 1998 [39] and have been widely exploited in different sensor applications.

The third way of insulating PPE chains has been reported by Bunz et al. (Table 4, entries 5, 6). The monomer 1,4-diiodo-2-methyl-5-(1′-hydroxyeth-2′-yl)benzene was treated with 20 equivalents of caprolactone in the presence of a tin(II) carboxylate to give a macromonomer (Scheme 3). Reaction of this macromonomer with acetylene gas in the presence of a Pd catalyst gave the desired grafted polymer in almost quantitative yields and with a high degree of polymerization (Table 4, entry 6). Zhu and Swager have recently reported grafted polymers by postfunctionalization of a suitable PPE [34].

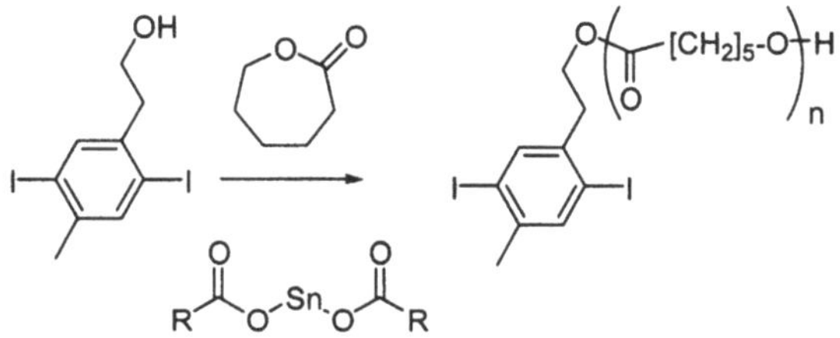

Scheme 3 Synthesis of the macromonomer

Other exploits have focused on the electronic manipulation of the PPEs via side chains (Table 4, entries 7–16) [36]. If a conjugated side chain is utilized, then the electronic properties of the resulting PPE are influenced. The synthesis of such cross-conjugated PPE–PPV derivatives involves the Pd-catalyzed coupling of a diethynylbenzene derivative with a 1,4-distyryl-2,5-diiodobenzene. The electronic properties of such PPEs are variable by changing the styryl side chain (Table 4, entries 7–12). The polymers are attractive as active layers in light-emitting diodes (LEDs) and thin-film transistors.

Water-soluble PPEs have been made by the suitable choice of ionic and highly hydrophilic side chains. These PPEs will play a role in bio-type sensing applications (*vide infra*). A clever example is Schanze's synthesis of a sulfonato-substituted PPE [40, 41]. The NaOH-promoted reaction of [1, 2]oxathiolane-2,2-dioxide with 2,5-diiodohydroquinone gives a bis-sulfonate that can be coupled (Pd-catalyzed) to 1,4-diethynylbenzene to form a water-soluble PPE (Scheme 4); Table 6, entry 1). In a similar approach, Schanze et al. [42] made a

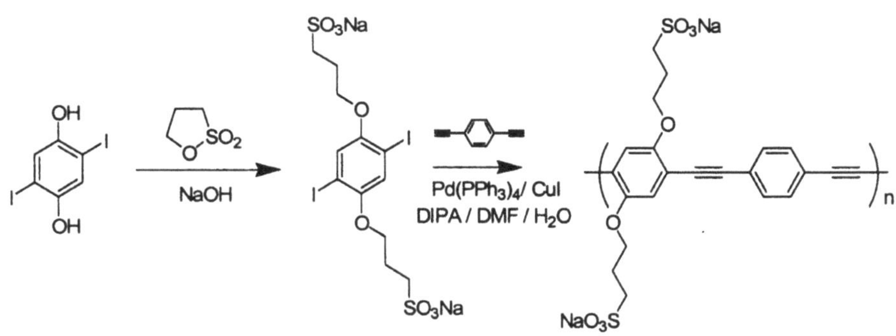

Scheme 4 Synthesis of a water-soluble PPE derivative by Schanze et al

Table 6 Water- soluble PPEs containing only *para*-phenylene units in the main chain

	Diyne	Dihalide	Catalyst	Polymer	Yield (%) M_n PDI
1[a]	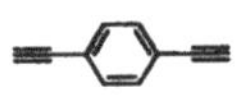	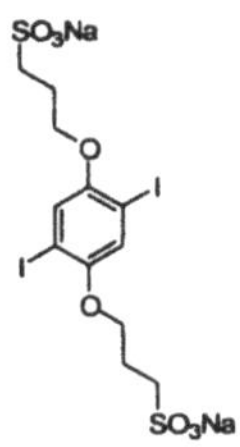	$Pd(PPh_3)_4$ CuI DIPA/DMF/ H_2O	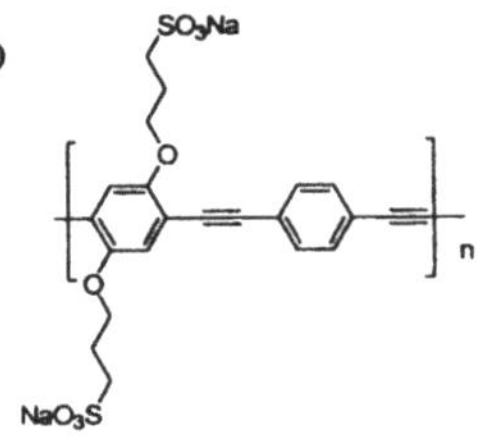	69% $M_w = 10^5$ according to ultrafiltration, No polydispersity available
2[b]	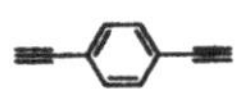	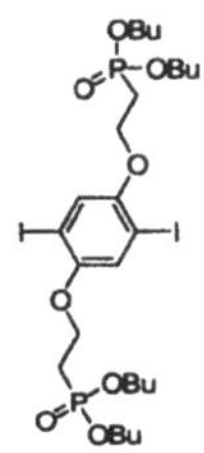	$Pd(PPh_3)_4$ CuI, DIPA, DMAC then Me_3SiBr in Et_3N followed by NaOH/MeOH	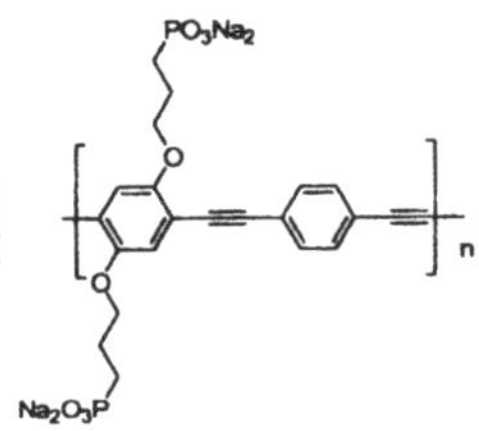	49% 55×10^3 1.98
3[c]	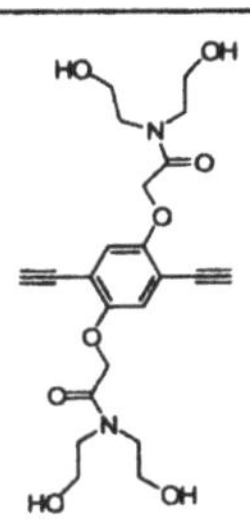	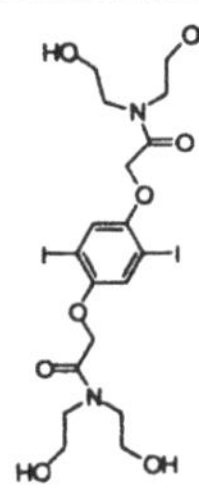	$(Ph_3P)_4Pd$ CuI, toluene, morpholine	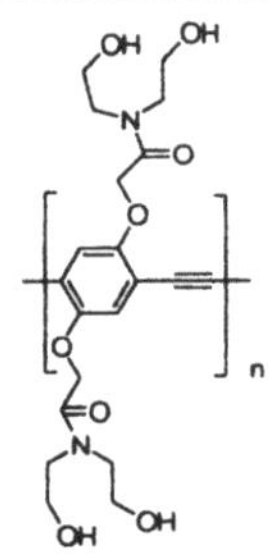	nd 32×10^3 1.40
4[c]	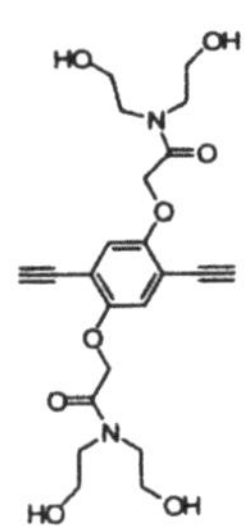	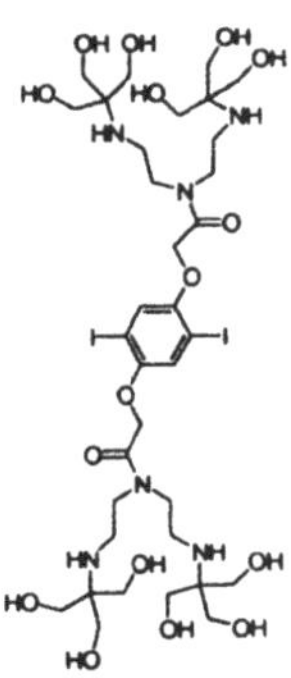	$(Ph_3P)_4Pd$ CuI, toluene, morpholine	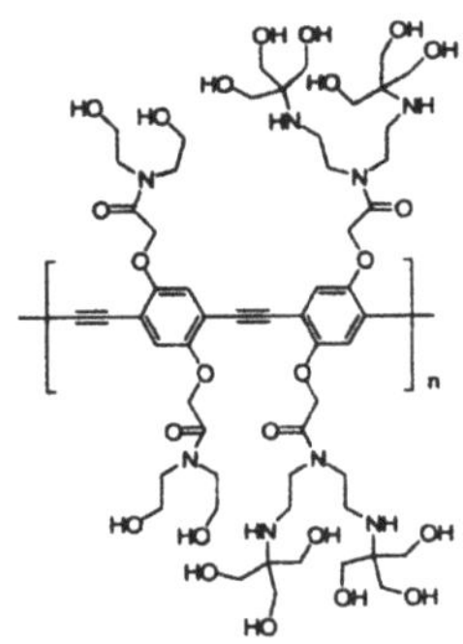	nd 13×10^3 1.90 only slightly soluble

Table 6 (continued)

	Diyne	Dihalide	Catalyst	Polymer	Yield (%) M_n PDI
5[d]	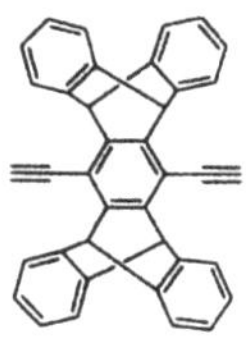	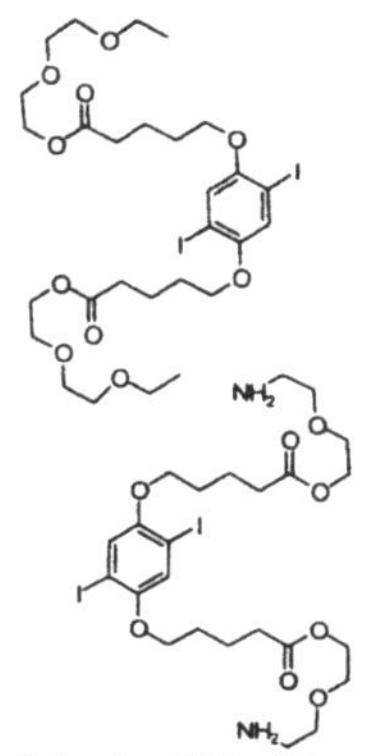 2:3 mix of diiodides	$(Ph_3P)_4Pd$ CuI, toluene, morpholine	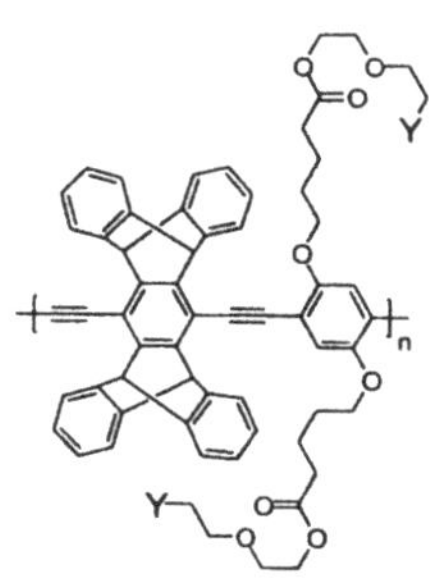 Y = $-NH_2$ or $-O-CH_2-CH_3$	67% 26×10^3 1.8 LiBr in DMF for GPC
6[e]	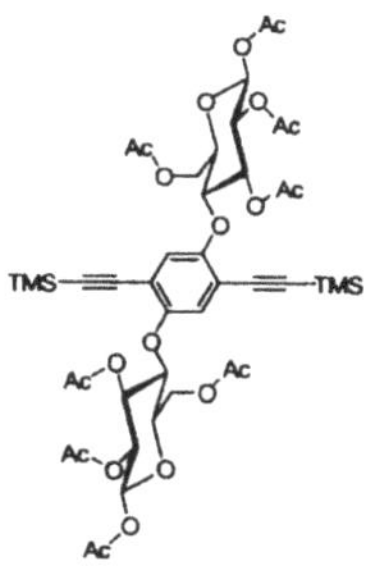	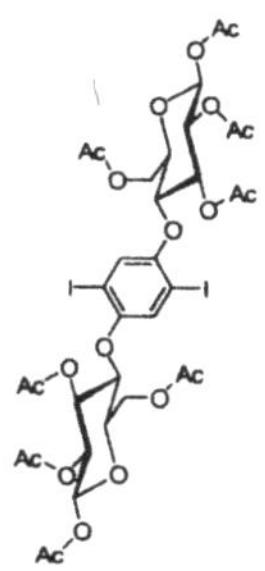	$(Ph_3P)_4Pd$ Ag_2O, THF 60 °C, then KOH/MeOH	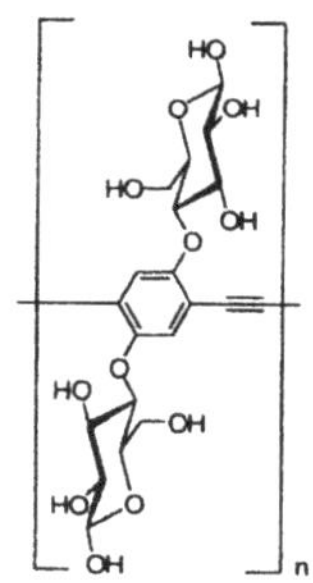	86% 5.0×10^3 1.4
7[e]	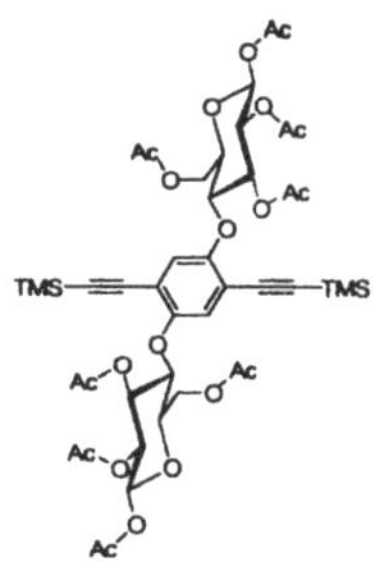	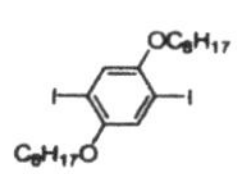	$(Ph_3P)_4Pd$ Ag_2O, THF 60 °C, then KOH/MeOH	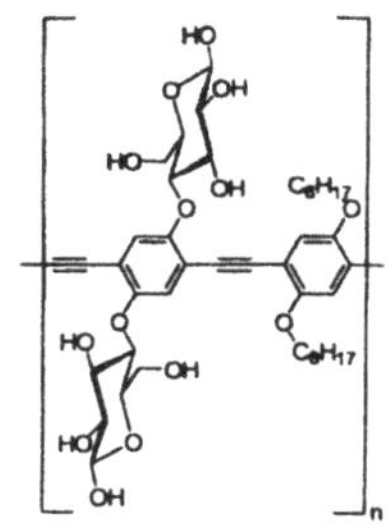	86% 20×10^3 1.7
8[e]	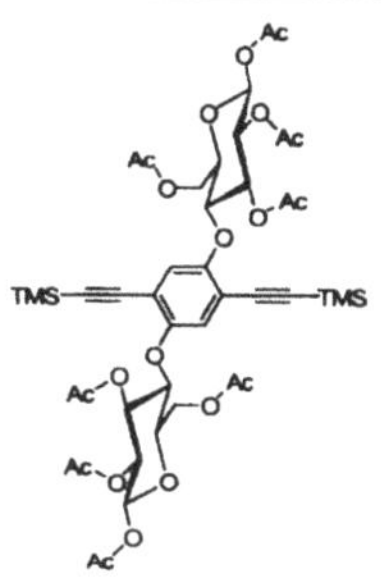	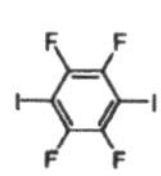	$(Ph_3P)_4Pd$ Ag_2O, THF 60 °C, then KOH/MeOH	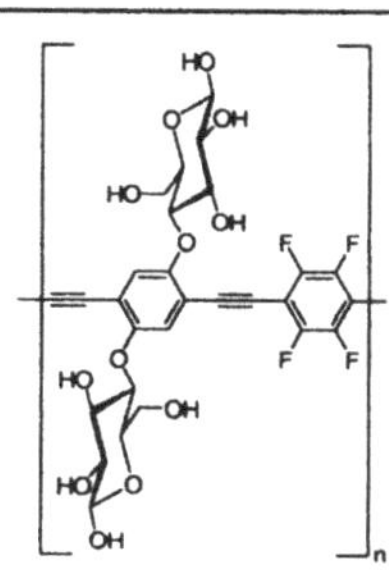	90% 4.2×10^3 3.5

Table 6 (continued)

	Diyne	Dihalide	Catalyst	Polymer	Yield (%) M_n PDI
9[e]	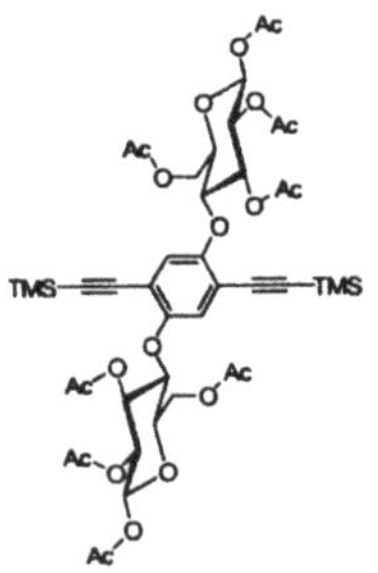	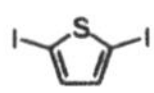	$(Ph_3P)_4Pd$ Ag_2O, THF 60 °C, then KOH/MeOH	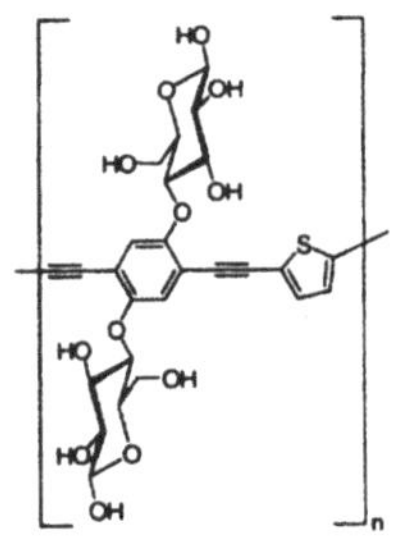	90% 2.6×10^3 6.4
10[e]		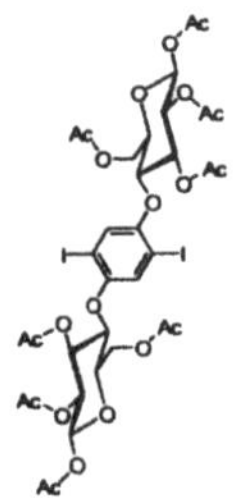	$(Ph_3P)_4Pd$ Ag_2O, THF 60 °C, then KOH/MeOH	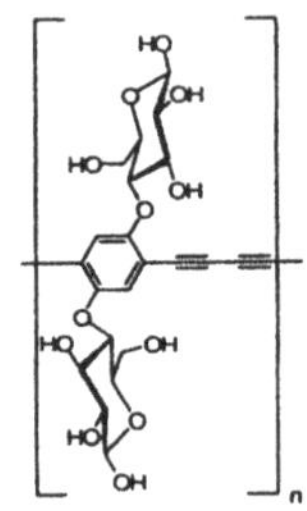	95% 12×10^3 2.4
11[f]	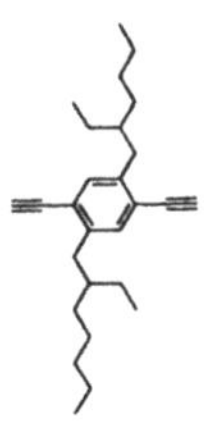	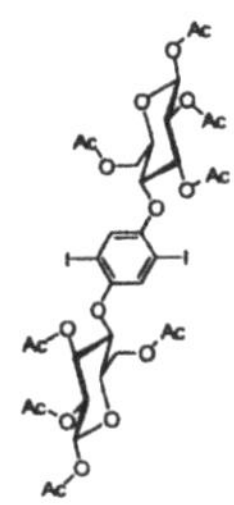	$(Ph_3P)_2PdCl_2$ CuI, piperidine toluene 50 °C 24h	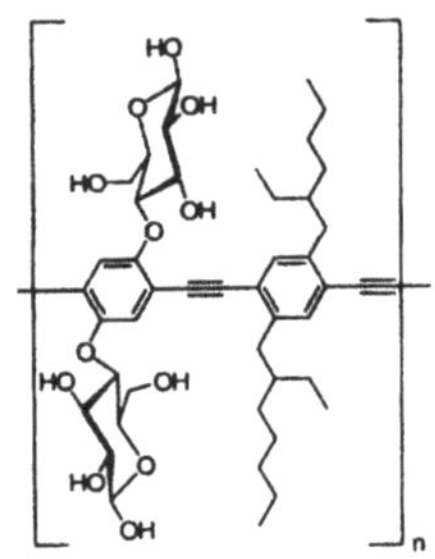	99% $P_n = 75$ 2.2

[a] Tan C, Pinto MR, Schanze KS (2002) Polym Prepr 43:126–127; Tan C, Pinto MR, Schanze KS (2002) Chem Commun 446–447.
[b] Pinto MR, Reynolds JR, Schanze KS, (2002) Polym Prepr 43:139–140; Pinto MR, Kristal BM, Schanze KS (2003) Langmuir 19:6523–6533.
[c] Kuroda K, Swager TM (2003) Chem Commun 26–27.
[d] Moon JH, Deans R, Krueger E, Hancock LF (2003) Chem Commun 104–105.
[e] Babudri F, Colangiuli D, DiLorenzo PA, Farinola GM, Omar OH, Naso F (2003) Chem Commun 130–131.
[f] Erdogan B, Wilson JM, Bunz UHF (2002) Macromolecules 35:7863–7864.

phosphonate monomer that furnished a phosphonate-substituted PPE (Table 6, entry 2). If charges are not desirable to make the PAE water soluble, it is possible to utilize a mini amine dendrimer (Table 6, entries 3, 4) as comonomer to induce water solubility [43, 44]. Alternatively, attached glucose residues, as shown by Bunz et al. and by Naso et al., can be utilized toward that end (Table 6, entries 6–11) [45, 46]. The chemistry and use of water-soluble PPEs and PAEs will experience growth in the near future due to the imminent usefulness of these materials in sensory and bio-type applications.

3.2
Main-Chain Manipulation

Main-chain manipulation offers an opportunity to dramatically change the electronic and physical properties of the PAEs. Popular approaches are the introduction of *meta* linkages into the polymers, the introduction of aromatic hydrocarbons other than benzene, the introduction of heterocycles, and the substitution of a fraction of the connecting alkyne groups by double bonds. The last strategy leads to polymers that are hybrids between PPEs and PPVs.

3.3
Meta PPEs and Related Compounds

Utilizing *meta*-substituted diynes and dihalides, Table 7 shows that these building blocks can be incorporated into kinked PAEs in a mix-and-match approach. While the molecular weights of the *meta* PPEs are usually not as high as those obtained for the *para* PPEs, they are decent. A common side reaction in the synthesis of the *meta* polymers is the formation of cyclic oligomers, as has been shown early on [47]. From a point of synthesis, it is remarkable that diazo groups as well as nitroxide radicals survive the Pd-catalyzed coupling conditions (Table 7, entry 3). An interesting development in this area is Arnt and Tew's recent contributions that allow water-soluble *meta* PPEs to be made utilizing a *t*-Boc protecting-group strategy [48, 49]. PPEs that have both *para* and *meta* linkages (Table 7, entries 4–8) have been made by Pang et al. and other groups [50–53], and these seem to be valuable as emitter materials in organic LEDs. They combine blue emission with facile processing, because these materials are not that rigid. They are significantly more soluble than similar *para* PPEs. The *meta* units at the same time interrupt conjugation and lead to the sought-after blue emission in the solid state.

3.4
PAEs Containing Aromatic Units Other Than Benzene in the Main Chain

In Table 8 all-hydrocarbon PAE backbones are shown, while in Table 9 PAEs with heterocyclic backbones are listed. Alkyne metathesis is a powerful tool to make poly(fluorenyleneethynylene)s (PFEs, Table 8, entries 1–7). The

Table 7 PPEs containing *meta*-phenylene units in the main chain

	Diyne	Dihalide	Catalyst	Polymer	Yield (%) M_n, PDI
1 [a]	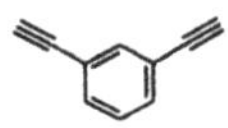	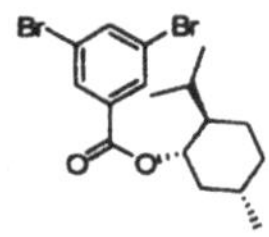	$(Ph_3P)_2PdCl_2$ PPh_3, CuI, NEt_3, THF	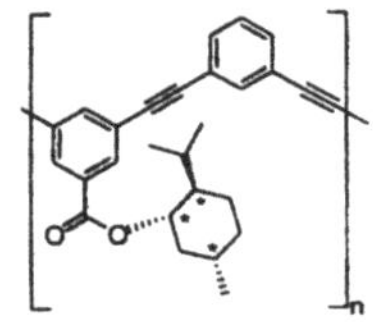	100 $1.6x10^3$ 1.9
2 [a]	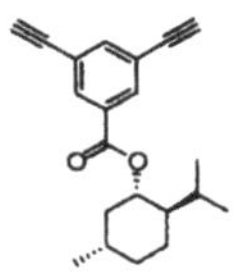	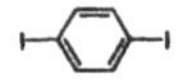	$(Ph_3P)_2PdCl_2$ PPh_3, CuI, NEt_3, THF	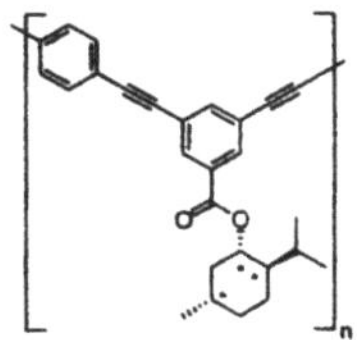	98 $5.3x10^3$ 1.6
3 [b]	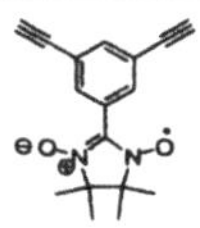	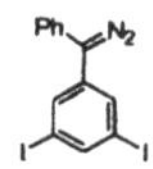	$(Ph_3P)_2PdCl_2$ CuI, NEt_3, pyridine, 3h	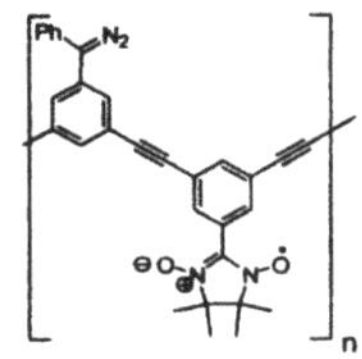	nd $15x10^3$ nd
4 [c]	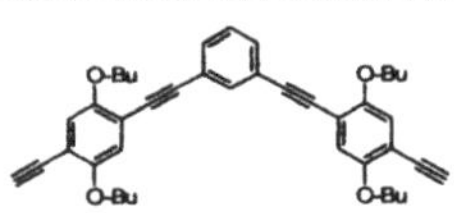	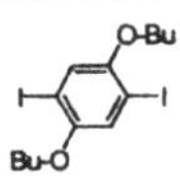	$(Ph_3P)_2PdCl_2$ CuI, NEt_3 toluene	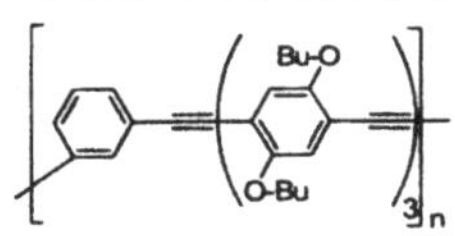	80 $55x 10^3$ 2.2
5 [c]	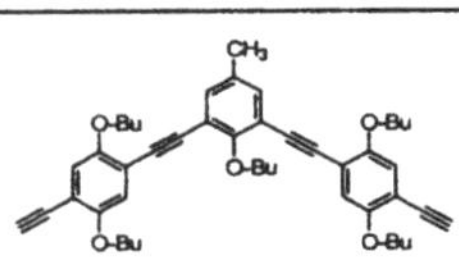	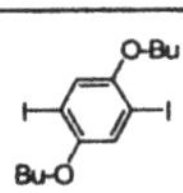	$(Ph_3P)_2PdCl_2$ CuI, NEt_3 toluene	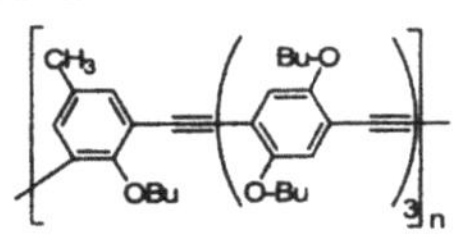	78 $20x 10^3$ 1.8
6 [d]	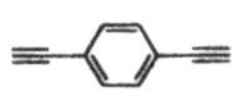	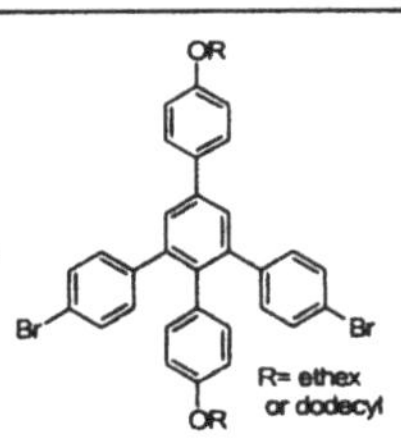	$(Ph_3P)_2PdCl_2$ CuI, NEt_3, PPh_3, toluene	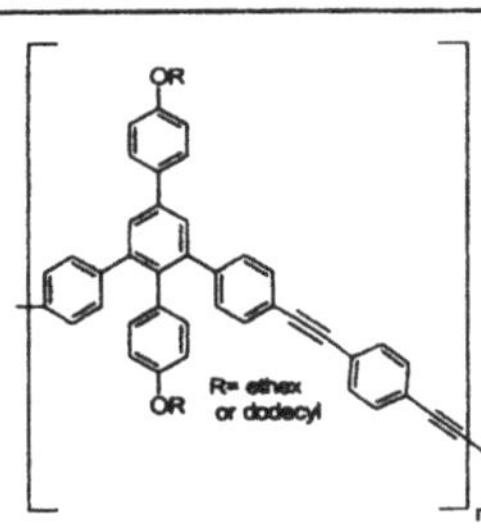	Ethex 82 $11x 10^3$ 2.1
7 [d]	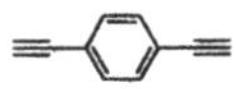				Dodec 90 $8.4x 10^3$ 2.6
8 [e]	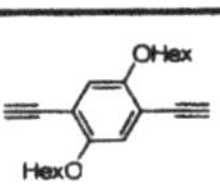	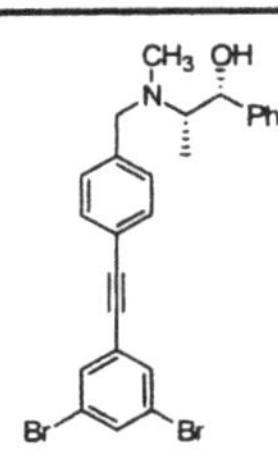	$(Ph_3P)_4Pd$ CuI, NEt_3, THF, 24 h reflux	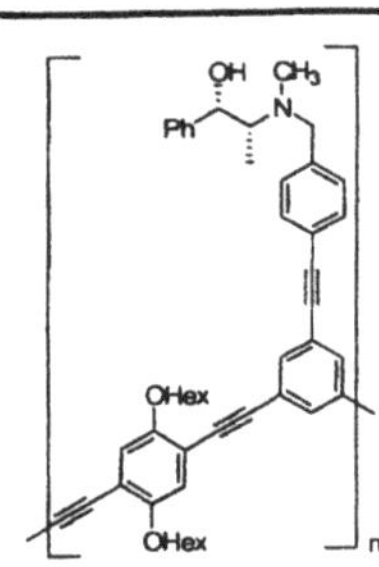	49 $4.9x10^3$ 1.44

Table 7 (continued)

	Diyne	Dihalide	Catalyst	Polymer	Yield (%) M_n, PDI
9[f]	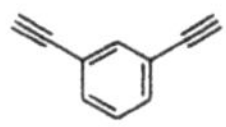	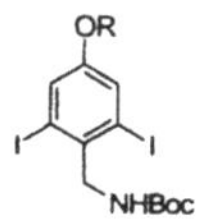	$(Ph_3P)_4Pd$ CuI, DIPA, PPh_3,toluene 10h, 65 °C then HCl/dioxane	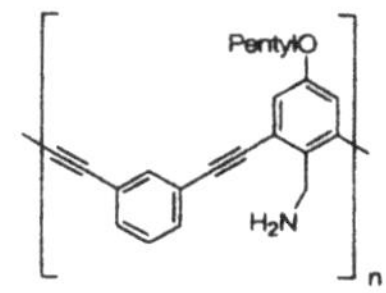	R= pentyl 52 $1.8x10^4$ 2.2
10[f]	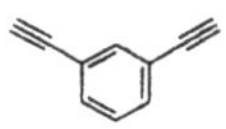	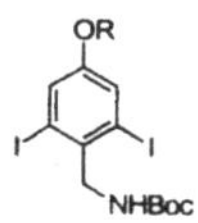	$(Ph_3P)_4Pd$ CuI, DIPA, PPh_3,toluene 10h, 65 °C then HCl/dioxane	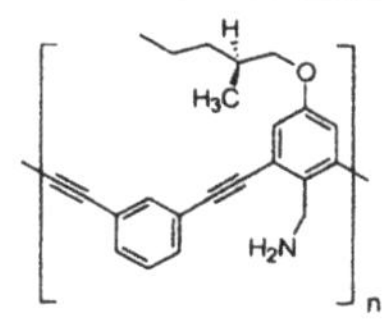	R = (S)-2-me-butyl $1.2x10^4$ 1.3
11[f]	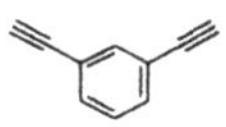	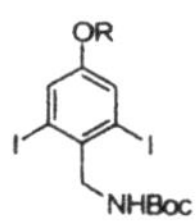	$(Ph_3P)_4Pd$ CuI, DIPA, PPh_3,toluene 10h, 65 °C then HCl/dioxane	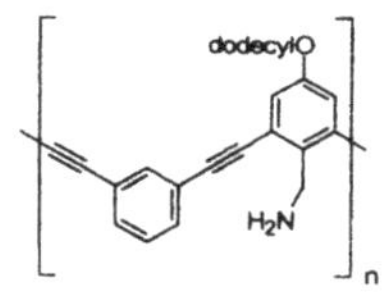	R= dodec 41 $1.0 x10^4$ 1.4
12[g]	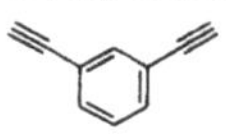	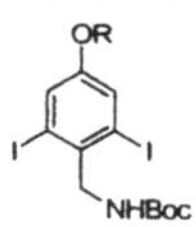	$(Ph_3P)_4Pd$ CuI, DIPA, PPh_3,toluene 10h, 65 °C then HCl/dioxane	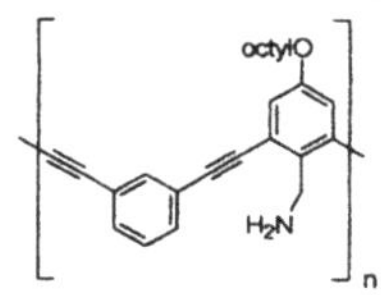	R = oct na $7.7 x10^3$ 1.3
13[h]	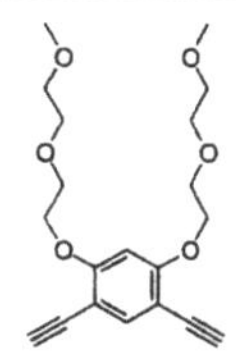	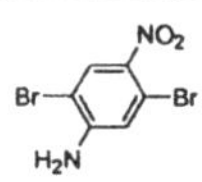	$(Ph_3P)_2PdCl_2$ CuI, NEt_3, THF	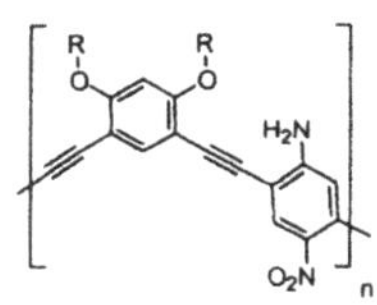	76 $1.7 x10^4$ 1.35

[a] Shinora KI, Aoki T, Kaneo T, Oikawa E, (2001) Polymer 42:351–355.
[b] Akita T, Koga N, (2001) Polyhedron 20:1475–1477.
[c] Chu Q, Pang Y, Ding L, Karasz FE (2002) Macromolecules 35:7569–7574.
[d] Spiliopoulos IK, Mikroyannidis JA (2002) J Polym Sci A 40:1449–1455.
[e] Sun C, Hu QS (2002) Polym Prepr 43:692–693.
[f] Arnt L, Tew GN (2002) J Am Chem Soc 124:7664–7665.
[g] Arnt L, Tew GN (2003) Langmuir 19:2404–2408.
[h] Koishi K, Ikeda T, Kondo K, Sakaguchi T, Kamada K, Tawa K, Ohta K (2001) Macromol Chem Phys 201:525–532.

Table 8 PAEs containing aromatic units (other than benzene) in the main chain

	Monomer 1	Monomer 2	Catalyst	Polymer	Yield (%) M_n PDI
1[a]	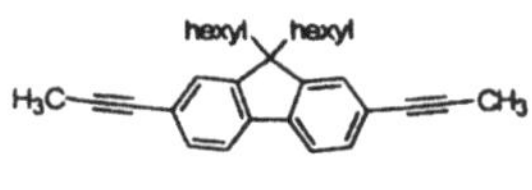		$Mo(CO)_6$ 4-chlorophenol 140 °C, 15 h	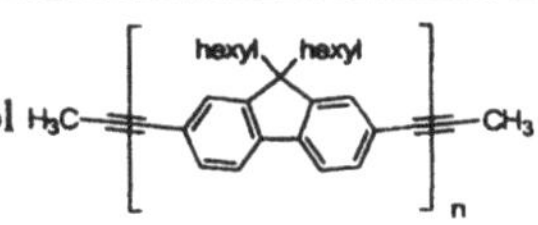	Hexyl 23, 14 $\times 10^3$ 4.2
2[a]	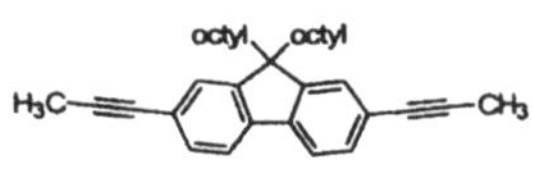		$Mo(CO)_6$ 4-chlorophenol 140 °C, 15 h	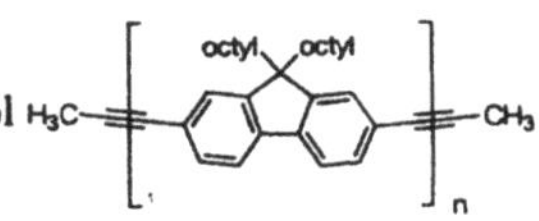	Octyl 74, 11 $\times 10^3$ 5.8
3[a]	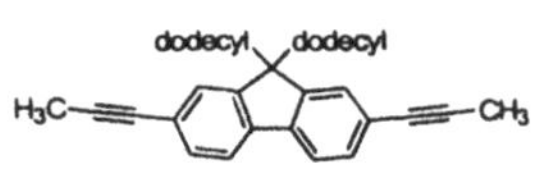		$Mo(CO)_6$ 4-chlorophenol 140 °C, 15 h	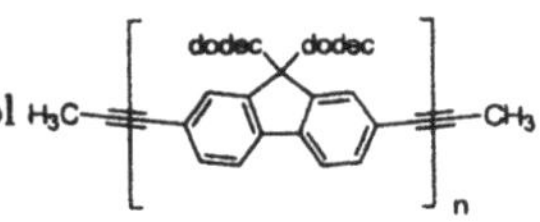	Dodecyl 69, 10 $\times 10^3$ 4.4
4[a]	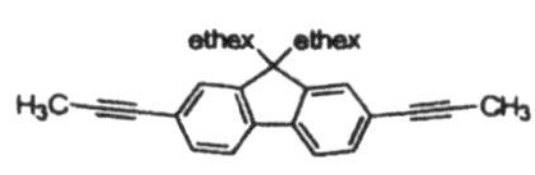		$Mo(CO)_6$ 4-chlorophenol 140 °C, 15 h	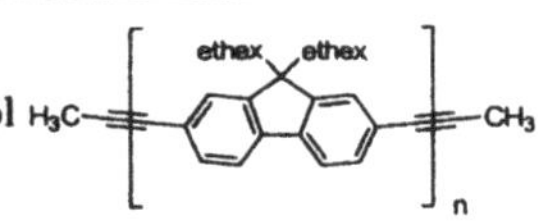	Ethylhexyl 34, 17 $\times 10^3$ 3.6
5[a]	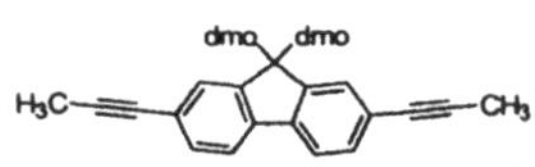 dmo = (S)-3,7-dimethyloctyl		$Mo(CO)_6$ 4-chlorophenol 140 °C, 15 h	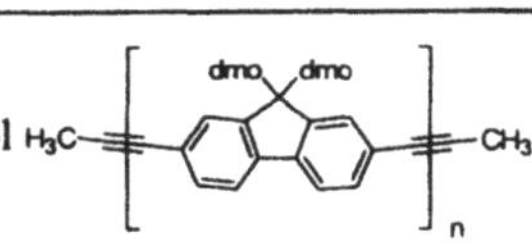	dmo 65, 10 $\times 10^3$ 5.4
6[b]	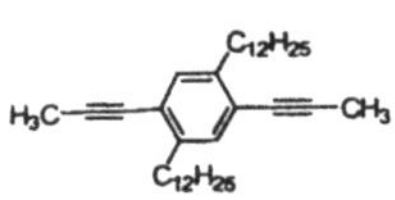 90%	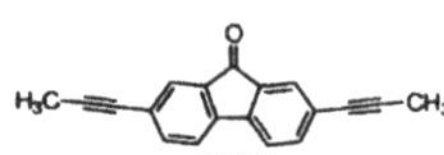 10%	$Mo(CO)_6$ 4-chlorophenol 140 °C, 15 h	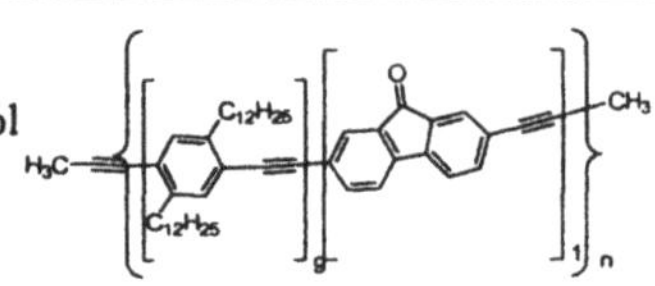	75, 14 $\times 10^3$ 8.8
7[b]	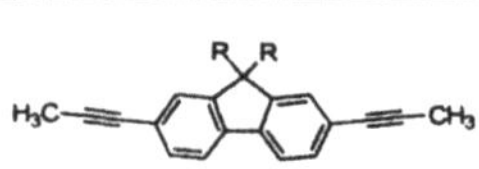 90 %	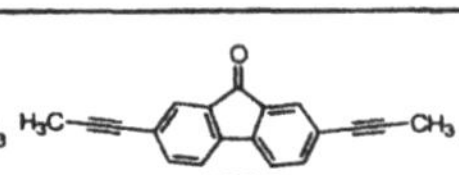 10%	$Mo(CO)_6$ 4-chlorophenol 140 °C, 15 h	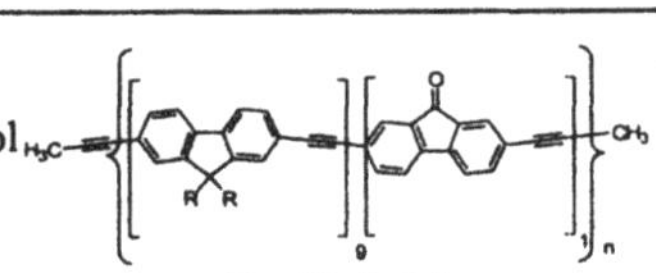 R = Dodecyl	90, 5 $\times 10^3$ 3.5
8[c]	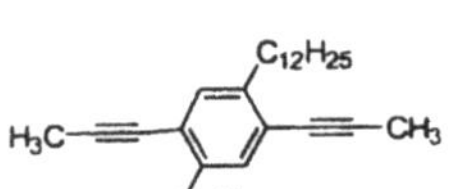 80%	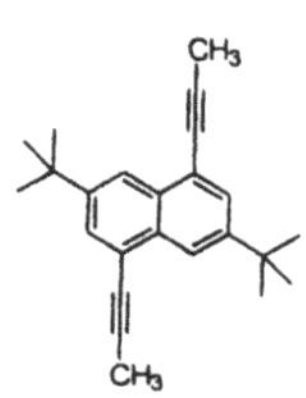 20%	$Mo(CO)_6$ 4-chlorophenol 140 °C, 15 h	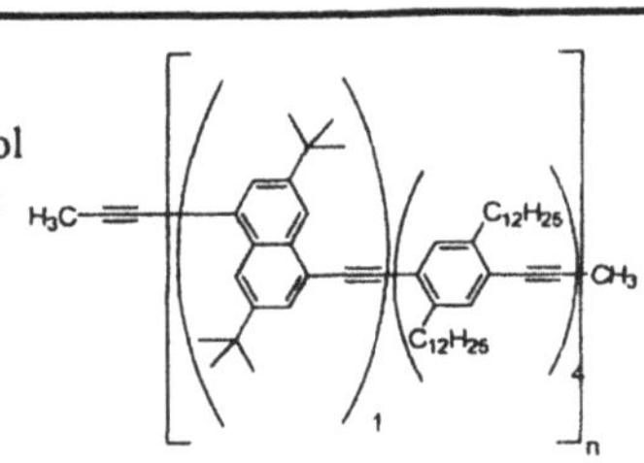	86 $P_n = 96$ 2.9

Table 8 (continued)

	Monomer 1	Monomer 2	Catalyst	Polymer	Yield (%) M_n PDI
9[d]	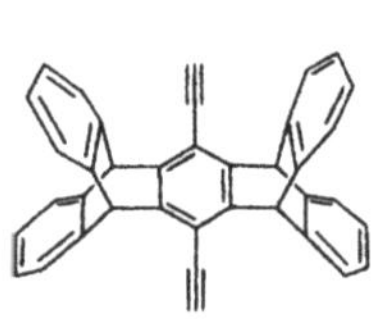	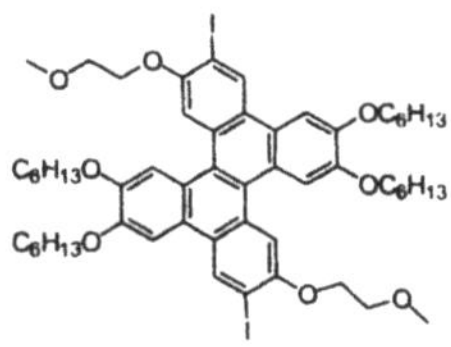	$(Ph_3P)_2PdCl_2$ CuI/toluene DIPA mix 60 °C 4 d	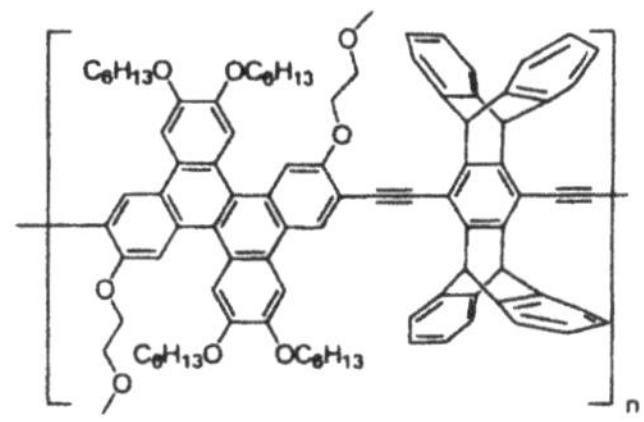	91 19 x 10^3 4.0
10[d]	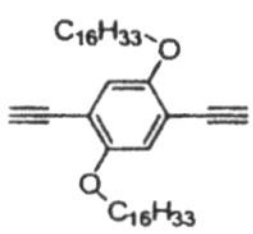	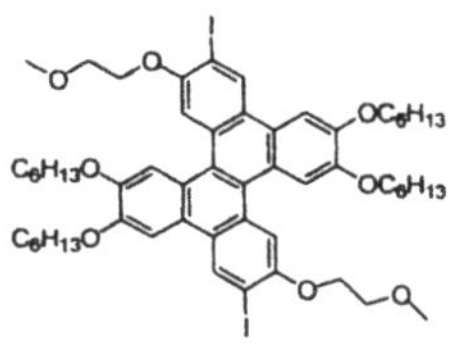	$(Ph_3P)_2PdCl_2$ CuI/toluene DIPA mix 60 °C 4 d	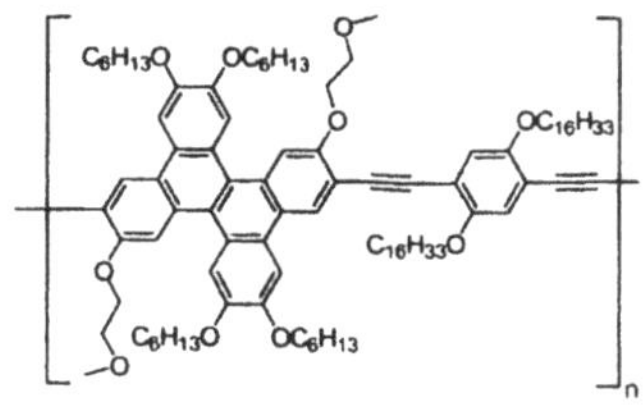	77 17 x 10^3 2.5
11[d]	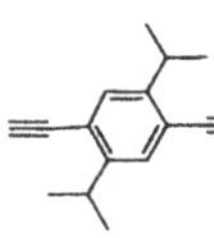	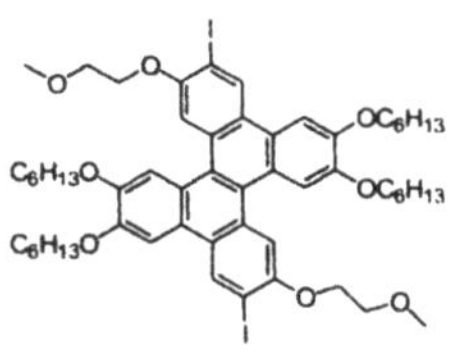	$(Ph_3P)_2PdCl_2$ CuI/toluene DIPA mix 60 °C 4 d	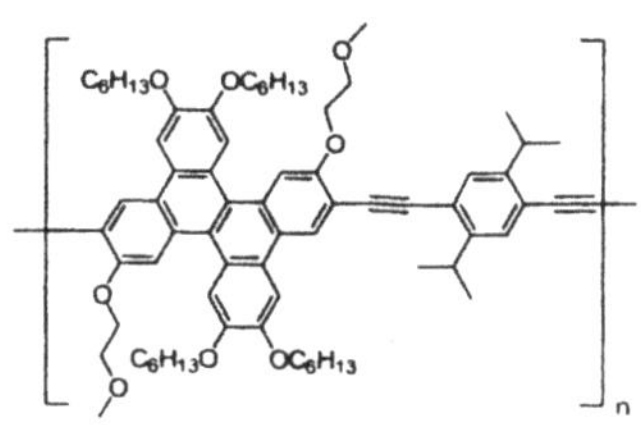	31 20 x 10^3 1.8
12[e]	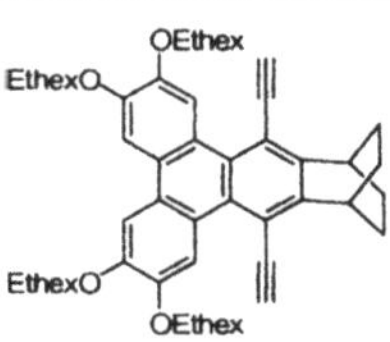	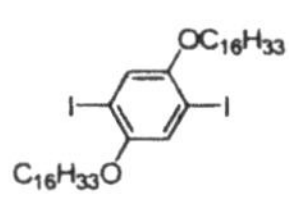	$(Ph_3P)_4Pd$ CuI/toluene DIPA mix	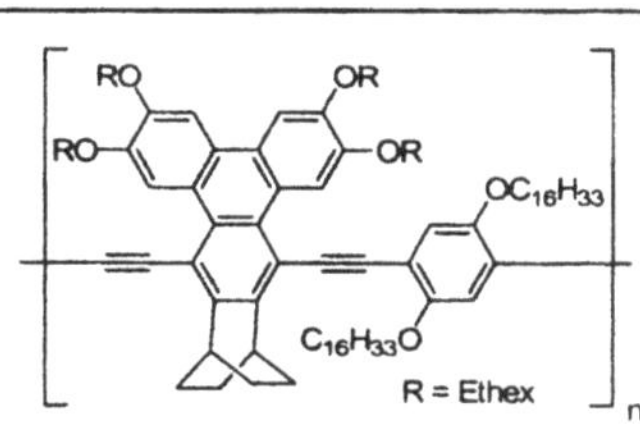	nd 45 x 10^3 2.5
13[e]	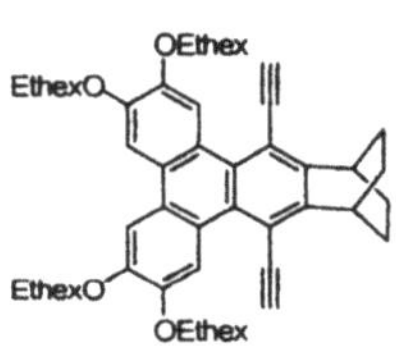	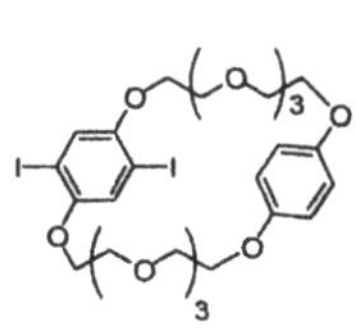	$(Ph_3P)_4Pd$ CuI/toluene DIPA mix	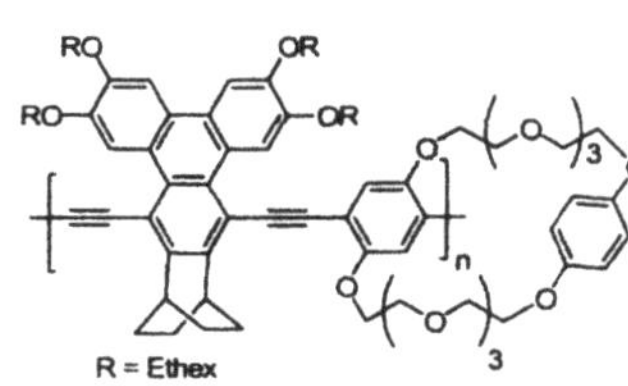	Nd 103 x 10^3 1.9

[a] Pschirer NG, Bunz UHF (2000) Macromolecules 33:3961–3963.
[b] Pschirer NG, Byrd K, Bunz UHF (2001) Macromolecules 33:8590–8591.
[c] Pschirer NG, Vaughn ME, Dong YB, zur Loye HC, Bunz UHF (2000) Chem Commun 85–86.
[d] Yamaguchi S, Swager TM (2001) J Am Chem Soc 123 12087–12088.
[e] Rose A, Lugmair CG, Swager TM (2001) J Am Chem Soc 123 11298–11299.

monomers are obtained by electrophilic iodination of fluorene, attachment of the solubilizing side chains via a nucleophilic phase-transfer alkylation, and Pd-catalyzed propynylation [54]. The entries 6–8 of Table 8 show that copolymers of poly(fluorenyleneethynylene) can be made by alkyne metathesis. Simple mixing of the monomers in the presence of $Mo(CO)_6$ and a suitable phenol suffices. Surprisingly, it is possible to incorporate dipropynylfluorenone into the formed PFEs, and if not more than 10 mol% of the fluorenone monomer is utilized, the obtained polymers show excellent molecular weights [55]. In the same approach it is possible to introduce naphthalene units into the PPE main chain by cometathesis of dipropynylnaphthalene and dipropynyldidodecylbenzene (Table 8, entry 8).

Regardless of whether the Pd-catalyzed coupling or alkyne metathesis is utilized to make PAEs, the critical step is the synthesis of the diiodoarene monomers. In this section some of the more interesting syntheses are showcased. The synthesis of dipropynyldi-*tert*-butylnaphthalene is shown in Scheme 5. Starting from naphthalene, Friedel–Crafts alkylation with 2-chloro-2-methylbutane gives a mixture of two di-*tert*-butylnaphthalenes that are separated by crystallization. Iodination of the correct isomer is followed by a Pd-catalyzed coupling of propyne to the diiodide to give the desired 1,5-dipropynyl-3,8-di-*tert*-butylnaphthalene [56] ready for ADIMET.

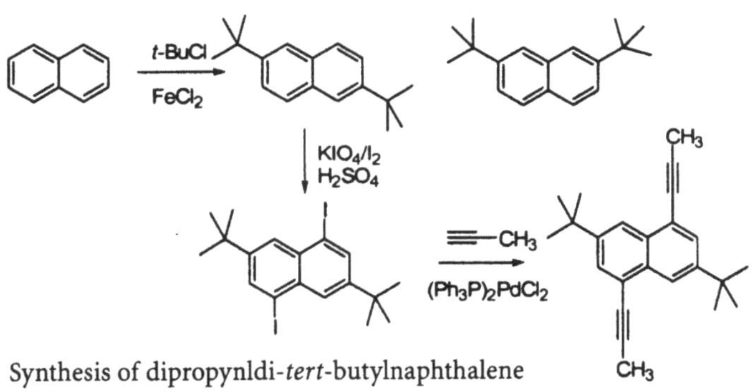

Scheme 5 Synthesis of dipropynldi-*tert*-butylnaphthalene

Scheme 6 shows Yamaguchi and Swager's synthesis of a decorated diiododibenzochrysene [57]. Starting from a suitable diphenyltolane precursor,

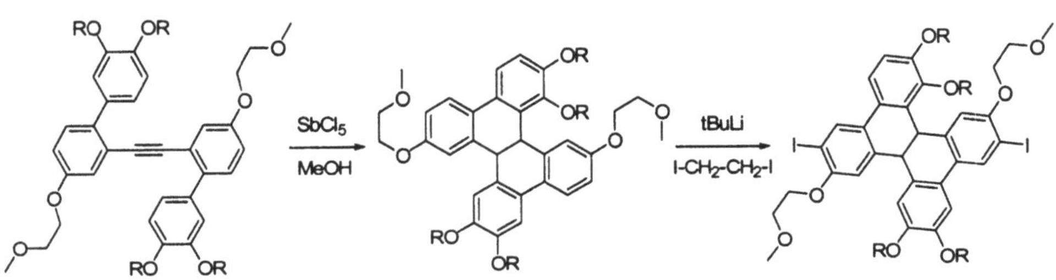

Scheme 6 Synthesis of a diiododibenzochrysene by Swager et al.

oxidative cyclization leads to the desired dibenzochrysene. Double *ortho* metalation by *tert*-BuLi is cleverly directed by the ethylene glycol ether to give the desired diiododibenzochrysene after workup with 1,2-diiodoethane. This monomer can then be coupled to diynes to give electronically tailored PAEs (Table 8, entry 9).

A second example from the same group is the synthesis of an elaborate diethynyltriphenylene derivative (Scheme 7; Table 8, entries 12, 13) [58]. Zn/Pd-promoted homocoupling of a 4-iodo-1,2-dialkoxybenzene furnishes the desired tetraalkoxybiphenyl, an electron-rich aromatic system. Iron trichloride-catalyzed Friedel–Crafts arylation of the biphenyl derivative with dimethoxybenzene furnishes an unsymmetrical triphenylene derivative. Deprotection, oxidation, and subsequent Diels–Alder reaction with cyclohexadiene is followed by catalytic hydrogenation and reoxidation. TMS–CC–Li attack on the quinone delivers the alkyne modules, treatment with $SnCl_2$ aromatizes the six-membered ring, while KOH in MeOH removes the TMS groups cleanly to give the elaborate monomer.

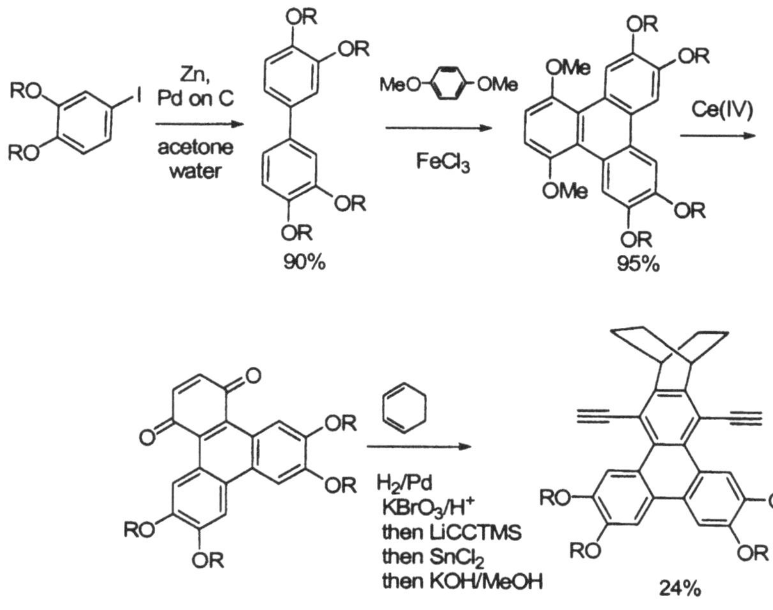

Scheme 7 Synthesis of a triphenylene diiodide by Swager et al.

The synthesis of the heterocyclic diiodides can be tricky (Scheme 8). Direct iodination of quinoline was reported by Kiamuddin et al. [59] to give 5,8-diiodoquinoline. Their procedure did not furnish any products according to Bunz et al., but classical electrophilic iodination leads to a single product that was identified not to be the 5,8-isomer but 3,6-diiodoquinoline, according to the X-ray crystal structure of a diethynylated derivative [60].

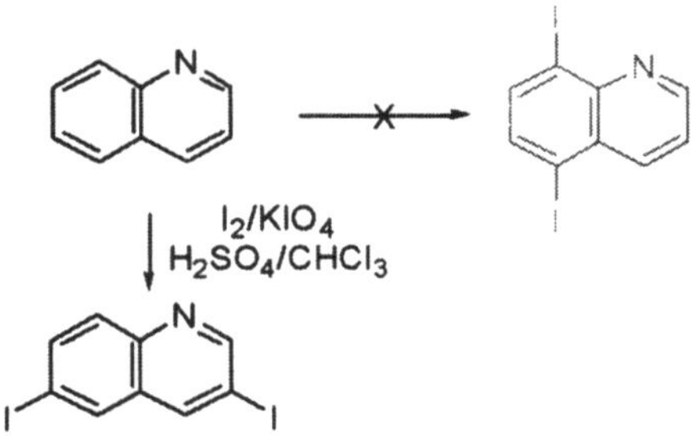

Scheme 8 Synthesis of 3,6-diiodoquinoline

Attempts to directly iodinate quinoxaline failed, and the synthesis of 2,3-diphenyl-5,8-dibromoquinoxaline is somewhat more involved (Scheme 9) [61]. Starting from *ortho*-phenylenediamine, reaction with $SOCl_2$ gives benzothiadiazole in high yield. Bromination in HBr furnishes 4,7-dibromobenzothiadiazole, which can be alkynylated or directly reduced [62]. Reduction of the dibromide with sodium borohydride leaves the halide substituents unmolested but opens the ring to furnish 1,4-dibromo-2,3-diaminobenzene. Reaction of this intermediate with a 1,2-dione furnishes a 2,3-disubstituted 5,8-dibromoquinoxaline. Pd-catalyzed alkynylation finishes the sequence off and removal of the TMS groups yields the desired 5,8-diethynylquinoxaline monomers (Table 9, entries 13, 14).

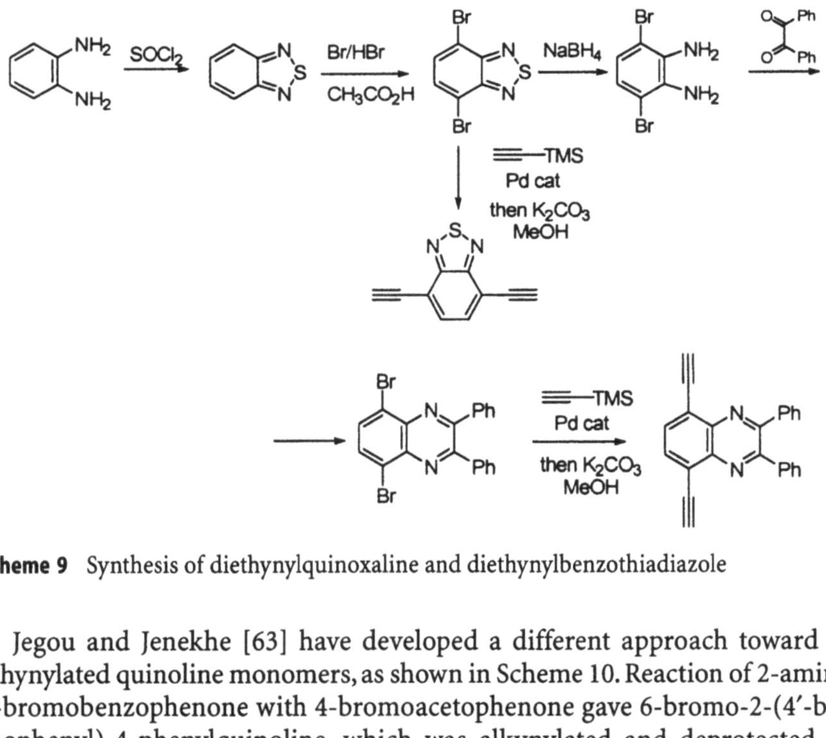

Scheme 9 Synthesis of diethynylquinoxaline and diethynylbenzothiadiazole

Jegou and Jenekhe [63] have developed a different approach toward diethynylated quinoline monomers, as shown in Scheme 10. Reaction of 2-amino-5-bromobenzophenone with 4-bromoacetophenone gave 6-bromo-2-(4′-bromophenyl)-4-phenylquinoline, which was alkynylated and deprotected ac-

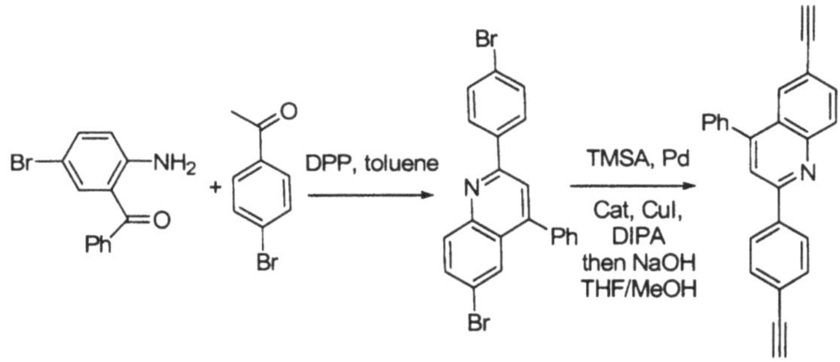

Scheme 10 Jenekhe's synthesis of an alkynylated quinoline monomer

cording to standard procedures. The dialkynylated monomer was coupled to a 2,5-dibromothiophene derivative to give the desired PAE (Table 9, entries 8,9).

The heterocyclic PAEs are useful for low-bandgap applications, as n-type semiconductors, and in sensory applications. Again, as long as the alkynylated or iodinated monomers are available, the synthesis of the corresponding PAE is not a problem, and either the Pd-catalyzed couplings or alkyne metathesis can be utilized toward that end.

Because of the usefulness of PPVs in electrooptical applications, there is a great interest in hybrid structures that contain structural elements of both the PPVs and the PPEs. Table 10 shows the hitherto made PPV–PPE derivatives. Three different types of reactions have been utilized to make these main-chain PPE–PPV hybrids. Alkyne metathesis with in situ catalysts formed from molybdenum hexacarbonyl and 4-trifluoromethylphenol utilizing suitably substituted dipropynylstilbenes cleanly gave PPE–PPV hybrids in high yields [64]. The degree of polymerization in these polymers, P_n=70–150, is quite significant (Table 10, entries 1–5). The double bonds do not react under the metathesis conditions and remain unmolested according to ^{1}H and ^{13}C NMR spectroscopies. The alkyne metathesis process gives easy access to PPE–PPV hybrids of high molecular weight and purity.

The second process that is utilized for making PPE–PPV hybrids is the Heck–Cassar–Sonogashira–Hagihara reaction and it works well for the synthesis of PPE–PPV hybrids (Table 10, entries 5, 17–19) [65–67]. The third method utilizes a Horner olefination reaction (Table 10, entries 6–16). An alkyne-containing dialdehyde is treated with a benzylic bisphosphonate in the presence of potassium *tert*-butoxide in toluene [68, 69]. The yield of the reaction is between 50 and 90% and the degree of polymerization of the obtained polymers varies in the range P_n=12–80 repeat units. The vinyl groups are formed predominantly in the *trans* configuration (92–96%). While the degree of polymerization is not competitive compared to the results obtained by alkyne metathesis, the flexibility of the approach allows the introduction of almost any functional bisaldehyde into these polymers. Several more dialdehydes (see Table 10, entry 12) have been utilized to generate complex PPE–PPV hybrids by this valuable method.

Table 9 PAEs containing heterocyclic moieties in the main chain

	Monomer 1	Monomer 2	Catalyst	Polymer	Yield (%) M_n, PDI
1[a]	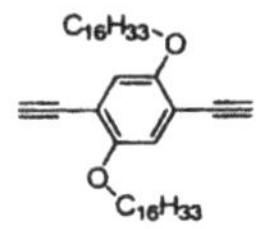	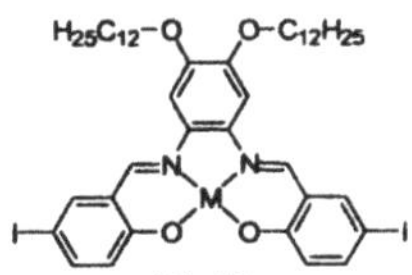 M : Zn	$(Ph_3P)_4Pd$, CuI, THF, DIPA, 70 °C 24 h	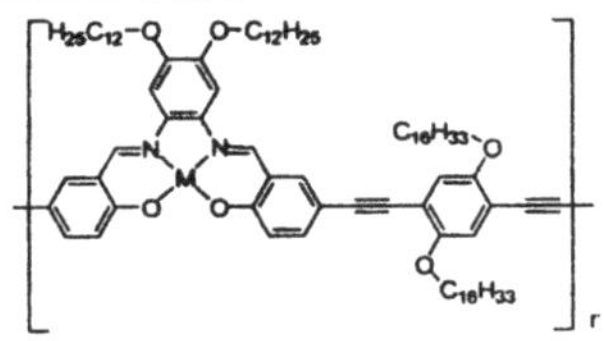	86 12.2 x 10^3 3.0
3[a]	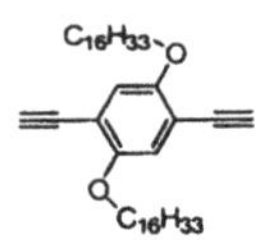	M : Ni	$(Ph_3P)_4Pd$, CuI, THF, DIPA, 70 °C 24 h	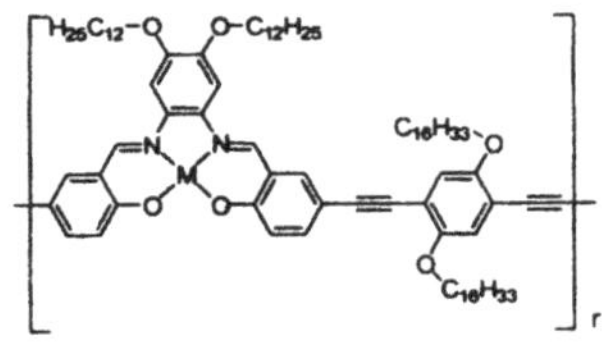	83 9.6 x 10^3 2.1
4[a]	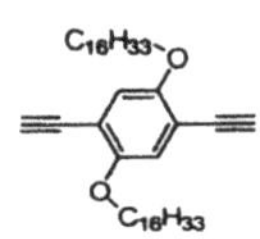	M : V=O	$(Ph_3P)_4Pd$, CuI, THF, DIPA, 70 °C 24 h	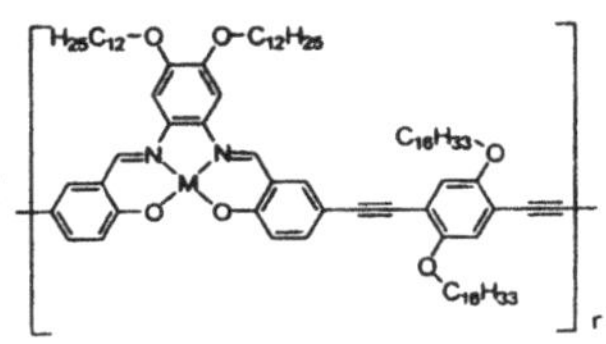	77 21.5 x 10^3 3.9
5[a]	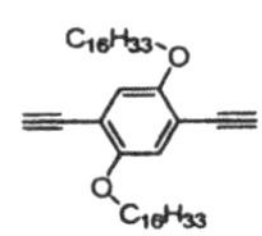	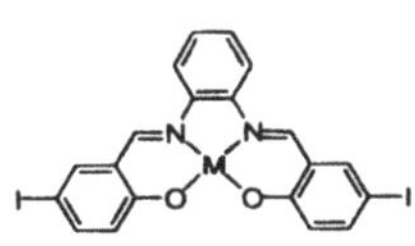 M = Zn	$(Ph_3P)_4Pd$, CuI, THF, DIPA, 70 °C 24 h	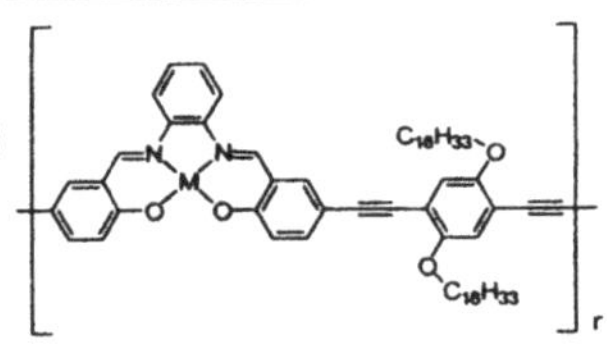	insol. NA NA Two more examples
6[b]	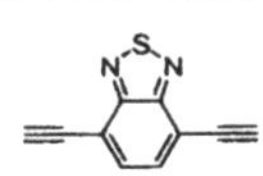	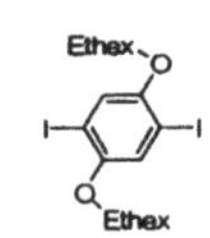	$(Ph_3P)_2PdCl_2$ PPh_3, CuI, piperidine	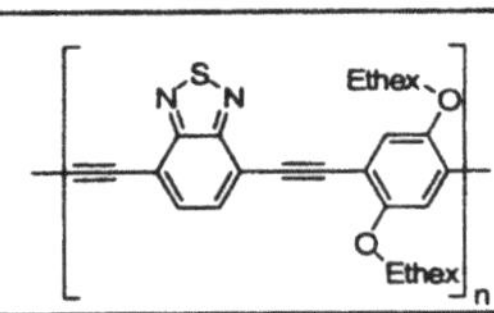	58 5.5 x 10^3 2.78
7[c]	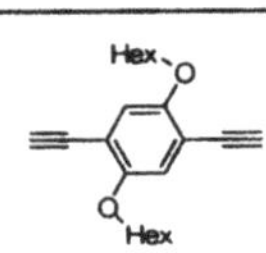	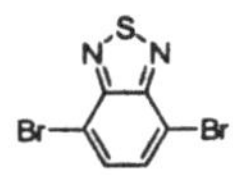	$(Ph_3P)_4Pd$, CuI, NEt_3 60 °C, 5 h	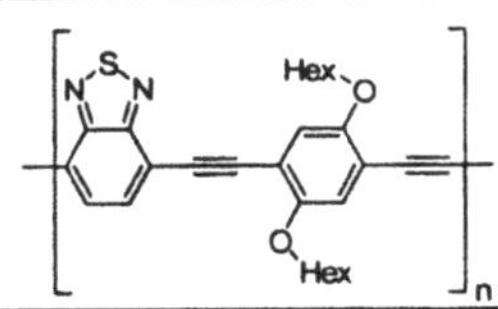	100 8.5 x 10^3 3.88
8[d]	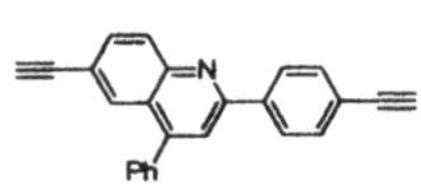	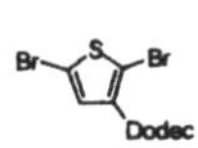	$(Ph_3P)_2PdCl_2$, CuI, THF, DIPA, 60 °C 72 h	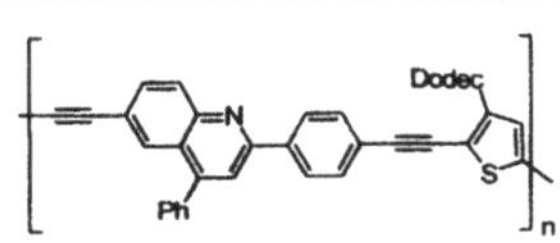	78 9.7 x 10^3 1.89
9[d]	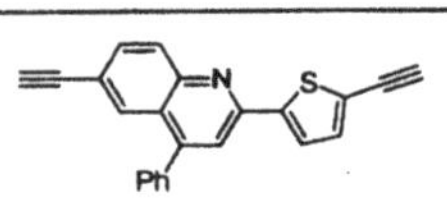	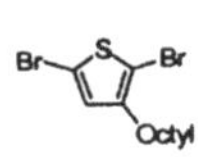	$Ph_3P)_2PdCl_2$, CuI, THF, DIPA, 60 °C 72 h	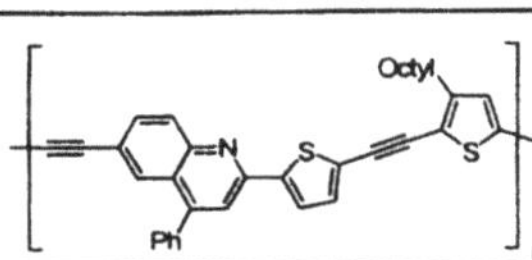	83 14.6 x 10^3 1.95

Table 9 (continued)

	Monomer 1	Monomer 2	Catalyst	Polymer	Yield (%) M_n, PDI
10[d]	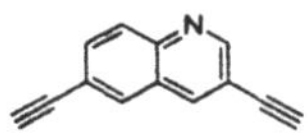	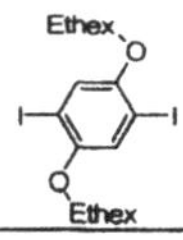	$(Ph_3P)_2PdCl_2$, CuI, THF, piperidine 23 °C, 48 h	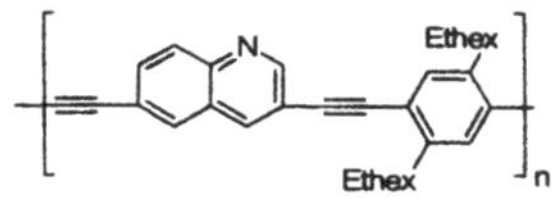	88 P_n = 95 3.1
11[d]	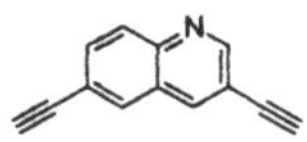	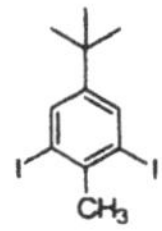	$(Ph_3P)_2PdCl_2$, CuI, THF, piperidine 23 °C, 48 h	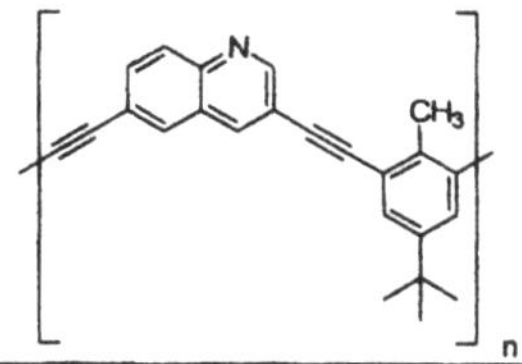	86 P_n = 11 1.8
12[d]	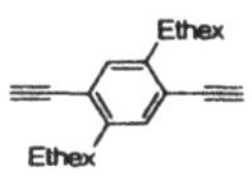	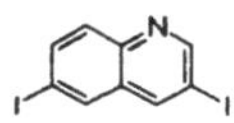	$(Ph_3P)_2PdCl_2$, CuI, THF, piperidine 23 °C, 48 h	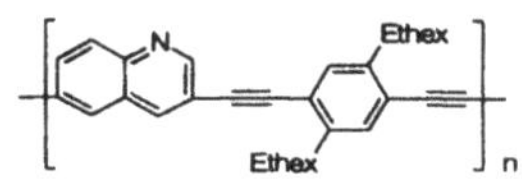	93 P_n = 37 3.3
13[e]	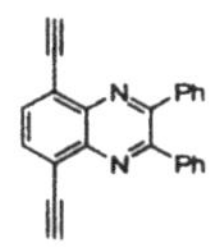	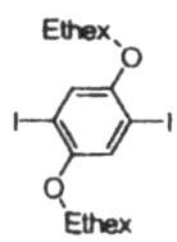	$(Ph_3P)_2PdCl_2$, CuI, THF, piperidine	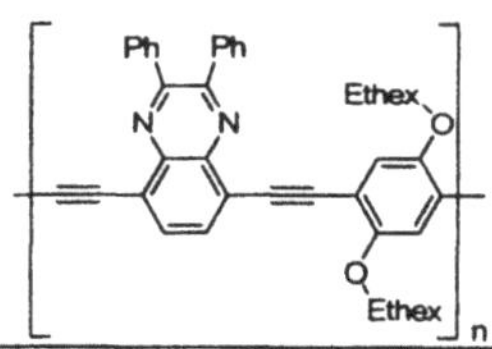	95% Pn = 110 3.03
14[e]	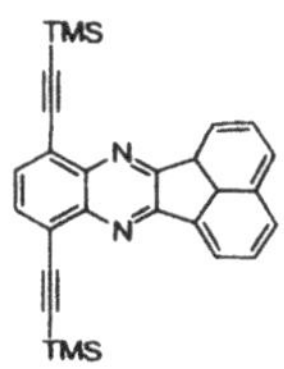	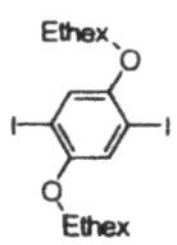	$(Ph_3P)_2PdCl_2$, CuI, THF, piperidine KOH/THF/ EtOH	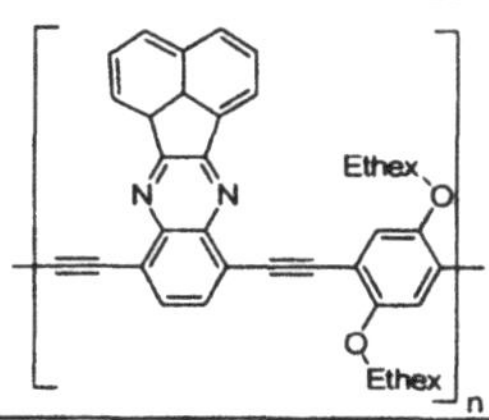	86% P_n = 36 1.44 sol. frac- tion
15[f]	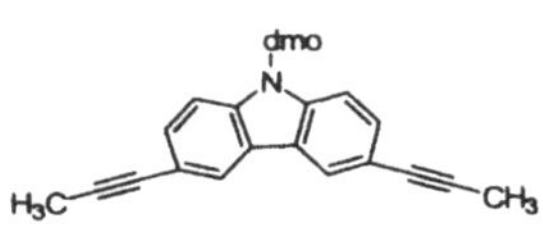		$Mo(CO)_6$ 4-chloro- phenol 140 °C, 24 h	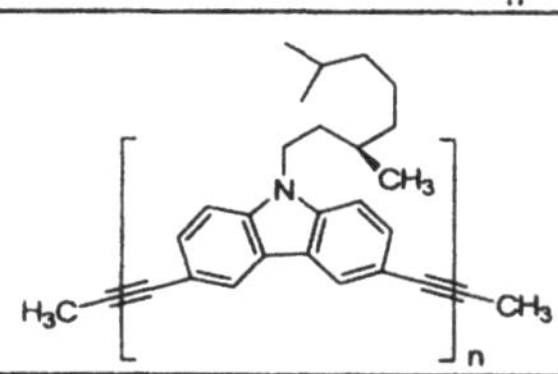	R = dmo 86 P_n = 31 3.1
16[f]	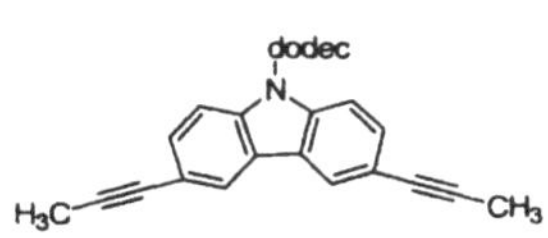		$Mo(CO)_6$ 4-chloro- phenol 140 °C, 24 h	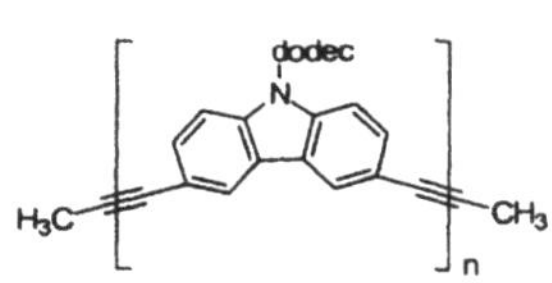	R = $C_{12}H_{25}$ 98 P_n = 24 2.3
17[f]	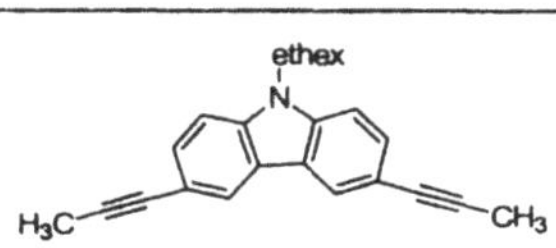		$Mo(CO)_6$ 4-chloro- phenol 140 °C, 24 h	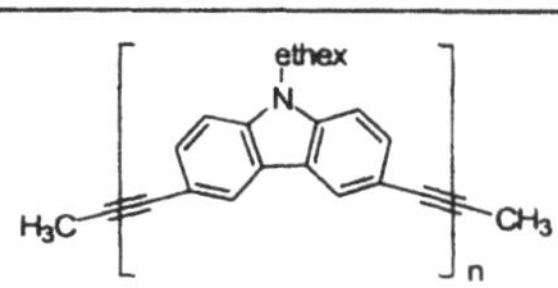	R = Ethex 94 P_n = 26 2.7

Table 9 (continued)

	Monomer 1	Monomer 2	Catalyst	Polymer	Yield (%) M_n, PDI
18 [f]	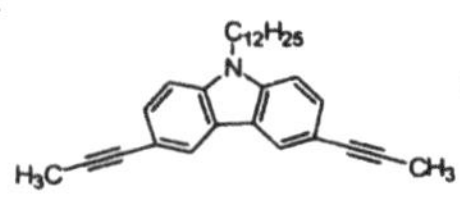	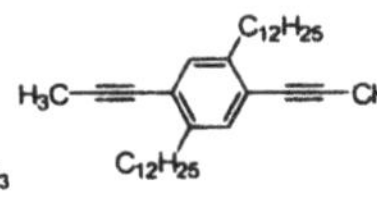	$Mo(CO)_6$ 4-chloro-phenol 140 °C, 24 h	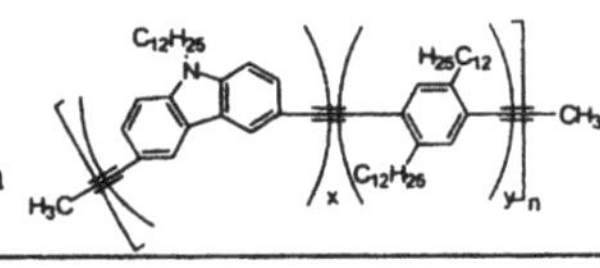	x= 0.2; y = 0.8 P_n = 93 3.0
19 [g]	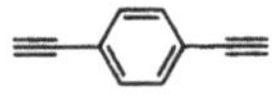	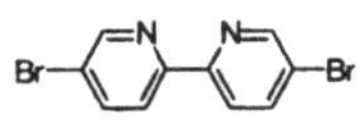	CuI $Pd(PPh_3)_4$ DIPA 80°C, 2-6.5d	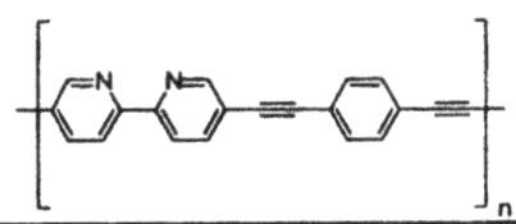	91 P_n = 4 nd
20 [h]	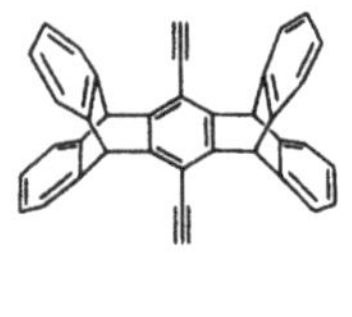	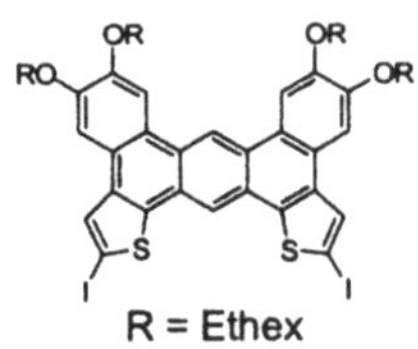	$Pd(PPh_3)_4$ CuI in 3:1 toluene/ DIPA 80°C, 2-6.5d	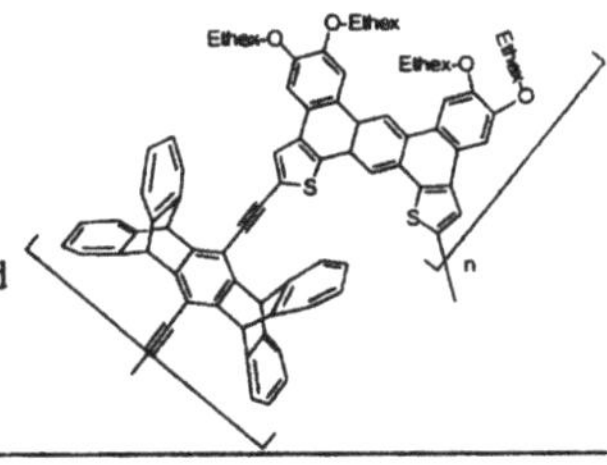	67 M_n = 16.5 x 10^3 nd
21 [h]	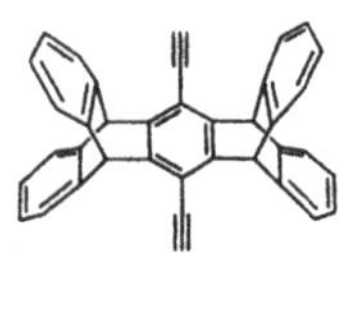	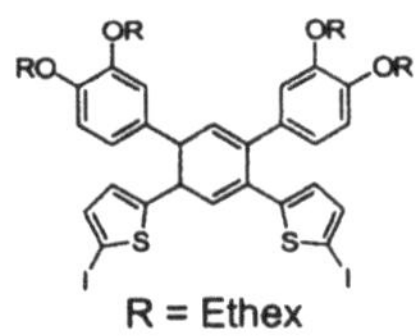	$Pd(PPh_3)_4$ CuI in 3:1 toluene/ DIPA 80°C, 2-6.5d	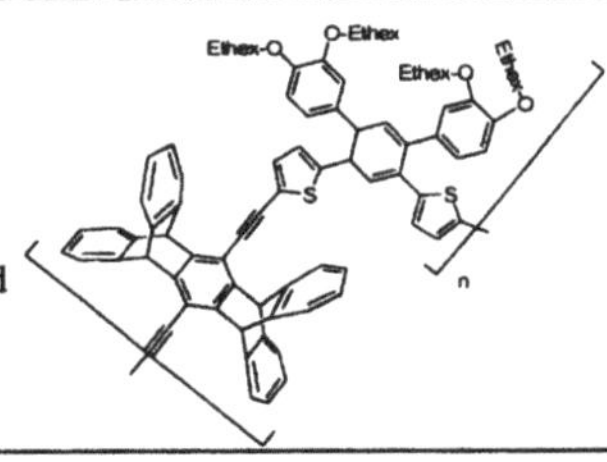	82 M_n = 21 x 10^3 nd
22 [h]	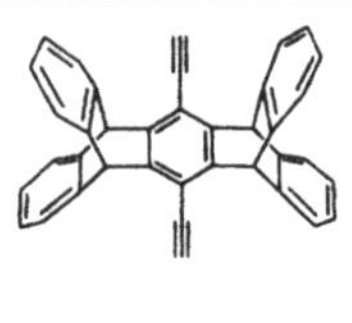	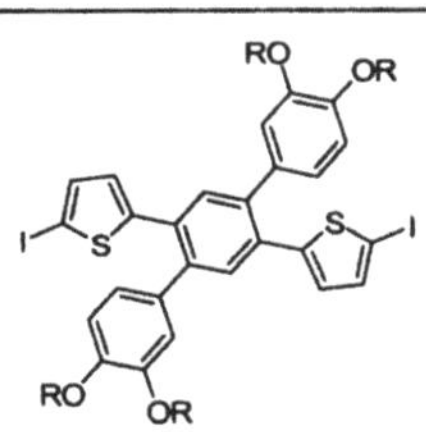	$Pd(PPh_3)_4$ CuI in 3:1 toluene/ DIPA 80°C, 2-6.5d	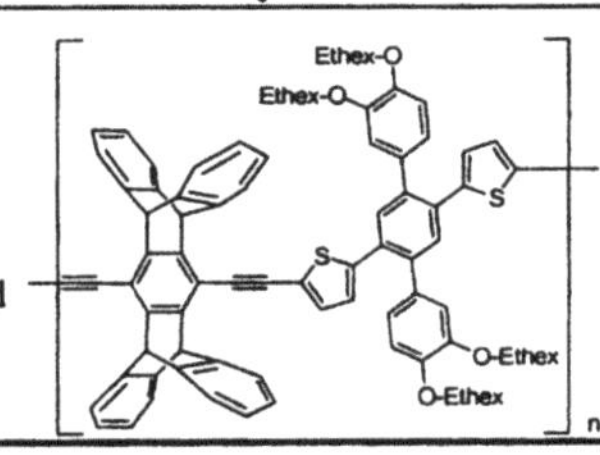	nd M_n = 15 x 10^3 nd
23 [h]	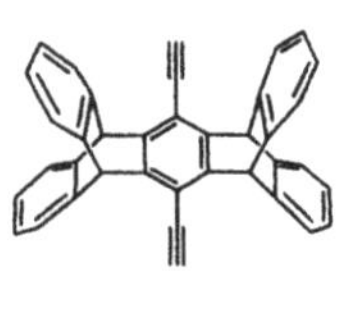	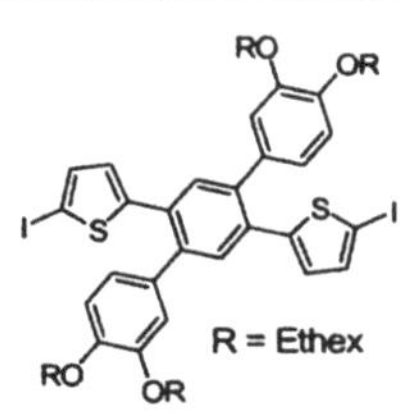	$Pd(PPh_3)_4$ CuI in 3:1 toluene/ DIPA 80°C, 2-6.5d	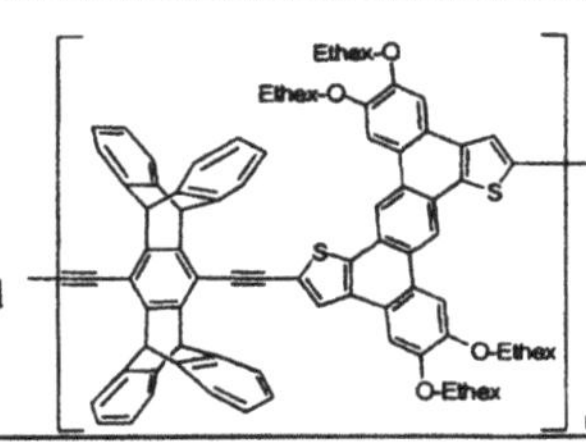	86 M_n= 20 x 10^3 nd
24 [i]	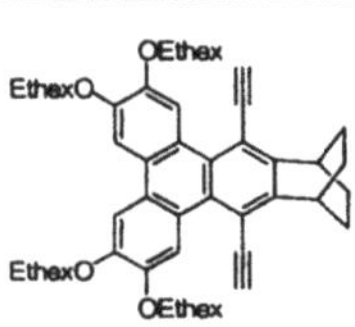	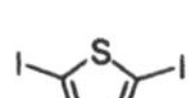	$Pd(PPh_3)_4$ CuI in 3:1 toluene/ DIPA 80°C, 2-6.5d	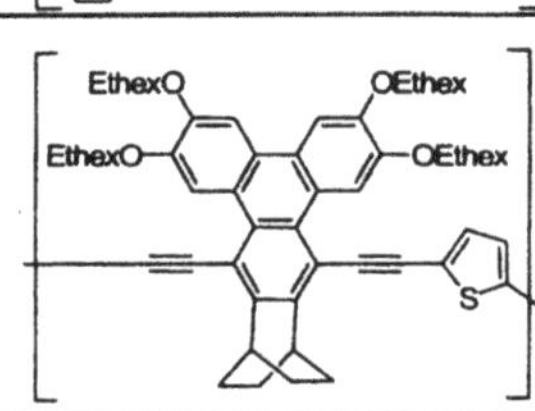	nd M_n= 11 x 10^4 1.3

Table 9 (continued)

	Monomer 1	Monomer 2	Catalyst	Polymer	Yield (%) M_n, PDI
25[j]	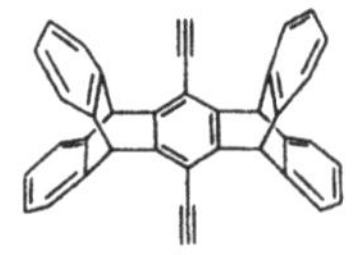	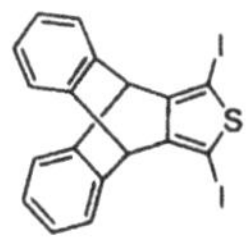	$Pd(PPh_3)_4$ CuI in 3:1 toluene/ DIPA 80°C, 2-6.5d	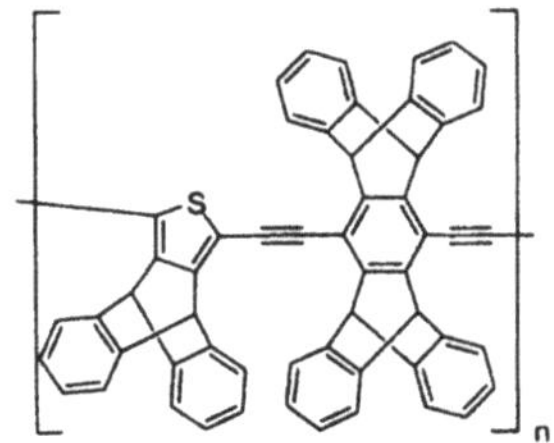	62 M_n= 15 x 10^3 3.4
26[k]	Acetylene gas	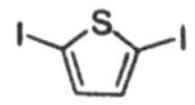	NEt_3 vapor $Pd(OAc)_2$, PPh_3, CuI	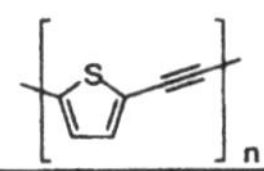	Insolubl e, nd
27[l]	H—≡—$SnBu_3$	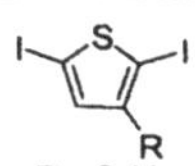 R= C_4H_9	$(Ph_3P)_4Pd$ dioxane reflux then LDA then reflux 2-8h	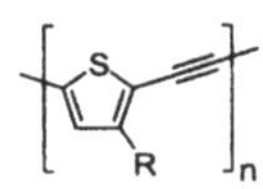	97 33 2.8
28[l]	H—≡—$SnBu_3$	R= $C_{16}H_{33}$			92 31 2.7
29[m]	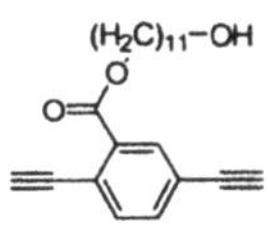	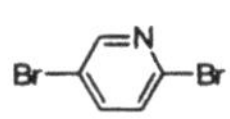	$PdCl_2$, CuI, NEt_3, PPh_3	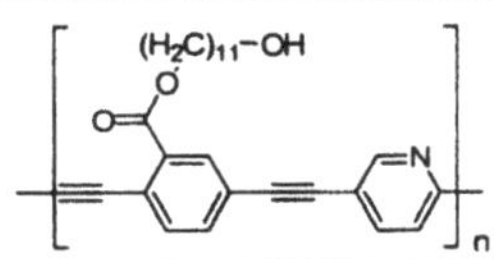	90% 15 x 10^3 3.0
30[m]	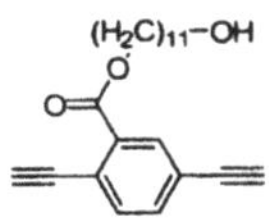	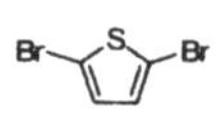	$PdCl_2$, CuI, NEt_3, PPh_3	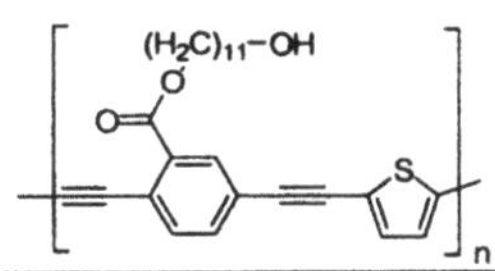	80-90% 9 x 10^3 2.2

a Leung ACW, Chong JH, Patrick BO, MacLachlan MJ (2003) Macromolecules 36:5051–5054.
b Bangcuyo CG, Evans U, Myrick ML, Bunz UHF (2001) Macromolecules 34:7592–7594.
c Morikita T, Yamaguchi I, Yamamoto T (2001) Adv Mater 13:1862–1864.
d Jegou G, Jenekhe SA (2001) Macromolecules 34:7926–7928.
e Bangcuyo CG, Rampey Vaughn ME, Quan LT, Angel SM, Smith MD, Bunz UHF (2002) Macromolecules 35:1563–1568; Bangcuyo CG, Ellsworth JM, Evans U, Myrick ML, Bunz UHF (2003) Macromolecules 36:546–548.
f Brizius G, Kroth S, Bunz UHF (2002) Macromolecules 35:5317–5319.
g Grummt UW, Birckner E, Klemm E, Egbe DAM, Heise B (2000) J Phys Org Chem 13:112–126.
h Tovar JD, Rose A, Swager TM (2002) J Am Chem Soc 124:7762–7769.
i Rose A, Lugmair CG, Swager TM (2001) J Am Chem Soc 123:11298–11299.
j Williams VE, Swager TM (2000) Macromolecules 33:4069–4073.
k McCaughey B, Costello C, Wang D, Hampsey JE, Yang Z, Li C, Brinker CJ, Lu Y (2003) Adv Mater 15:1266–1269.
l Pizzoferrato R, Berliocchi M, DiCarlo A, Lugli P, Venanzi M, Micozzi A, Ricci A, LoSterzo C (2003) Macromolecules 36:2215–2223.
m Arias Marin E, LeMoigne J, Maillou T, Guillon D, Moggio I, Geffroy B (2003) Macromolecules 36:3570–3579.

Table 10 Mixed PPE–PPV hybrids

Ref	Monomer 1 Monomer 2	Catalyst	Polymer	Yield (%) P_n or M_n PDI
1[a]	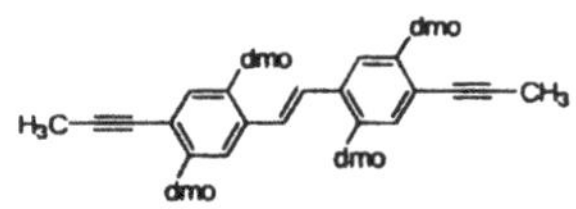	$Mo(CO)_6$ 4-chloro-phenol, *o*-dichloro-benzene	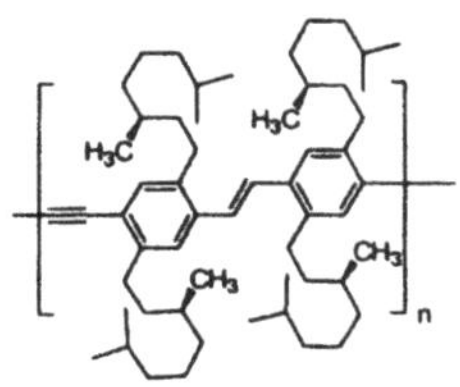	3,7-dmo 78 152 3.43
2[a] *a*	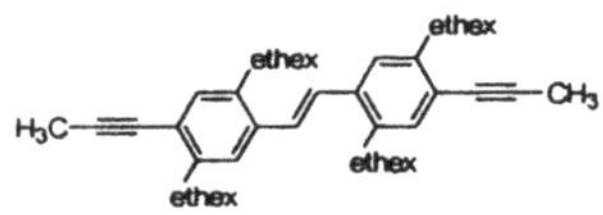	$Mo(CO)_6$ 4-chloro-phenol, *o*-dichloro-benzene	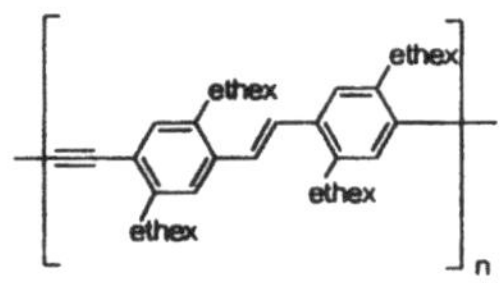	Ethex 92 77 2.09
3[a] *a*	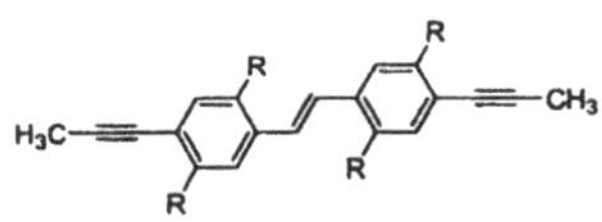	$Mo(CO)_6$ 4-chloro-phenol, *o*-dichloro-benzene	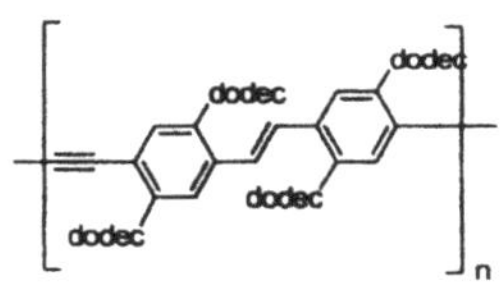	Dodecyl 96 685 4.29
4[a] *a*	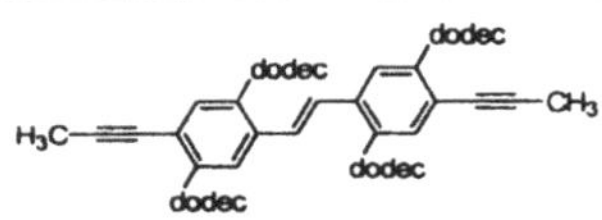	$Mo(CO)_6$ 4-chloro-phenol, *o*-dichloro-benzene	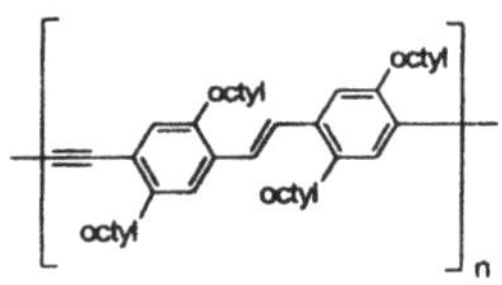	Octyl 98 100 2.54
5[b] b	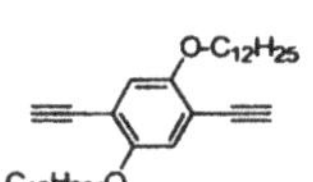 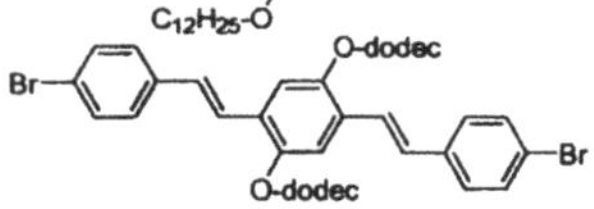	$(Ph_3P)_4Pd$ CuI, DIPA, PPh_3, Toluene	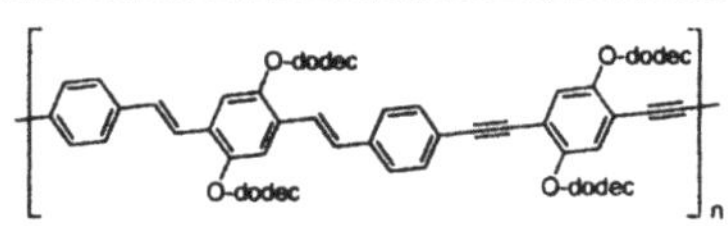	26 8 3.4
6[b] b	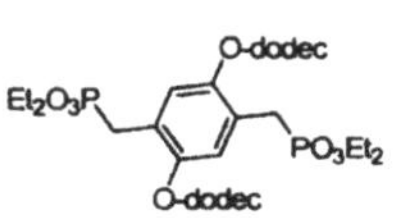 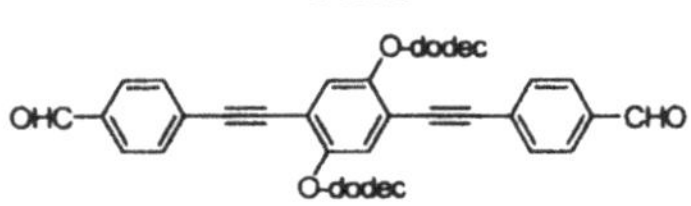	KO*t*Bu toluene	several other PPEVs of the same structrure with different side chains were reported via the Horner approach. See ref. b)	66 77 6.2

Table 10 (continued)

Ref	Monomer 1 Monomer 2	Catalyst	Polymer	Yield (%) P_n or M_n PDI
7 [c] c	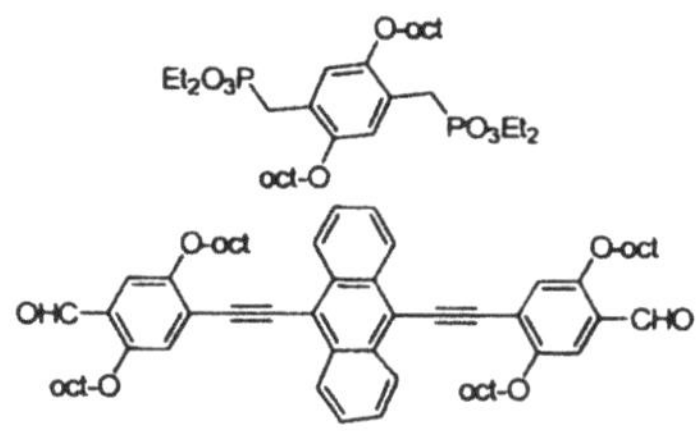	KO*t*Bu toluene	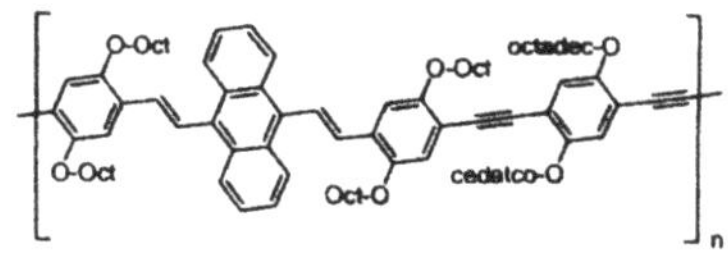	87 16 2.0
8 [c]	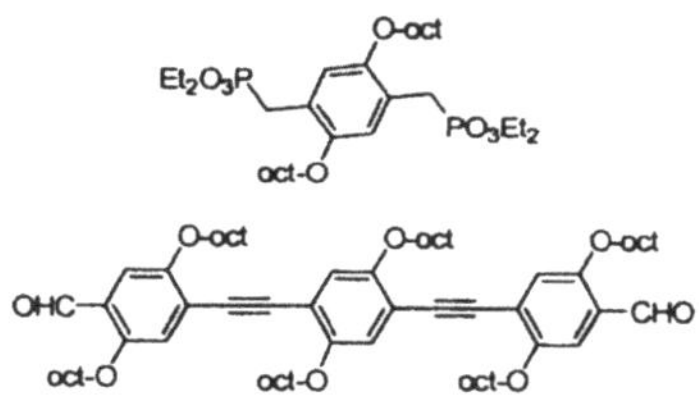	KO*t*Bu toluene	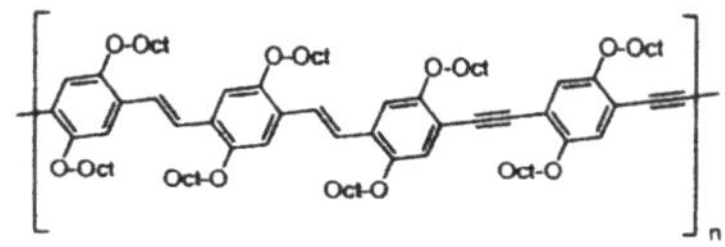	88 12 2.2
9 [c]	several other PPEVs of the same structrure with different side chains but with similar degree of polymerization were reported via the Horner approach. See ref. c)			
10 [d] d	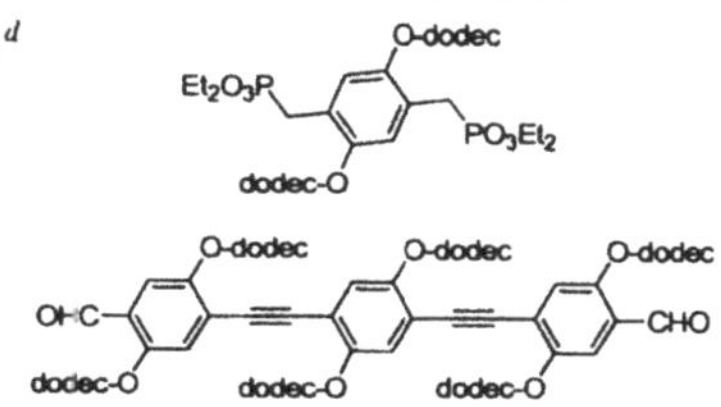	KO*t*Bu toluene	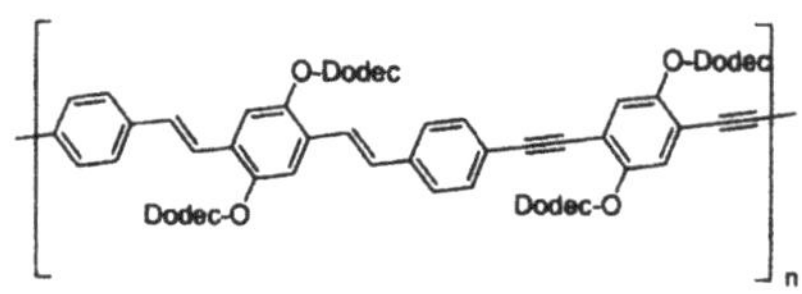	Not reported 77 6.2
11 [d] d	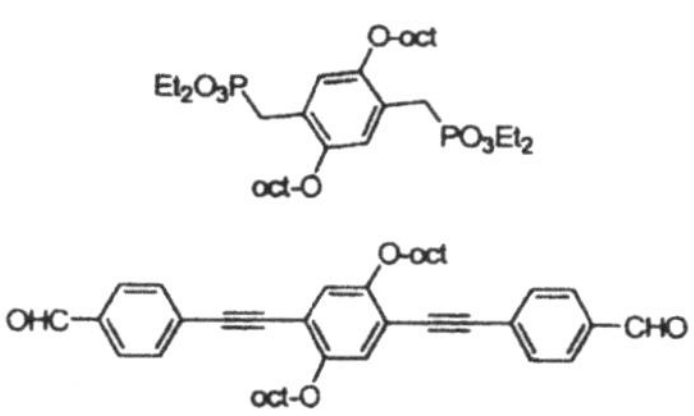	KO*t*Bu toluene	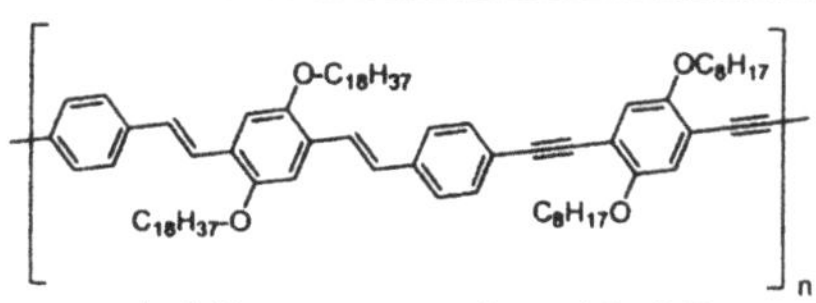 nematic LC; more examples with different side chains are reported.	69 42 3.1
12 [e] e	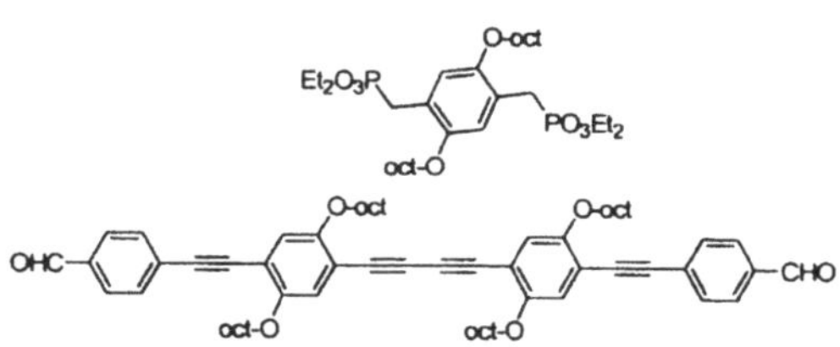		KOtBu/toluene, Horner method	54 12 2.2

Table 10 (continued)

Ref	Monomer 1 Monomer 2	Catalyst	Polymer	Yield (%) P_n or M_n PDI
	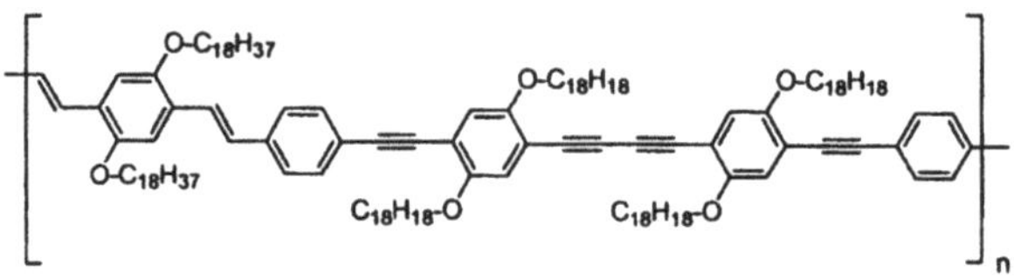		Polymer	
13[f] f	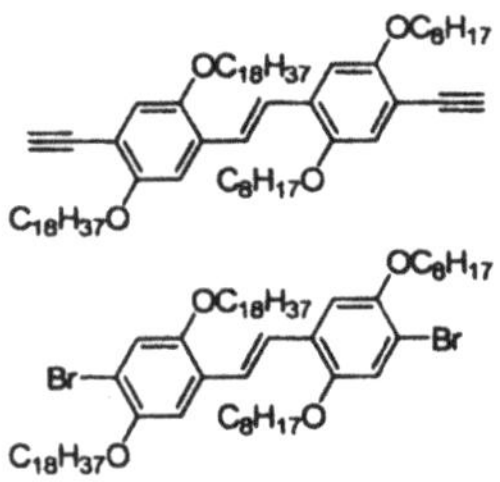	$(Ph_3P)_4Pd$ CuI, DIPA, PPh_3, toluene	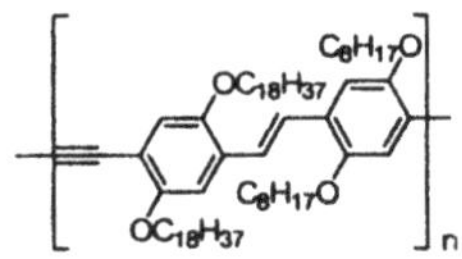	85% 8 4.0
14[f]	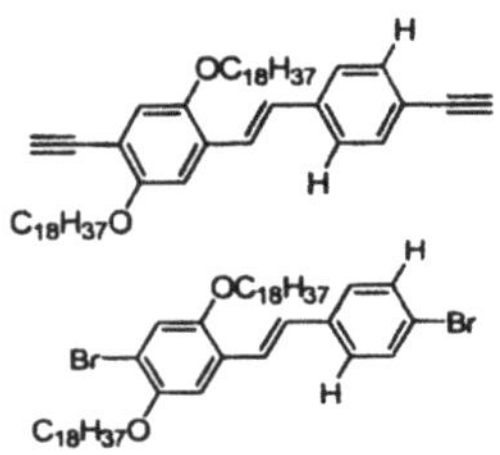	$(Ph_3P)_4Pd$ CuI, DIPA, PPh_3, toluene	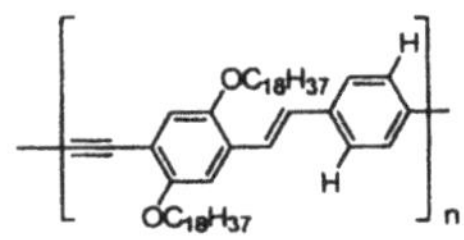	74% 8 4.4
15[f]	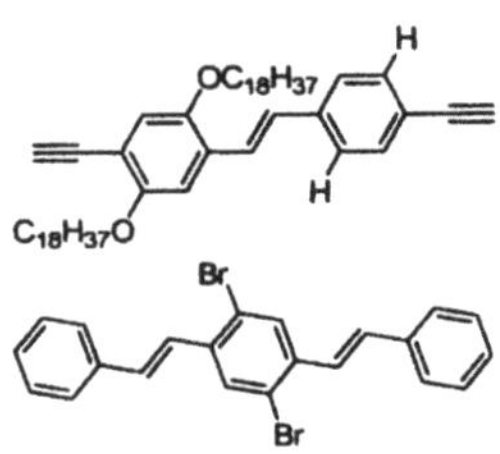	$(Ph_3P)_4Pd$ CuI, DIPA, PPh_3, toluene	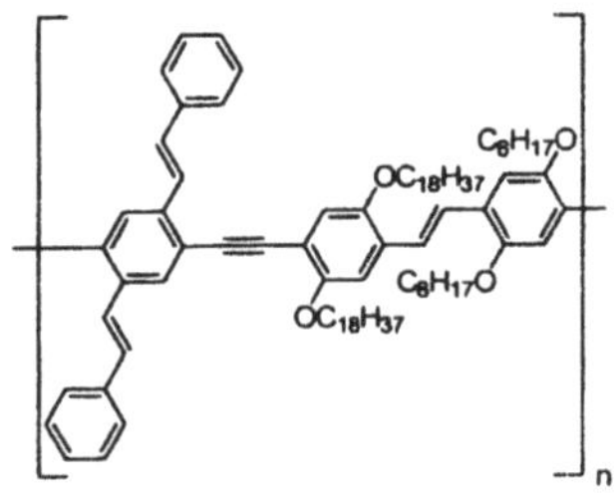	81 14 4.7
16[g]	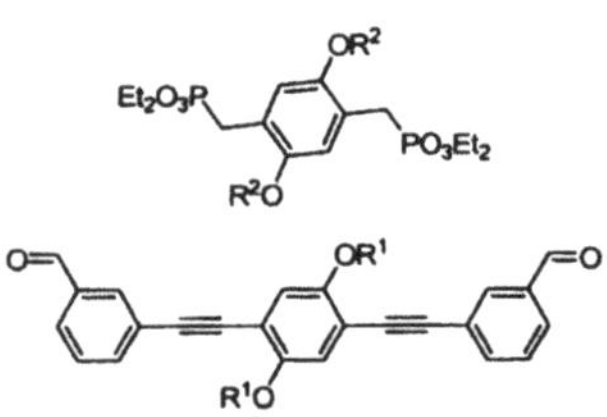 Monomer 1 R^2 = hexyl, Monomer 2 R^1 = hexyl	KOtBu/toluene, Horner method	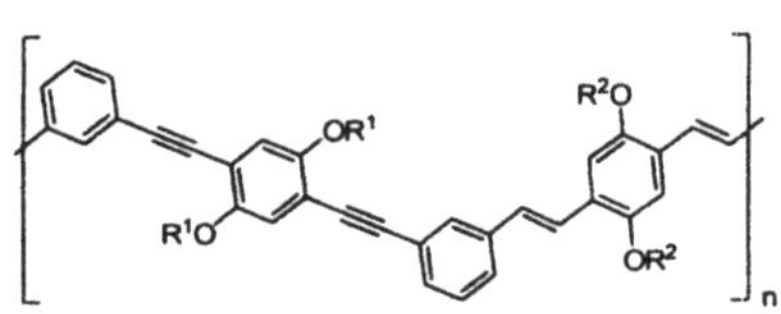	82 73 1.36

Table 10 (continued)

Ref	Monomer 1 Monomer 2	Catalyst	Polymer	Yield (%) P_n or M_n PDI
17[g]	$R^1 = R^2$ = dodecyl		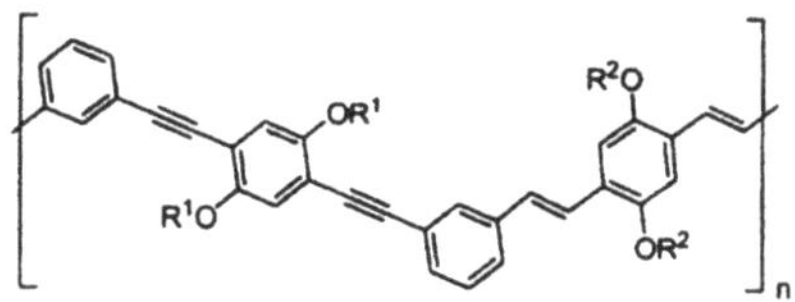	85 53 1.40
18[g]	R^1 = dodecyl; R^1= hexyl		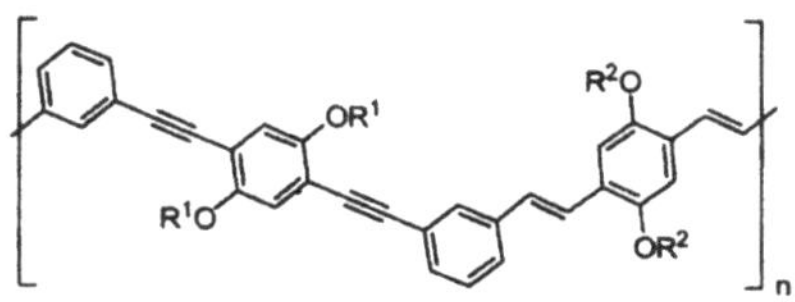	80 27 1.72
19[g]	R^1 = hexyl; R^1= dodecyl *trans* content of the double bonds is 92-96%		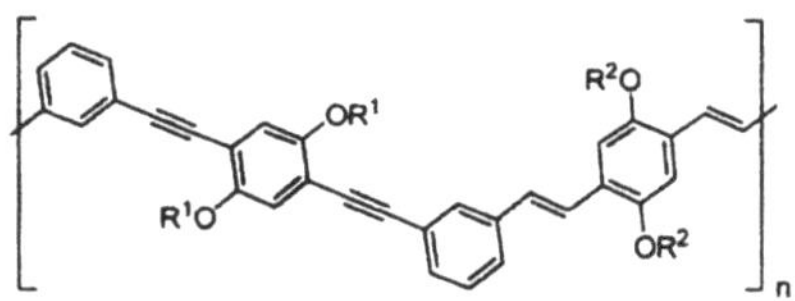	88 76 1.59
20[h]	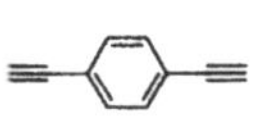 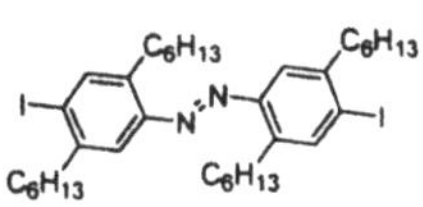	$(Ph_3P)PdCl_2$ 1 mol% CuI, NEt_3, DMF	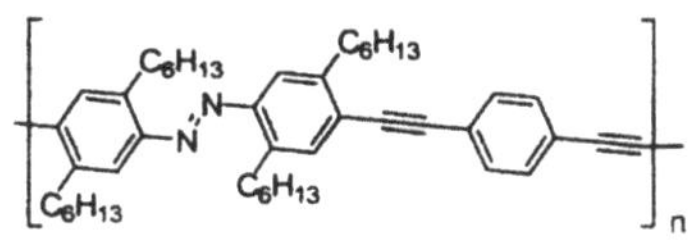	100 33 5.4
21[h]	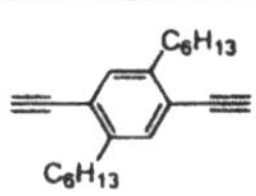 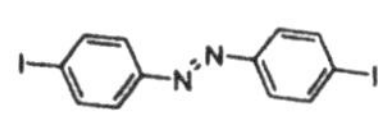	$(Ph_3P)PdCl_2$ 1 mol% CuI, NEt_3, DMF	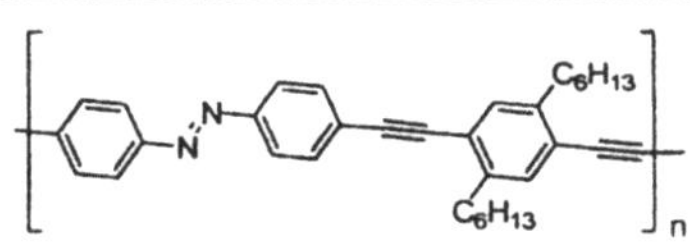	100 11 2.7
22[h]	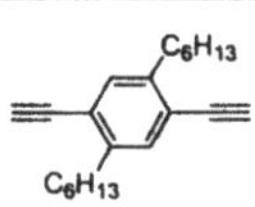 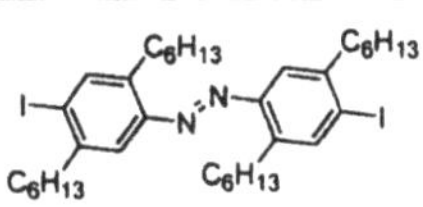	$(Ph_3P)PdCl_2$ 1 mol% CuI, NEt_3, DMF	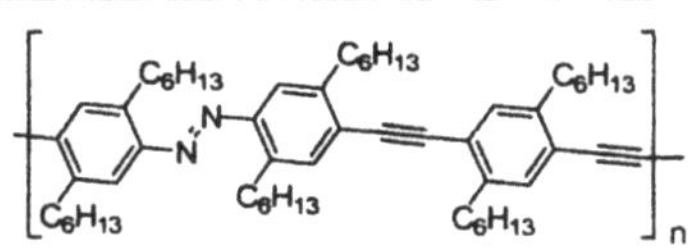	100 15 3.83

[a] Brizius G, Pschirer NG, Steffen W, Stitzer K, zur Loye HC, Bunz UHF (2000) J Am Chem Soc 122:12435–12440.
[b] Egbe DAM, Tillmann H, Birckner E, Klemm E (2001) Macromol Chem Phys 202:2712–2726.
[c] Egbe DAM, Bader C, Nowotny J, Gunther W, Klemm E (2003) Macromolecules 36:5459–5469.
[d] Egbe DAM, Roll CP, Birckner E, Grummt UW, Stockmann R, Klemm E (2002) Macromolecules 35:3825–3837.
[e] Egbe DAM, Birckner E, Klemm E (2002) J Polym Sci A 40:2670–2679.
[f] Egbe DAM, Bader C, Klemm E, Ding L, Karasz FE, Grummt UW, Birckner E (2003) Macromolecules 36:9303–9312.
[g] Chu Q, Pang Y, Ding L, Karasz FE (2003) Macromolecules 363848–3853.
[h] Izumi A, Nomura R, Masuda T (2001) Macromolecules 31:4342–4347.

A paper by Meijer et al. described the synthesis and electronic properties of a series of PPE–PPV hybrid oligomers. The approach is shown in Scheme 11 [70]. Starting from a brominated stilbene, reaction with a series of diethynylbenzenes furnished pentameric PE–PV hybrids in yields around 80%. The materials are used as emitters in LEDs and other electrooptical devices. It was noted that fibrous aggregates formed upon melting and recooling of thin films of these pentameric PE–PV hybrids.

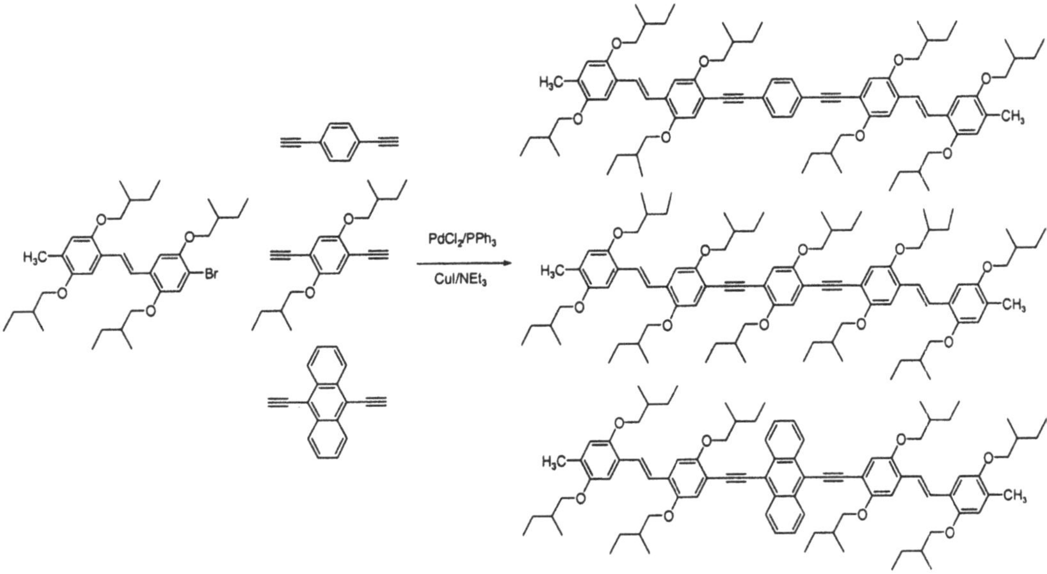

Scheme 11 Synthesis of PE–PV hybrids. From: Schenning APHJ, Tsipis AC, Meskers SCJ, Beljonne D, Meijer EW, Bredas JL (2002) Chem Mater 14:1362–1368

3.5
Synthesis of PAEs Utilizing a Combinatorial Approach: Caveat Emptor

Combinatorial synthesis with high-throughput screening is a powerful approach for the discovery of new materials, new catalysts, and biologically active compounds [71–75]. In a clever exploit of the high-throughput approach, Lavastre et al. [76] explored a library of highly fluorescent PAEs. Scheme 12 shows the combinations of different monomers that were utilized in this approach. According to the Lavastre group, the polymers B10, D10, G10, and H10 were highly fluorescent in the solid state. In solution many more polymers showed significant fluorescence including the polymers G11 and G12. And while it appears reasonable that dialkyfluorene-containing monomers will give rise to highly fluorescent solid-state materials, the notion that polymers G11 and G12 should be fluorescent was questioned by McLachlan et al. [77]. They prepared a series of polymers including G11 and G12 and found that these compounds were nonfluorescent in solution and in the solid state. They attributed the strong fluorescence of G11 and G12, reported by Lavastre et al., to

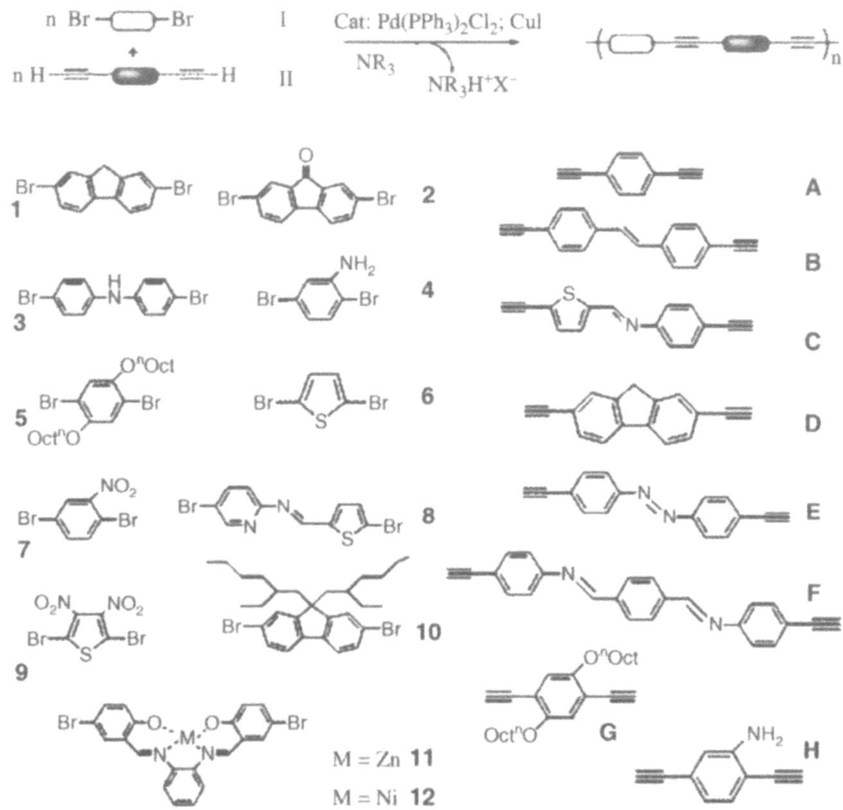

Scheme 12 Combinatorial approach for the synthesis of a wide variety of PAEs. From: Lavastre O, Illitchev I, Jegou G, Dixneuf PH (2002) J Am Chem Soc 124:5278–5279

the presence of residual dibromosalphene monomer; this example may serve as a caveat.

Lavastre's approach is valid, but the use of (a) less reactive brominated arenes and (b) the obvious lack of solubilizing groups restricts the value of this approach. Perhaps *all* of the 96 polymers would have to be made in a conventional approach and compared to the polymers obtained by the high-throughput method to draw further conclusions. In addition, the fluorescence was measured in situ, in a THF/diisopropylamine mixture in the presence of the Pd–Cu catalyst. That approach may have an intrinsic problem too, because it is known that PAEs are solvatochromic, and as a consequence, the obtained fluorescence data might be biased. However, with some tweaking this is a powerful method to obtain a cornucopia of different structures quickly and without too much synthetic effort, if the monomers are easily available.

3.6 Nonconjugated PAE Derivatives

A noteworthy development is the use of the Heck–Cassar–Sonogashira–Hagihara coupling in the synthesis of nonconjugated AE-containing polymers. Table 11 shows hitherto reported examples. The introduction of nonconjugated segments into PAEs is of interest, because it (a) enhances the solubility and processability of the formed polymers, and (b) in these polymers the chromophores are well defined and do not suffer from differing conjugation lengths etc. In addition, these polymers can fold either in solution or in the solid state, which makes them interesting objects for structural studies. A subset (Table 11,

Table 11 Non-conjugated PAEs

	Diyne	Dihalide	Catalyst	Polymer	Yield (%)	M_n	PDI	Ref
1	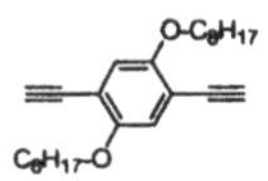	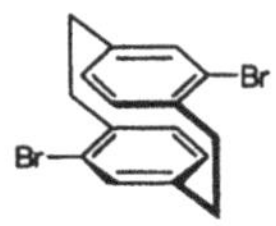	$(Ph_3P)_2PdCl_2$ PPh_3 CuI HN^iPr_2 THF Reflux 48 h	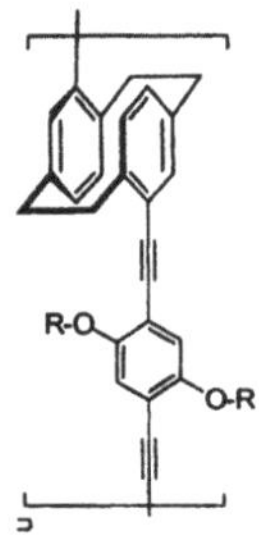	73	3.1 x10^3	1.7	[a]
2	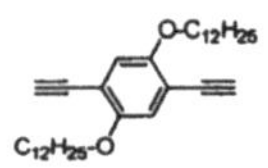	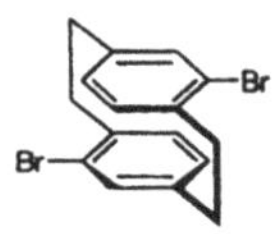	$(Ph_3P)_2PdCl_2$ PPh_3 CuI HN^iPr_2 THF Reflux 48 h		98	8.0 x10^3	1.8	[a]
3	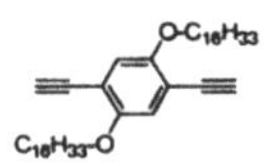	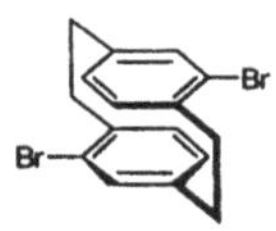	$(Ph_3P)_2PdCl_2$ PPh_3 CuI HN^iPr_2 THF Reflux 48 h		70	5.9 x10^3	1.5	[a]
4	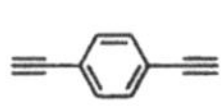	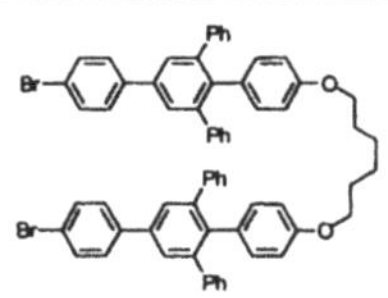	$(Ph_3P)_2PdCl_2$ PPh_3 CuI NEt_3 70°C 48h	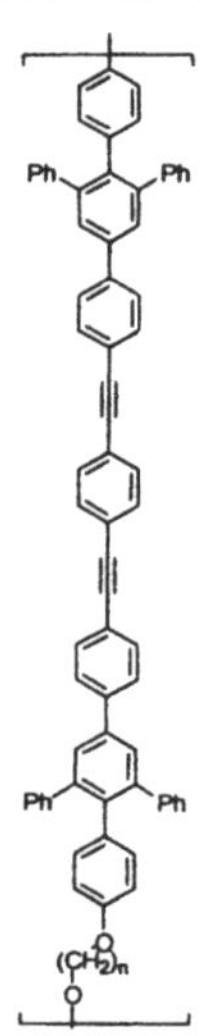	71	3.3 x10^3	2.5	[b]
5	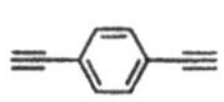	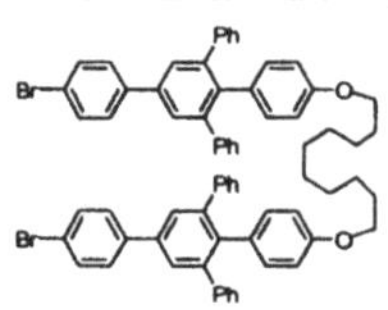	$(Ph_3P)_2PdCl_2$ PPh_3 CuI NEt_3 70°C 48h		89	5.9 x10^3	2.1	[b]

Table 11 (continued)

	Diyne	Dihalide	Catalyst	Polymer	Yield (%)	M_n	PDI	Ref
6	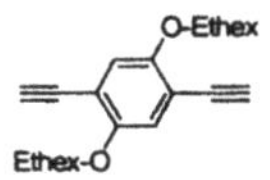	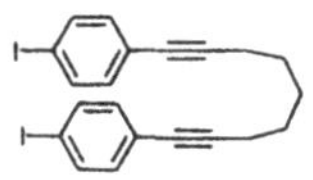	$(Ph_3P)_4Pd$ CuI HN^iPr_2 Toluene 75°C 20 h	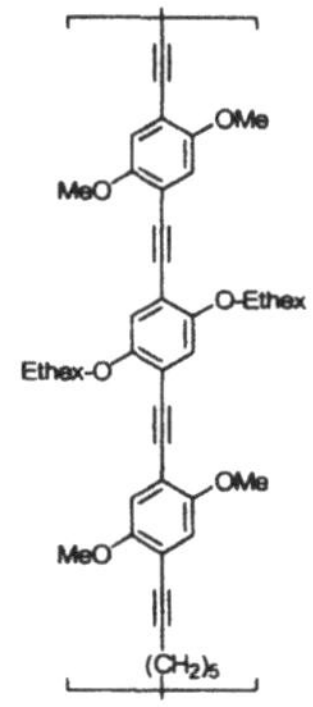	70	9.3 $x10^3$	na	c
	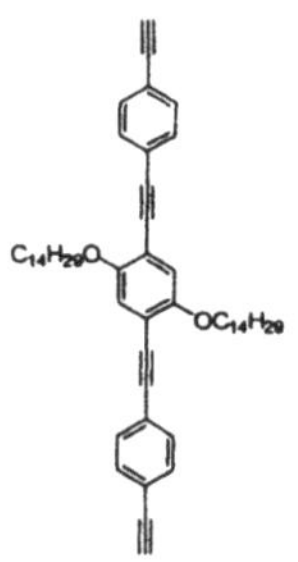	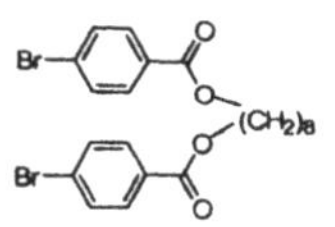	$PdCl_2$ PPh_3 $Cu(OAc)_2$ NEt_3 THF	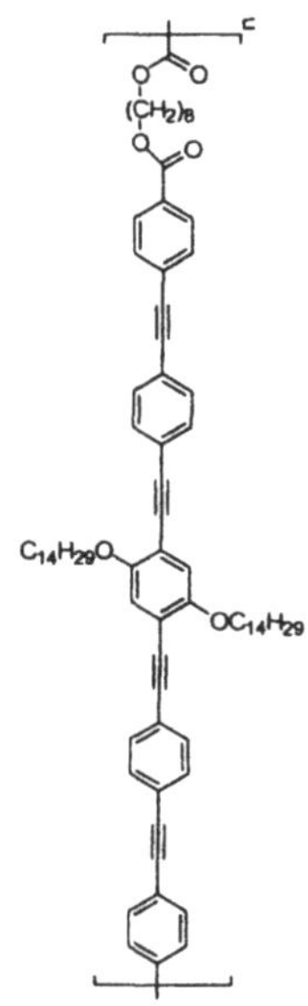	>100%	na	na	d

[a] Morisaki Y, Chujo Y (2002) Macromolecules 35:587–589.
[b] Spiliopoulos IK, Mikroyannidis JA (2001) Macromolecules 34:5711–5718.
[c] Dellsperger S, Dötz F, Smith P, Weder C (2000) Macromol Chem Phys 201:192–198.
[d] Moggio I, LeMoigne J, Arias Marin E, Issautier D, Thierry A, Comoretto D, Dellepiane G, Cuniberti C (2001) Macromolecules 34:7091–7099.

entries 1–3) of these syntheses deals with polymers that show "through-space" interactions of cyclophane-containing PAEs. These are made by the Pd-catalyzed coupling of dibromocyclophane to a suitable diethynylbenzene derivative. To obtain processable polymers the dialkynylbenzene has to carry solubilizing side chains such as octyloxy, dodecyloxy, or hexadecyloxy. Yields are acceptable in most of the reported cases and the obtained molecular weights are moderate to good, perhaps due to the formation of cyclic oligomers.

The shown topologies suggest that there are many more possibilities in this area. Equipping the tethers with recognition elements and utilizing two or more different conjugated "phores" should give rise to the formation of highly chromic materials with attractive sensory and electrooptical properties. The size of the utilized phore will influence the physical and electronic properties of the nonconjugated PAE.

4 Polymer-Analogous Reactions of PAEs

4.1 Reduction of the Main Chain

An attractive challenge is the postfunctionalization of PPEs to generate novel polymers in a single step. The alkyne groups in the PPEs are the natural point of attack. The simplest reaction would be the reduction of the C≡C triple bonds of PPEs into C–C single bonds. While this transformation appears trivial on paper it is not easily done in the laboratory. Attempts to reduce PPEs by catalytic hydrogenation utilizing H_2 and Pd/charcoal at temperatures up to 120 °C and with pressures up to 10 atm of H_2 failed; the starting material was isolated. Only if much more forceful hydrogenation conditions were applied was the reduction of the triple bonds possible. Utilizing Wilkinson's catalyst at 350 °C and under a H_2 pressure of 500 bar in a steel autoclave, dialkyl PPE's alkyne groups were fully hydrogenated (Scheme 13) [78, 79]. At lower temperatures and lower H_2 pressures partial hydrogenation was observed. According to ^{13}C NMR experiments, the formed polymer was inhomogeneous and contained segments with –C≡C–, –CH=CH–, and CH_2–CH_2 groups in a random fashion. These polymers therefore could not be well characterized.

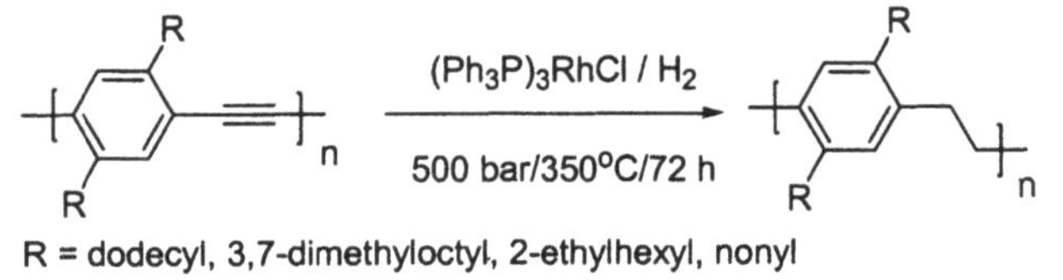

Scheme 13 Reduction of dialkyl PPEs utilizing catalytic hydrogenation

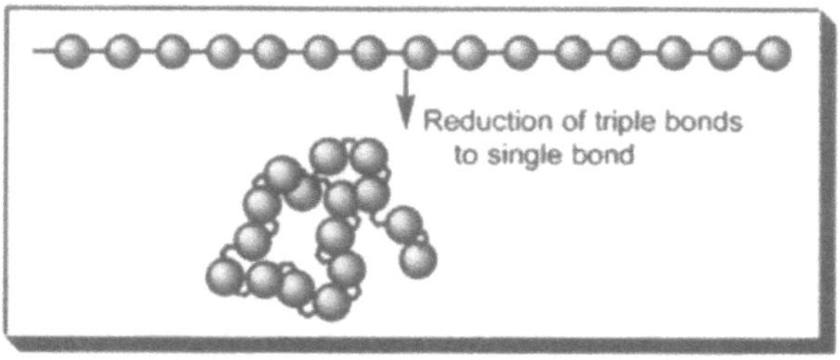

Fig. 2 Reduction of PPEs to poly(dialkyl *p*-xylylene)s: turning rods into coils

The sluggish reaction of the alkyne bonds in PPEs is surprising, because the triple bond in tolane can be hydrogenated under much milder conditions. The adjoining alkyl groups seem to protect the PPE backbone efficiently, apparently by a steric effect. In addition, the embedding of the tolane units into a conjugated chain will change the electronics via a shift of the frontier orbitals. This may further attenuate their reactivity. The decreased reactivity of the PPEs under the hydrogenation conditions seems to be yet another example of the high stability of PPEs under ambient conditions as well as under heating.

Catalytic hydrogenation was applied to transform a series of rigid-rod PPEs of different molecular weights into coiled and floppy polymers of identical degree of polymerization and of very similar molecular weight (Fig. 2). That approach allowed investigation of the decrease of apparent molecular weights when transforming a rigid rod into a floppy coil [79]. The apparent molecular weight of a series of didodecyl PPEs of different molecular weights was determined by gel-permeation chromatography (GPC) vs. polystyrene. The samples were then hydrogenated and again their molecular weight was determined by GPC. By correlating the "before" and "after" molecular weight it was concluded that there is a scaling factor of 1.4 by which the PPEs' molecular weight appears to be overestimated. These data are in qualitative agreement with the results obtained by Cotts et al. [80] in an earlier light scattering study addressing the question of flexibility and rigidity of PPEs.

While the catalytic reduction of PPEs is powerful, it is necessary to work at high temperatures and with a stainless-steel autoclave, which may not always be practical. Weder et al. reported an alternative way to reduce dialk*oxy*-substituted PPEs [81]. Treating the PPEs with a mixture of tosylhydrazide and tripropylamine at elevated temperatures gave the reduced species in high yield and excellent purity (Scheme 14). The purported intermediate is diimine, formed by elimination of Ar–SO_2–H from the hydrazide. Diimine reduces the triple bonds to give the fully reduced polymer. Attempts to utilize the diimine method to reduce dialkyl PPEs, however, failed perhaps due to the more massive steric shielding by the alkyl groups in these polymers.

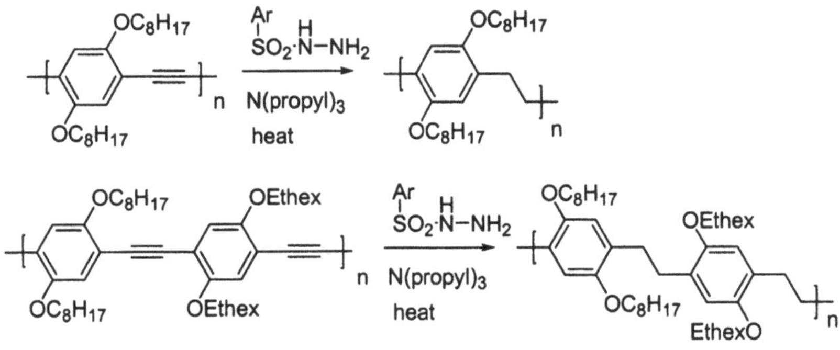

Scheme 14 Reduction of dialkoxy PPEs by Weder et al. (tosylhydrazide method)

4.2 Organometallic Cross-linking

An approach to functionalize the backbone of the PPEs was pioneered by Weder et al. [82–84]. Alkyne groups are used as ligands for organometallic fragments. Reaction of a dialkoxy PPE with an excess of (Ph–CH=CH$_2$)$_2$Pt$_2$Cl$_4$ furnished a complex (Scheme 15) in which a significant fraction of the triple bonds were complexed by a –Pt$_2$(Ph–CH=CH$_2$)Cl$_4$ unit. If more PPE is added, the polymer can be cross-linked and an organometallic gel network will be obtained. Alternatively, if a spin-cast solution is utilized, an insoluble but clear film forms. Both the cross-linked gel and the insoluble film can be redissolved upon addition of an excess of styrene, suggesting that the complexation of the triple bonds is reversible. An attractive application of this concept was realized by treating a solution of PPE with (Ph–CH=CH$_2$)$_3$Pt (Scheme 16). Thin-film samples of the Pt(0)–PPE network showed very high charge-carrier mobilities (1.5×10^{-2} cm^2 V^{-1} s^{-1}), the highest hitherto observed for a disordered conjugated polymer. The authors contend that these Pt-cross-linked materials could lead to new higher-performance semiconducting devices [85, 86].

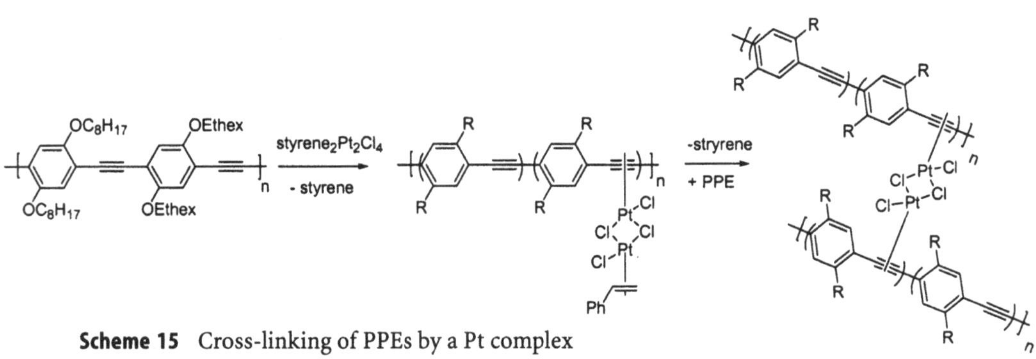

Scheme 15 Cross-linking of PPEs by a Pt complex

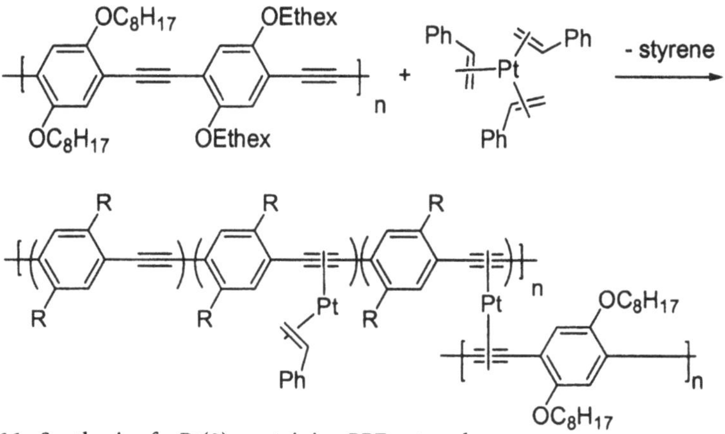

Scheme 16 Synthesis of a Pt(0)-containing PPE network

4.3 Side-Chain Postfunctionalization

An efficient postfunctionalization scheme would be a powerful approach to synthesize PPEs from a common precursor. The chemical diversity would be introduced late in the synthetic process, so that different PPEs would be available by a one-step procedure utilizing a preformed PPE backbone with a chemical handle attached. Wagner and Nuyken [87] have made an effort in that direction (Scheme 17). Starting from a PPE that contains a benzylic hydroxyl group, reaction with a mixture of PPh_3 and tetrabromomethane led to the clean conversion of the benzylic hydroxyl group into a benzylic bromide. This approach allows attaching of a nucleophilic group to the side chain by way of an S_N2-type reaction of the benzylic bromide. While the authors did not report any specific reactions, a cornucopia of new derivatives can be imagined utilizing their platform and performing S_N2 reactions.

A somewhat related approach for the attachment of polystyrene side chains onto preformed PPEs has been reported by Swager et al. (Scheme 18) [35]. Starting from a hydroxyalkyl-substituted PPE, reaction with bromoacetyl bromide furnishes an intermediate bromoester. Controlled radical polymerization with styrene furnishes a graft copolymer in which polystyrene side chains are attached to every second repeating unit of the PPE. Due to the presence of the ester linkage, the polystyrene side chains can be removed by hydrolysis. Surprisingly, there are no large molecular weight differences (GPC) between the grafted polymer and its "stripped" congener.

A last example of a postfunctionalized PPE was reported by Bunz et al. A grafted PPE made by polymerization of a macromonomer was treated with a biotin carboxylic acid chloride in the absence or in the presence of triethy-

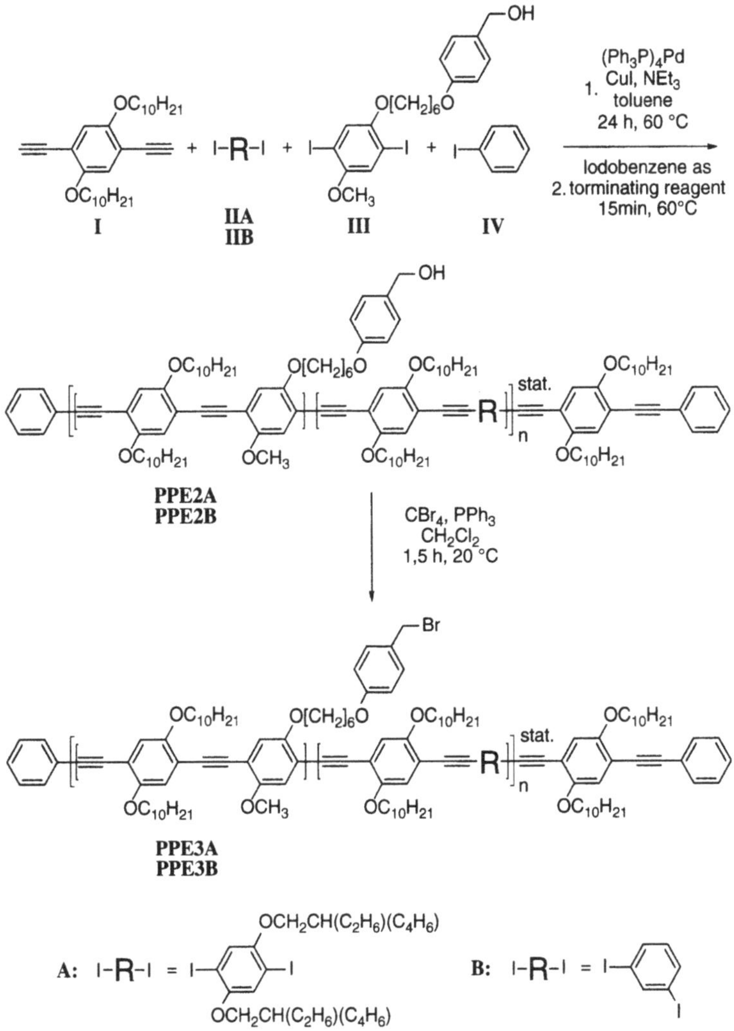

Scheme 17 Benzyl bromide-functionalized PPEs

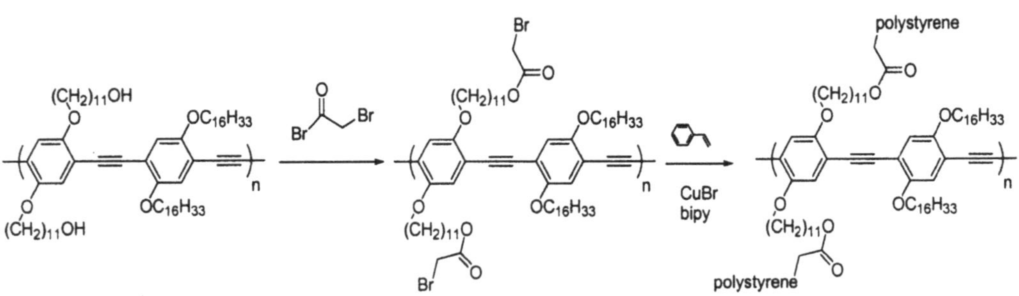

Scheme 18 Synthesis of a graft copolymer by Swager et aI. utilizing a postfunctionalization approach

lamine (Scheme 19) [8]. In the absence of triethylamine 5–7% of the repeating units were biotinylated, while in the presence of triethylamine up to 20% of the repeating units were biotin-decorated. The biotinylated PPEs were successfully utilized as a sensitive detection platform for streptavidin or avidin. A similarly biotinylated PPE was described by Whitten et al. who utilized this material as a FRET-based sensing platform for single-stranded DNA [88].

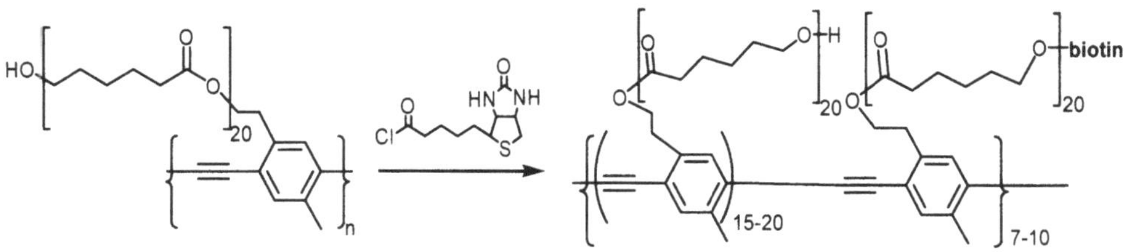

Scheme 19 Synthesis of a biotin-substituted PPE

5
Summary and Conclusions

The synthesis of novel PAEs has been comprehensively described in this article, covering the literature from 1999 to the end of 2003. Compared to the first decade of PPE chemistry (1990–1999) attention has shifted from the synthesis of simple dialkyl and dialkoxy PPEs (the focus of the last comprehensive PAE review [1]) to PAEs that show either sophisticated macromolecular architectures, water solubility, or are reversibly cross-linked, such as Weder's Pt- and Pd-connected PPEs. The increase in complexity leads to a host of new and attractive applications of these robust and chemically almost indestructible polymer backbones. Noteworthy is the recent trend to water-soluble and bioactive or biocompatible and biochromic PAEs. Such PAEs should find important applications in sensory devices and other biorelated diagnostic tools. Overall, PAEs have graduated from a synthetic curiosity into sophisticated and useful materials with a broad range of properties. The synthetic methods to make the conjugated backbone, i.e., form the aryleneealkynylene units, have not changed much and are either the Pd-catalyzed coupling of the Heck–Cassar–Sonogashira type or alkyne metathesis utilizing simple mixtures of $Mo(CO)_6$ and acidic phenols at enhanced temperatures. In the future we will see PAEs in different applications that will range from microstructured materials and semiconductor devices to explosive and infectious disease detection.

Acknowledgments The author acknowledges generous funding for the PAE projects from the National Science Foundation (CHE 0138659, DMR 0138948), the Petroleum Research Funds, the Department of Energy (DE-FG02-03ER46029), The Georgia Institute of Technology, The University of South Carolina, and the National Institute of Health. I am grateful to the gifted and enthusiastic coworkers and collaborators I have had over the last 10 years. Particularly important for the success of the PAE project were (and are): Dr. L. Kloppenburg, Dr. N.G. Pschirer, Dr. W. Steffen, Dr. D. Song, Dr. P.-H. Ge, Dr. G. Brizius, Dr. T. Miteva, C.G. Bangcuyo, J.N. Wilson, Y.-Q. Wang, B. Englert, Dr. I.-B. Kim, J. Leech, a series of excellent undergraduates from the University of South Carolina, Prof. M. Srinivasarao (PTFE, Gatech), Prof. U. Scherf (Univ. Wuppertal), Prof. D. Neher (Univ. Potsdam), Prof. C. Murphy (USC), Prof. M. Myrick (USC), Prof. R. Adams, (USC), Prof. H-C zur Loye (USC), Prof. M. Garcia-Garibay (UCLA), Prof. R. Schroeder (Bowling Green), Dr. V. Enkelmann, (MPIP Mainz), and last but not least the USC staff crystallographer Dr. M. Smith (USC).

References

1. Bunz UHF (2000) Chem Rev 100:1605–1644
2. Scherf U (1999) Top Curr Chem 201:163–222
3. Enkelmann V (1984) Adv Polym Sci 63:91–136
4. Halkyard CE, Rampey ME, Kloppenburg L, Studer-Martinez SL, Bunz UHF (1998) Macromolecules 31:8655–8659
5. Miteva T, Palmer L, Kloppenburg L, Neher D, Bunz UHF (2000) Macromolecules 33:652–654
6. Kim IB, Bunz UHF (unpublished results)
7. Lavigne JJ, Broughton DL, Wilson JN, Erdogan B, Bunz UHF (2000) Macromolecules 36:7409–7412
8. Wilson JN, Wang YQ, Lavigne JJ, Bunz UHF (2003) Chem Commun 1626–1627
9. Hotta S, Rughooputh DDV, Heeger AJ, Wudl F (1987) Macromolecules 20:212–215
10. Giesa R, Schulz RC (1990) Macromol Chem Phys 191:857–867
11. Thorand S, Krause N (1998) J Org Chem 63:8551–8553
12. Moroni M, LeMoigne J, Luzzati S (1994) Macromolecules 27:562–571
13. Zhou Q, Swager TM (1995) J Am Chem Soc 117:12593–12602; Steiger D, Smith P, Weder C (1997) Macromol Rapid Commun 18:643–649
14. McQuade TD, Kim J, Swager TM (2000) J Am Chem Soc 122:5885–5886
15. Nguyen P, Zheng YA, Agcos L, Lesley G, Marder TB (1994) Inorg Chim Acta 220:289–296
16. Hundertmark T, Littke AF, Buchwald SL, Fu GC (2000) Org Lett 2:1729–1731
17. Frederic P, Patt J, Hartwig JF (1995) Organometallics 14:3030–3039
18. Brizius G, Bunz UHF (2002) Org Lett 4:2829–2831
19. Grela K, Ignatowska J (2002) 4:3747–3749
20. Schrock RR, Clark DN, Sancho J, Wengrovius JH, Rocklage SM, Pedersen SF (1982) Organometallics 1:1645–1651
21. Fürstner A, Guth O, Rumbo A, Seidel G (1999) J Am Chem Soc 121:11108–11113
22. Blackwell JM, Figueroa JS, Stephens FH, Cummins CC (2003) Organometallics 22:3351–3353
23. Zhang W, Kraft S, Moore JS (2004) J Am Chem Soc 126:329–335
24. Wang YQ, Erdogan B, Wilson JN, Bunz UHF (2003) Chem Commun 1624–1625
25. Moon JH, Swager TM (2002) Macromolecules 35:6086–6089
26. Swager TM, Gil CJ, Wrighton MS (1995) J Phys Chem 99:4886–4893
27. Francke V, Mangel T, Mullen K (1998) Macromolecules 31:2448–2453
28. Huang WY, Matsuoka S, Kwei TK, Okamoto Y, Hu XS, Rafailovich MH, Sokolov J (2001) Macromolecules 34:7809–7816
29. Sato T, Jian D-L, Aida T (1999) J Am Chem Soc 121:10658–10659
30. Masuo S, Yoshikawa H, Asahi T, Masuhara H, Sato T, Jiang DL, Aida T (2003) J Phys Chem B 107:2471–2479

31. Masuo S, Yoshikawa H, Asahi T, Masuhara H, Sato T, Jiang DL, Aida T (2002) J Phys Chem B 106:905–909
32. Masuo S, Yoshikawa H, Asahi T, Masuhara H, Sato T, Jiang DL, Aida T (2001) J Phys Chem B 105:2885–2889
33. Jiang DL, Sato T, Aida T (2001) Chinese J Polym Sci 19:161–166
34. Zhu ZG, Swager TM (2003) J Am Chem Soc 125:9942–9943
35. Breen CA, Deng T, Breiner T, Thomas EL, Swager TM (2003) J Am Chem Soc 125:9670–9671
36. Wilson JN, Windscheif PM, Evans U, Myrick ML, Bunz UHF (2002) Macromolecules 35:8681–8683
37. Erdogan B, Wilson JM, Bunz UHF (2002) Macromolecules 35:7863–7864
38. Yang JS, Swager TM (1998) J Am Chem Soc 120:11864–11873
39. Yang JS, Swager TM (1998) J Am Chem Soc 120:5321–5322
40. Tan C, Pinto MR, Schanze KS (2002) Polym Preprints 43:126–127
41. Tan C, Pinto MR, Schanze KS (2002) Chem Commun 446–447
42. Pinto MR, Kristal BM, Schanze KS (2003) Langmuir 19:6523–6533
43. Kuroda K, Swager TM (2003) Chem Commun 26–27
44. Moon JH, Deans R, Krueger E, Hancock LF (2003) Chem Commun 104–105
45. Erdogan B, Wilson JM, Bunz UHF (2002) Macromolecules 35:7863–7864
46. Babudri F, Colangiuli D, DiLorenzo PA, Farinola GM, Omar OH, Naso F (2003) Chem Commun 130–131
47. Haley MM, Pak JJ, Brand SC (1999) Top Curr Chem 201:81–130
48. Arnt L, Tew GN (2002) J Am Chem Soc 124:7664–7665
49. Arnt L, Tew GN (2003) Langmuir 19:2404–2408
50. Chu Q, Pang Y, Ding L, Karasz FE (2003) Macromolecules 36:3848–3853
51. Chu Q, Pang Y, Ding L, Karasz FE (2002) Macromolecules 35:7569–7574
52. Li JA, Pang Y (2004) Synth Met 140:43–48
53. Chu QH, Pang Y (2003) Macromolecules 36:4614–4618
54. Kloppenburg L, Jones D, Bunz UHF (1999) Macromolecules 32:4194–4203
55. Pschirer NG, Byrd K, Bunz UHF (2001) Macromolecules 33:8590–8591
56. Pschirer NG, Marshall AR, Stanley C, Beckham HW, Bunz UHF (1999) Macromol Rapid Commun 21:493–495
57. Yamaguchi S, Swager TM (2001) J Am Chem Soc 123:12087–12088
58. Rose A, Lugmair CG, Swager TM (2001) J Am Chem Soc 123:11298–11299
59. Kiamuddin M, Haque ME, Pak J (1966) J Sci Ind Res 9:30–34
60. Bangcuyo CG, Rampey-Vaughn ME, Quan LT, Angel SM, Smith MD, Bunz UHF (2002) Macromolecules 35:1563–1568
61. Bangcuyo CG, Ellsworth JM, Evans U, Myrick ML, Bunz UHF (2003) Macromolecules 36:546–548
62. Bangcuyo CG, Evans U, Myrick ML, Bunz UHF (2001) Macromolecules 34:7592–7594
63. Jegou G, Jenekhe SA (2001) Macromolecules 34:7926–7928
64. Brizius G, Pschirer NG, Steffen W, Stitzer K, zur Loye H-C, Bunz UHF (2000) J Am Chem Soc 122:12435–12440
65. Egbe DAM, Bader C, Nowotny J, Gunther W, Klemm E (2003) Macromolecules 36:5459–5469
66. Egbe DAM, Roll CP, Birckner E, Grummt UW, Stockmann R, Klemm E (2002) Macromolecules 35:3825–3837
67. Egbe DAM, Tillmann H, Birckner E, Klemm E (2001) Macromol Chem Phys 202:2712–2726
68. Egbe DAM, Bader C, Klemm E, Ding L, Karasz FE, Grummt UW, Birckner E, (2003) Macromolecules 36:9303–9312

69. Egbe DAM, Birckner E, Klemm E (2002) J Polym Sci A 40:2670–2679
70. Schenning APHJ, Tsipis AC, Meskers SCJ, Beljonne D, Meijer EW, Bredas JL (2002) Chem Mater 14:1362–1368
71. Murer P, Lewandowski K, Svec F, Frechet JMJ (1998) Chem Commun 2559–2560
72. Brochini S, James K, Tangpasuthadol V, Kohn J (1997) J Am Chem Soc 119:4553–4554
73. Jandeleit B, Schaefer DJ, Powers TS, Turner HW, Weinberg WH (1999) Angew Chem 38:2494–2532
74. Lavastre O, Morken J (1999) Angew Chem 38:3163–3165
75. Bäuerle P (2004) Nachr Chem 52:19–24
76. Lavastre O, Illitchev I, Jegou G, Dixneuf PH (2002) J Am Chem Soc 124:5278–5279
77. Leung ACW, Chong JH, Patrick BO, MacLachlan MJ (2003) Macromolecules 36:5051–5054
78. Marshall AR, Bunz UHF (2001) Macromolecules 34:4688–4690
79. Ricks HL, Choudry UH, Marshall AR, Bunz UHF (2003) Macromolecules 36:1424–1425
80. Cotts PM, Swager TM, Zhou Q (1996) Macromolecules 29:7323–7328
81. Beck JB, Kokil A, Ray D, Rowan SJ, Weder C (2002) Macromolecules 35:590–593
82. Huber C, Bangerter F, Caseri WR, Weder C (2001) J Am Chem Soc 124:9978–9979
83. Huber C, Kokil A, Caseri WR, Weder C (2002) Organometallics 21:3817–3818
84. Kokil A, Huber C, Caseri WR, Weder C (2003) Macromol Chem Phys 205:40–45
85. Kokil A, Shiyanovskaya I, Singer KD, Weder C (2002) J Am Chem Soc 124:9978–9979
86. Kokil A, Shiyanovskaya I, Singer KD, Weder C (2003) Synth Met 138:513–517
87. Wagner M, Nuyken O (2003) Macromolecules 36:6716–6721
88. Kushon SA, Bradford K, Marin V, Suhrada C, Armitage BA, McBranch D, Whitten D (2003) Langmuir 19:6456–6464

Adv Polym Sci (2005) 177: 53–90
DOI 10.1007/b101375

Organometallic PAEs

Elisabeth Klemm (✉) · Thomas Pautzsch · Lars Blankenburg

Institut für Organische Chemie und Makromolekulare Chemie,
Friedrich-Schiller-Universität Jena, Humboldtstraße 10, 07743 Jena, Germany
Elisabeth.Klemm@uni-jena.de

Abstract This review describes recent results in the field of poly(aryleneethynylene)s (PAEs) that contain metal ions in the polymer backbone, or in the polymer side chain. This work is focused primarily on polymers possessing ligands of metal complexes as part of the aryleneethynylene chain. PAEs with porphyrinylene in the backbone have also been addressed. Synthetic routes toward the polymers, as well as their photochemical, photophysical, and electrochemical properties, are presented. Monodisperse oligo(phenyleneethynylene)s with terminal metal complexes or with a ferrocene and thiol at each end are mentioned.

Keywords Sonogashira polycondensation · Hybrid polymers · Polypyridine ligands · Copolymers · Macromolecular metal complexes

Abbreviations

^{13}C NMR	Carbon nuclear magnetic resonance
CV	Cyclic voltammetry
dba	Dibenzylideneacetone (ligand)
dppf	1,1′-Bis(diphenylphosphino)ferrocene
E^o	Oxidation potential $[(E_{pa}+E_{pc})/2]$
E_g^{opt}	Bandgap energy (optically measured)
E_{ox}	Oxidation potential
E_{pa}	Anodic peak potential
E_{pc}	Cathodic peak potential
fac	Facial
Fc	Ferrocene
FTIR	Fourier-transform infrared
GPC	Gel-permeation chromatography
^{1}H NMR	Proton nuclear magnetic resonance
HOMO	Highest occupied molecular orbital
LUMO	Lowest unoccupied molecular orbital
MLCT	Metal-to-ligand charge transfer
M_n	Number-average molecular weight
M_w	Weight-average molecular weight
NLO	Nonlinear optics
OPE	Oligo(phenyleneethynylene)
PAE	Poly(aryleneethynylene)
P_n	Degree of polymerization
PPE	Poly(phenyleneethynylene)
ppH	2-Phenylpyridine
SCE	Saturated calomel electrode
UV	Ultraviolet
Vis	Visible
ε	Extinction coefficient
$\lvert n_2 \rvert$	Nonlinear refraction index
Φ_{fl}	Fluorescence quantum yield
λ_{max}	Wavelength of maximum absorption
)))	Sonication

1 Introduction

Over the last three decades, there has been a great focus on the investigation of the chemical and physical properties of π-conjugated polymers, culminating in the award of the Nobel Prize for chemistry in the year 2000 to Alan J. Heeger, Alan G. MacDiarmid, and Hideki Shirakawa, who were all recognized for the discovery and development of conductive polymers [1–5]. π-Conjugated oligomers and polymers are of interest due to their unique optical and electronic properties, which enable them to be used as the active medium in optical, electronic, optoelectronic, and chemical sensing devices.

Among the π-conjugated polymers, poly(*p*-phenylenevinylene)s (PPVs) have attracted particular attention since the 1990 report of Friend et al. on

Table 1 Important review articles covering the same, related, or adjacent areas

Entry	Author	Citation	Title
1	Bunz UHF	Chem Rev (2000) 100:1605–1644	Poly(aryleneethynylene)s: synthesis, properties, structures, and applications
2	Kingsborough RP, Swager TM	Prog Inorg Chem (1999) 48:123–231	Transition metals in polymeric π-conjugated organic frameworks
3	Long NJ	Angew Chem Int Ed Engl (1995) 34:21–38	Organometallic compounds for nonlinear optics – the search for en-light-enment
4	Yamamoto T	Macromol Rapid Commun (2002) 23:583–606	π-Conjugated polymers with electronic and optical functionalities: preparation by organometallic polycondensation, properties, and applications
5	Kraft A, Grimsdale AC, Holmes AB	Angew Chem Int Ed (1998) 37:403–428	Electroluminescent conjugated polymers – seeing polymers in a new light
6	Giesa R	JMS – Rev Macromol Chem Phys (1996) C36:631–670	Synthesis and properties of conjugated poly(aryleneethynylene)s
7	Balzani V, Juris A, Venturi M	Chem Rev (1996) 96:759–833	Luminescent and redox-active polynuclear transition metal complexes
8	Schubert US, Eschbaumer C	Angew Chem Int Ed (2002) 41:2892–2926	Macromolecules containing bipyridine and terpyridine metal complexes: towards metallo-supramolecular polymers
9	Long NJ, Williams CK	Angew Chem Int Ed 2003, 42:2586–2617	Metal alkynyl sigma complexes: synthesis and materials
10	Ley KD, Schanze KS	Coord Chem Rev (1998) 171:287–307	Photophysics of metal–organic π-conjugated polymers
11	Rehan M	Acta Polym (1998) 49:201–224	Organic/inorganic hybrid polymers
12	Ziessel R	Synthesis (1999) 11:1839–1865	Making new supermolecules for the next century: multipurpose reagents from ethynyl-grafted oligopyridines
13	Martin RE, Diederich F	Angew Chem Int Ed (1999) 38:1350–1377	Linear monodisperse π-conjugated oligomers: model compounds for polymers and more
14	Tour JM	Acc Chem Res (2000) 33:791–804	Molecular electronics: synthesis and testing of components
15	Bunz UHF	Acc Chem Res (2001) 34:998–1010	Poly(*p*-phenyleneethynylene)s by alkyne metathesis

organic LEDs [6,7]. However, poly(phenyleneethynylene)s (PPEs), having similar properties to those of the PPV-type polymers, have received much less attention [8]. Furthermore, there has been little attention paid to π-conjugated hybrid polymers containing both organic and inorganic components [9, 10]. The situation is different, however, for defined monodisperse metal-containing oligomers. A multitude of compounds of this kind have been synthesized and characterized by the research groups of Tour, Ziessel, Diederich, and Schanze (Table 1). Table 1 lists a selection of literature on defined oligomeric arylene ethynylenes and other review articles related to the topic discussed here.

The introduction of transition metal ions into π-conjugated polymers provides enormous opportunities to fine-tune the physical properties of the resulting materials. From the strong interaction between transition metal complexes and conductive polymer backbones, new photophysical and electrochemical properties are expected. The resulting new materials should possess a wide range of interesting characteristics, such as photorefractive effects, photoconductivity, and novel redox properties [11–15]. Ruthenium(II)-polypyridine complexes are among the most studied molecules because they are chemically very stable, and they have numerous redox states and strong photoemission under photoexcitation [16] and electrochemical excitation [17]. The photochemical, photophysical, and electrochemical aspects of these studies have been reviewed [18–20]. In addition to the early interest in water splitting [21], ruthenium(II)-tris(2,2′-bipyridine) compounds have been exploited for artificial photosynthesis [22]. The strong specific photoluminescence and electroluminescence of these compounds are used for molecular recognition [23,24] and for chemical analysis [25,26].

The polymers derived from ruthenium(II)-polypyridine complexes have demonstrated promising potential for application in solar energy conversion, sensors, polymer-supported electrodes, nonlinear optics, photorefraction, and electroluminescence [27–32].

2 Scope and Structure

2.1 Main Structural Principles of Macromolecular Metal Complexes

Macromolecular metal complexes can be classified into three main categories, taking into consideration the manner of binding of a metal compound to suitable macroligands [33] (Fig. 1). Type 1 metal complexes are those with the metal ion or metal chelate at a macromolecular chain, network, or surface. One possible approach to synthesize such polymers is using the polymerization of vinyl-substituted metal complexes.

Type 2 metal complexes have the macromolecular metal complex as part of a polymer chain via the metal or the ligand. Different methods for their prepa-

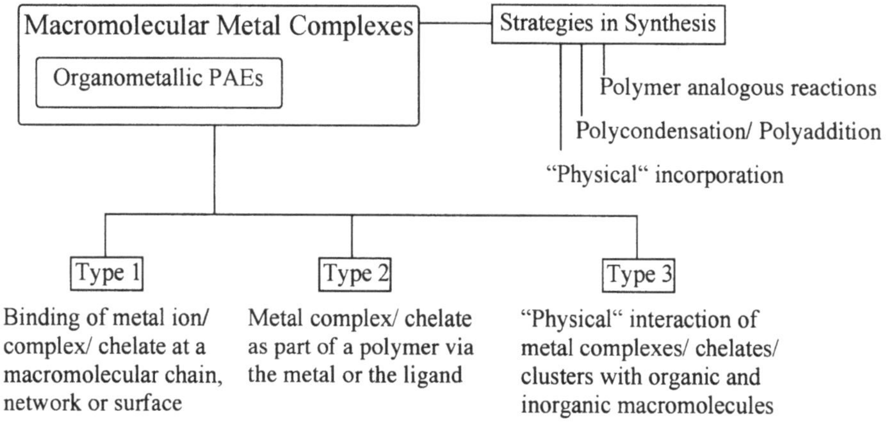

Fig. 1 Synthetic routes for macromolecular metal complexes

ration, such as the reaction of a bifunctional ligand with the metal compound, or the reaction of a bifunctional metal complex with another bifunctional reagent, are given. Type 2 macromolecular metal complexes are described in Sect. 3.

Type 3 metal complexes involve the "physical" interaction of a metal complex, chelates, or metal cluster with an organic polymer or inorganic high molecular weight compound. The preparation of type 3 compounds differs from those of type 1 and type 2, as they are ultimately achieved through the use of adsorption, deposition by evaporation, microencapsulation, and various other methods.

2.2
Structures of Organometallic PAEs

In this review, well-defined metal-containing PAEs are described whose primary structure is represented by one of the schematic drawings A–C and E shown in Fig. 2. In contrast to the structures shown in the A–C systems, E has a conjugated phenyleneethynylene with metal chelates as end groups. PAEs containing metal complex as side groups (D) have, up to now, not been described in the literature. The classes of compounds such as metal-bridged alkynes, the poly(metallayne)s, and polymer carbyne complexes (structures G and H) do not in fact represent PAEs.

An interesting report by Bunz, also on the topic of organometallic PAEs, can be found in the literature and addresses work in this field from 1995 to mid-1999 [10]. Our article will focus on developments in organometallic PAEs from 1997 to 2003, but also includes earlier significant references where necessary for discussion.

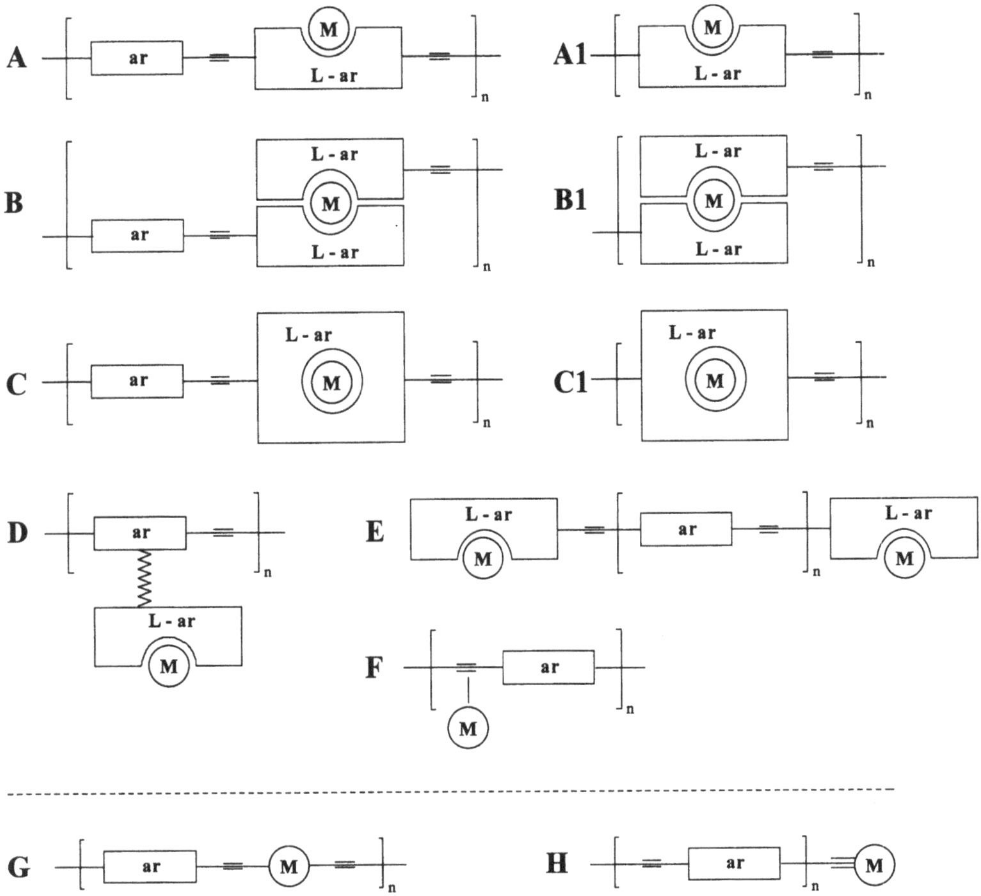

M = (transition) metal; ar = (hetero) arene bridging unit; L - ar = complexing ligand - (hetero) arene

Fig. 2 Schematic drawings of metal-containing PAE structures (see text)

3
Metal Complexes as Part of a PAE Chain

3.1
2,2′-Bipyridyl–Ru(II) Complexes in the PAE Main Chain

The Pd-catalyzed coupling reaction is a useful method of preparing organometallic heteroaryleneethynylene copolymers with alternating structure. To synthesize polymers with a defined high value of metal chelate units in the main chain, Klemm et al. utilized the Sonogashira cross-coupling reaction of the isomeric (4,4′- or 5,5′-dibromo-2,2′-bipyridine)-bis(2,2′-bipyridine) or (4,4′-*tert*-butyl-2,2′-bipyridine)ruthenium(II) complexes **5–8** (Scheme 2) [34–36] and diethynylarenes (phenylene, anthracenylene, 2,5-dialkoxyphenylene,

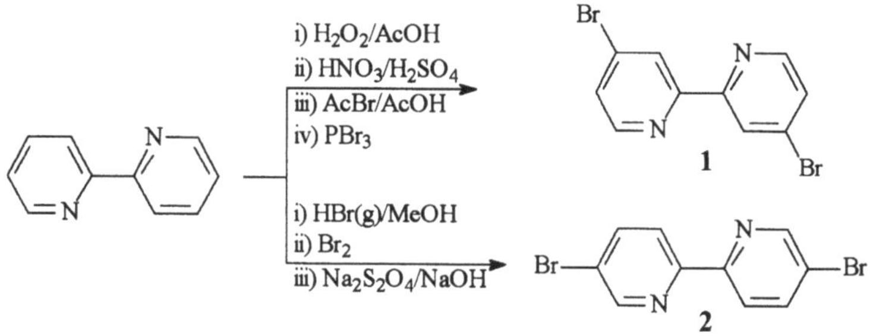

Scheme 1 Synthesis of 4,4′-or 5,5′-dibromo-2,2′-bipyridine

pyridinylene, *N*-phenyl-*N,N*-diphenyleneamine with different substituents) in a step-growth polymerization. The dibromo-2,2′-bipyridine isomers were synthesized by bromination of the dihydrobromide (5,5′-dibromo-2,2′-bipyridine, (**1**)) in a glass autoclave [37]. The preparation by the *N*-oxide route [38] forms the 4,4′-dibromo-2,2′-bipyridine (**2**) in good yield (Scheme 1).

Further complexation with (2,2′-bipyridine)$_2$RuCl$_2$*0.5H$_2$O (**3**) or (4,4′-*tert*-butyl-2,2′-bipyridine)$_2$RuCl$_2$*toluene (**4**) for 12 h in a mixture of ethanol/water (2:1) followed by precipitation gives the dark red ruthenium complexes **5–8** [39–41]. Recrystallization from EtOH/H$_2$O (2:1) yielded the pure products on the gram scale (Scheme 2).

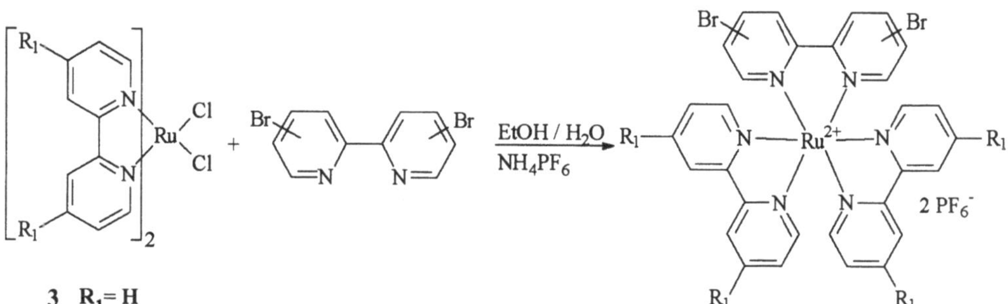

3 R$_1$= H
4 R$_1$= ***tert*-butyl**
1/2
5–8

Scheme 2 Synthesis of the dibromoruthenium(II) complexes

In order to obtain the *tert*-butyltriphenylamine derivative, the reaction according to Hartwig et al. and Buchwald et al. [42–46] (diphenylamine, *tert*-butylbromobenzene) was carried out, followed by bromination and ethynylation (Scheme 3).

The new polymers **9–28** were synthesized by the Pd-catalyzed reaction of (4,4′- or 5,5′-dibromo-2,2′-bipyridine)-bis(4,4′-*tert*-butyl-2,2′-bipyridine or 2,2′-bipyridine)ruthenium(II) complexes **5–8** and diethynylarenes (Table 2) in a step-growth polycondensation mechanism (Scheme 4). The typical reaction conditions used for the synthesis of the polymers involved stirring the argon-

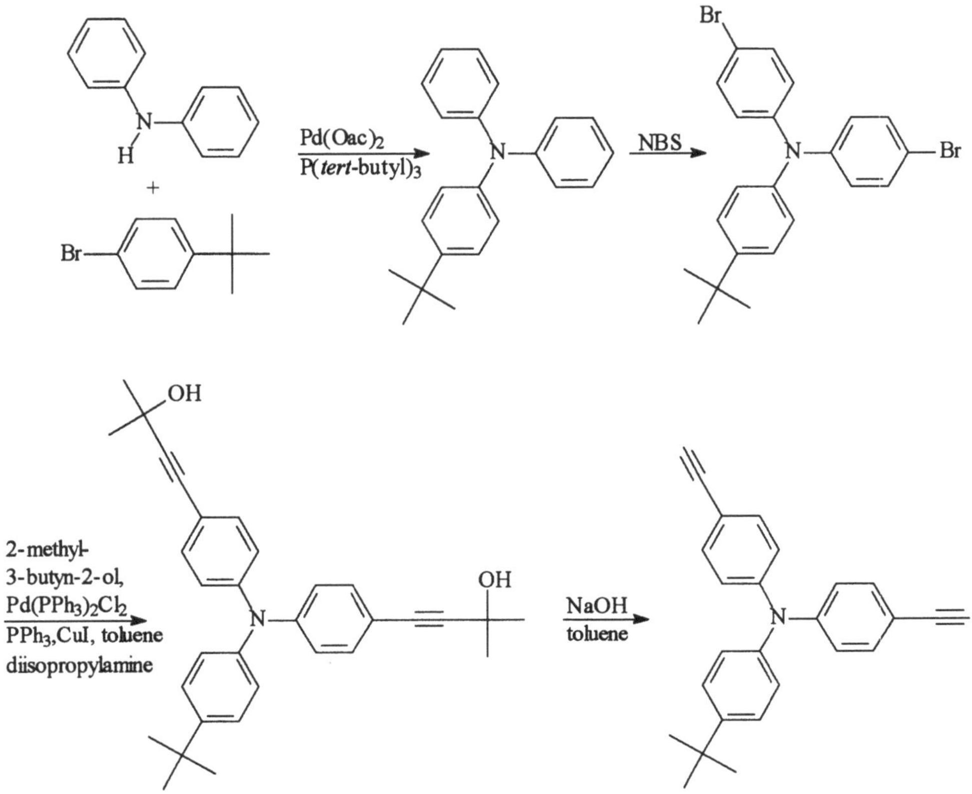

Scheme 3 Synthetic route to a diethynyl triphenylamine derivative

degassed mixture of the monomers and catalysts (4 mol% of $Pd(PPh_3)_4$ and 4 mol% of CuI) in acetonitrile/diisopropylamine for 72–96 h at a temperature of 75 °C. The polymers were isolated by dropwise addition to methanol and further precipitation with NH_4PF_6. Soxhlet extraction with methanol was then performed, followed by dissolution in acetone and dropwise addition to vigorously stirred *n*-hexane. This process then gave the pure red polymer with yields above 65–85%, pointing to a very high conversion. The polymers are all soluble in acetone, acetonitrile, DMF, and DMSO, with the exception of polymer **14**.

Moreover, it is remarkable that a simple change of the substituents in the ligands (*tert*-butyl rather than hydrogen) leads to a significant difference in the solubility of the polymers. Whereas polymer **14** (R_1 at the ligands is H), for example, is absolutely insoluble in all common organic solvents, polymer **11** (R_1=H, R_2=$C(CH_3)_3$) is completely soluble in acetone, acetonitrile, and DMSO. The polymers substituted with *tert*-butyl groups are also soluble in cyclohexanone. All the soluble polymers can be converted into transparent thin films (thickness between 150 nm and 3 µm). This is important for application in optical devices.

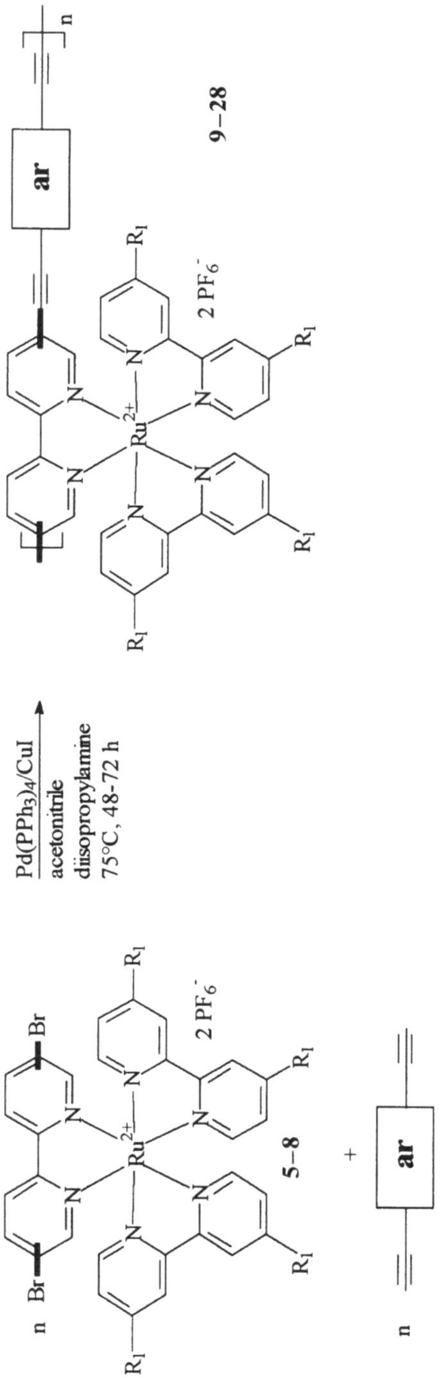

Scheme 4 Synthesis of the (2,2′-bipyridinylene)Ru(II) PAEs

Table 2 Structures and properties of 2,2′-bipyridinylene–Ru(II) polymers

polymer	ruthenium isomer	R_1-ligand	aromatic alkyne resp. ≡-**ar**-≡ (diethynylarenes)	$\overline{M}_n$ [a]	$\overline{M}_w$ [a]	P_n	$\overline{M}_n$ (Br) [b]	$\lambda_{max\,1}$ [c] [nm]	ε [c] $[M^{-1}.cm^{-1}]$	E_{ox1} [d]
9	5,5′-	*tert*-butyl	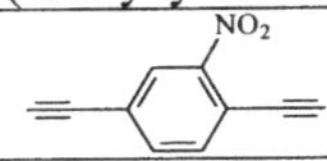	5800	16700	8	9700	434	17000	
10	5,5′-	*tert*-butyl	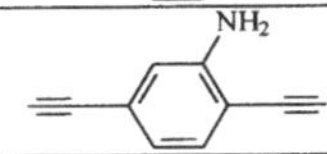	16400	97300	30	4800	450	55600	
11	5,5′-	*tert*-butyl	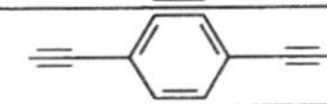	19400	74800	26	66600	418	82400	
12	4,4′-	*tert*-butyl	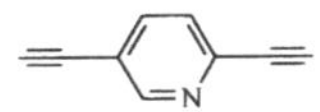	-	-	-	15700	422 495	17600 15700	
13	5,5′-	*tert*-butyl		-	-	-	8100	433	15900	
14	5,5′-	H	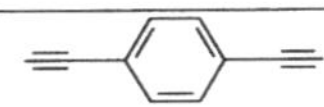	insoluble		-	-	-	-	
15	4,4′-	H	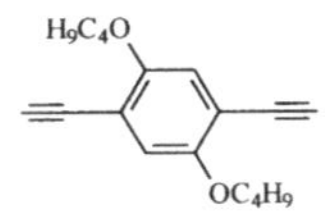	6900	15700	10	9800	426 494	43900 37400	1.34
16	5,5′-	H		16600	47500	34	31000	456	73800	
17	4,4′-	*tert*-butyl		25300	84900	39	13100	433 507	49800 38000	1.23
18	5,5′-	*tert*-butyl		22600	57700	32	80600	453	50900	1.27
19	5,5′-	H	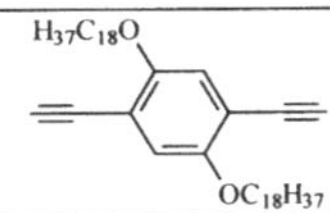	18800	62800	34	14600	456	66200	
20	4,4′-	*tert*-butyl	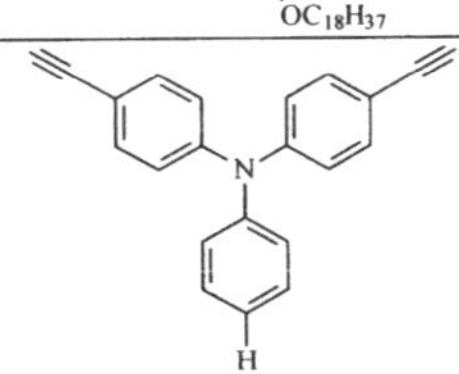	6200	17000	9	10700	426 499	47300 33100	
21	5,5′-	*tert*-butyl		16100	55400	23	9600	462	57400	1.25 1.10

Table 2 (continued)

polymer	ruthenium isomer	R_1-ligand	aromatic alkyne resp. ≡-**ar**-≡ (diethynylarenes)	$\overline{M}_n$ [a]	$\overline{M}_w$ [a]	P_n	$\overline{M}_n$ (Br) [b]	$\lambda_{max\,1}$ [c] [nm]	ε [c] [$M^{-1}.cm^{-1}$]	E_{ox1} [d]
22	4,4′-	*tert*-butyl	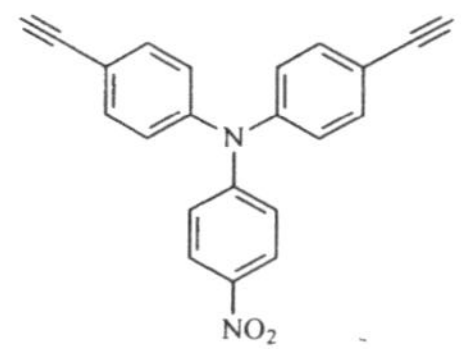	6300	16000	8	29600	415	60800	1.33
								496	32000	1.22
23	5,5′-	*tert*-butyl		19600	109900	32	17400	436	72600	1.26
										1.26
24	4,4′-	*tert*-butyl	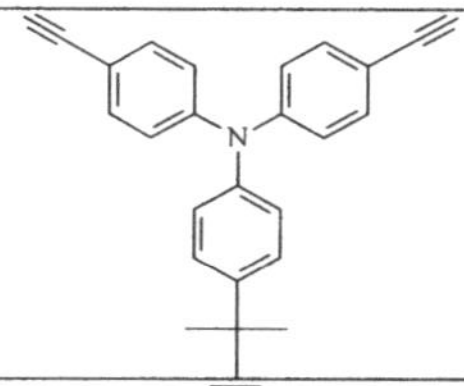	-	-	-	32000	428	19900	1.24
								500	40000	1.07
25	5,5′-	*tert*-butyl		-	-	-	16600	471	60400	1.25
										1.05
26	5,5′-	*tert*-butyl	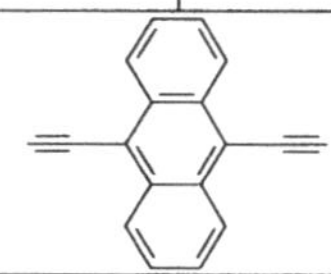	3200	9300	5	4400			
27	4,4′-	*tert*-butyl	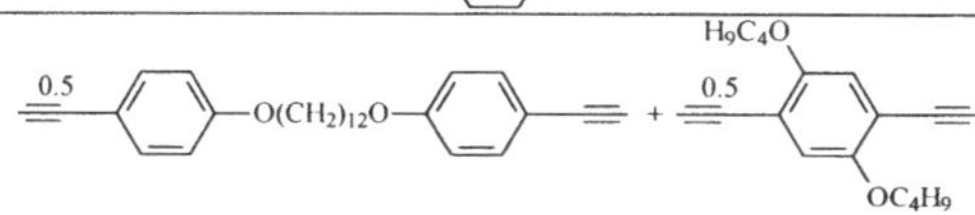	4500	23000	8	5600	425	30900	
								496	25400	
28	5,5′-	*tert*-butyl		14500	73400	17	6500	421	52900	

[a] GPC (DMSO, 70 °C, 0.1 M $LiPF_6$, pullulane and dextrane standard).
[b] Calculated from elemental analysis, one bromine per chain end.
[c] Acetonitrile solution.
[d] $Ag/AgCl/CH_3CN$ reference and Pt electrode, 0.25 M Bu_4NPF_6, standard ferrocene, potentials referenced to SCE ($E_{1/2}$=420 mV (SCE) for Fc/Fc^+).
Refs.: Pautzsch T, Klemm E (2002) Macromolecules 35:1569; Pautzsch T, Blankenburg L, Hager M, Egbe DAM, Klemm E (2003) Proc SPIE 5215; Pautzsch T, Blankenburg L, Klemm (2003) Polym Sci A Polym Chem 42:722.

The chemical structure of the polymers was confirmed by ^{1}H NMR and elemental analysis, and spectroscopically characterized in comparison with monodisperse low molecular weight model compounds. Scheme 5 outlines the approach to the model compounds. Model compounds **31–34** were synthesized by complexation of the ruthenium-free model ligands **29/30** with **3/4**. The model ligands were synthesized in toluene/diisopropylamine, in a similar fashion as the polycondensation using $Pd(PPh_3)_4$ and CuI as catalyst (Sonogashira reaction) [34, 47–49].

The M_n determination of the rigid polymers, such as the organometallic PAEs, using GPC methods (polystyrene standard) is not trivial. Trials conducted with the model compounds yield unreasonable and nonreproducible results. A good correctness can be obtained with a NOVEMA 300 (PSS, Mainz) column and pullulan/dextran standards. DMSO with $LiPF_6$ at 75 °C was used as the mobile phase. Comparison of the real molecular weight of model compounds (1,000, 2,000, and 4,000 g/mol) with the molecular weights obtained by GPC proved that GPC methods led to the underestimation of the molecular weights by a factor of 2–2.4. When compared to the results of end group determination, the molecular weights are overestimated by a factor of approximately 2.7. The 4,4′-connected polymers have lower maxima in the molecular weight distribution when compared with the 5,5′-substituted polymers.

UV/Vis absorption spectra of the polymers and the model complexes show four absorption maxima in acetonitrile. The absorption maxima in the visible region (around 450 and 440 nm, respectively) are similar to those of $Ru(bpy)_3^{2+}$, and therefore correspond to the metal-to-ligand charge-transfer (MLCT) band of ruthenium(II) complex units. The high molar absorptivity can therefore be explained by the fact that the MLCT band is likely buried under the considerably more intense ligand-centered π–π^* transition.

It is remarkable that the polymers containing the 4,4′-diethynylbipyridylene structure give a significant redshift in the MLCT absorption spectra compared to the linear polymers. A possible explanation for this observation may be attributed to the longer conjugation found in the 4,4′-isomer in comparison to the 5,5′-isomer (including the ligands in the conjugation). A similar interpretation is given for phenanthroline-based PAEs containing Ru(II) [34].

It is interesting to consider the effects that the inserted side groups in the arylene units have on absorption behavior. One can observe that an exchange of the amino (**10**) with nitro (**9**) groups leads to a hypsochromic shift and a reduction in extinction. The introduction of pyridine (**13**) shows the same result. Substituting triphenylamine segments into this framework leads to absorption behavior that is similar to that of amino- or alkoxy-substituted phenylene units. The triphenylamine has an electron-donating effect, like the aforementioned side groups. Noteworthy is that substitution in the 4″-position at the external triphenylamine phenylene with an electron-withdrawing group ($-NO_2$) (**23**) leads to the expected hypsochromic shift, while *tert*-butyl substitution gives a small bathochromic shift (**20–25**).

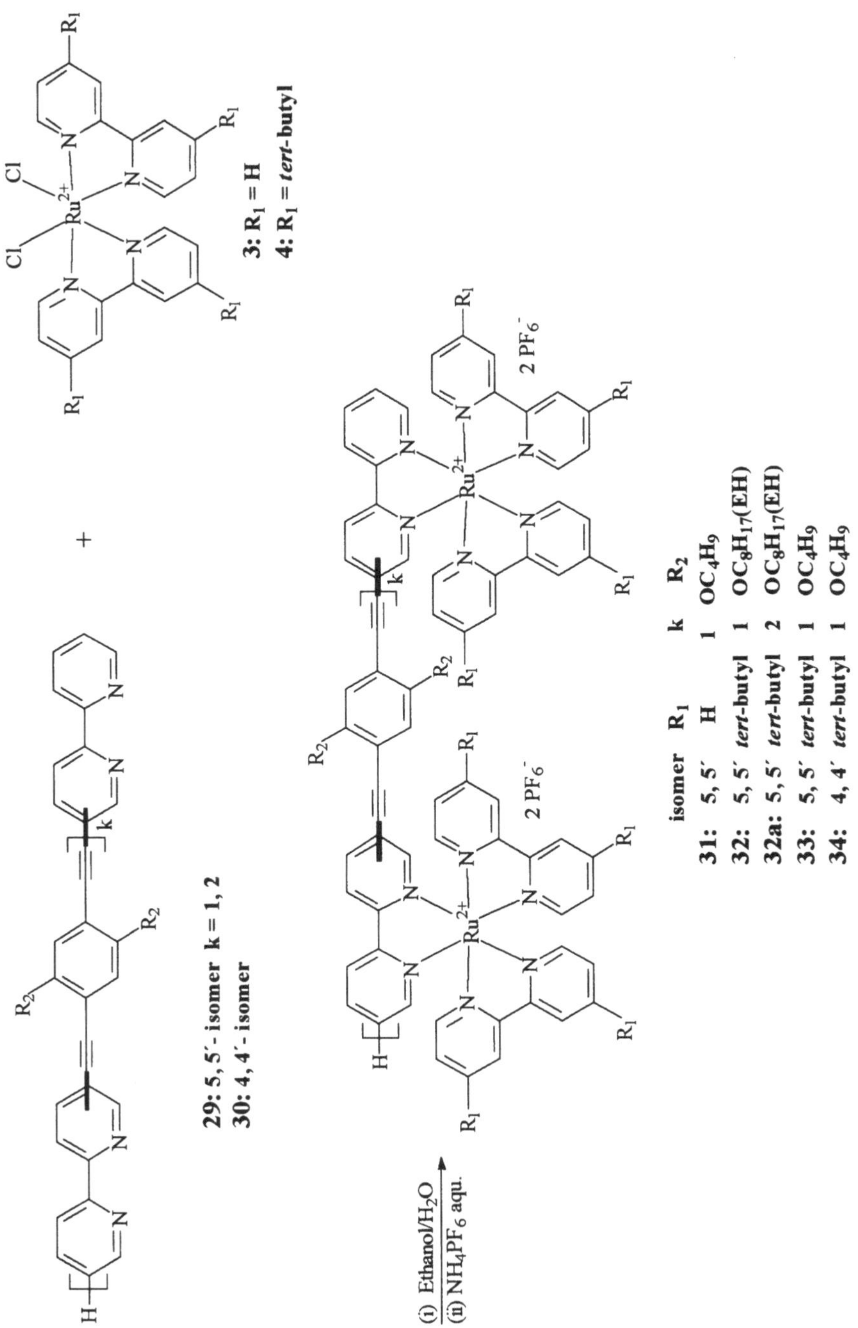

Scheme 5 Synthetic route to Ru(II) model compounds

All of the ruthenium polymers show emission when excited at λ_{max} (absorption). A large Stokes shift and a small quantum yield characterize the emission behavior; the luminescence quantum yield of the polymers is 1%. Thermogravimetric analyses in air indicate high thermal stability of the polymers, with thermal decomposition starting at approximately 290 °C. The polymers have no glass transition temperature.

The polymers can be oxidized by differential pulse polarography. Their oxidation is metal-centered and leads to Ru(III) compounds. The potential is located around +1.26 V (SCE). It can be stated that the polymers which contain the triphenylamine structure units in the main chain show, as expected, an additional peak caused by the amine nitrogen. Substitution at the triphenylamine by electron-donating substituents lowers these potentials to 1.05 V (**25**), whereas acceptor substituents cause an increase of the oxidation potential (**23**).

The third-order nonlinearity of the new compounds has been investigated by time-resolved degenerate four-wave mixing (Rentsch et al.). The measurements were performed in solutions at 1,047 nm with picosecond time resolution, and on thin films at 800 nm with femtosecond time resolution. The observed ultrafast response (less than 14 ps/180 fs), together with the possibility that these compounds can form thin films for waveguides, makes these polymers promising candidates for use as fast all-optical switching devices. $|n_2|$ values as high as 4.9×10^{-13} cm^2/W were measured [31, 50].

3.2 1,10-Phenanthrolyl–Ru(II) Complexes in the PAE Main Chain

The rigid framework of 1,10-phenanthroline ligands is an attractive feature for the construction of molecular assemblies. Although possessing favorable metal binding properties, they have rarely been employed for this purpose [51–55]. This was the motivation for Tor et al. to investigate the synthesis and spectral properties of 3,8-bis(arylethynyl)-1,10-phenanthrolines (**36**) and their Ru(II) complexes (**37**) [56]. The ligands were formed by Pd-catalyzed cross-coupling reactions using 3,8-dibromo-1,10-phenanthroline (**35**) and substituted phenylacetylenes under sonication at room temperature (Scheme 6).

The most intensive transition of the 1,10-phenanthroline structure is polarized along the 3 to 8 position [57]. So Tor et al. proposed that extension of conjugation along this axis can lead to derivatives with emission in the visible region. The substitution at the phenylacetylenes allowed for the study of the influence of the ligands (**36**) on the absorption behavior of the Ru(II) complexes (**37**). For complexation K_2RuCl_5 was used and the complexes were obtained as [Ru(phenanthroline ligand)$_3$]$^{2+}$(PF_6^-)$_2$ (**37**). This novel series of highly conjugated 3,8-bis(arylethynyl)-1,10-phenanthroline ligands and their complexes demonstrate that it is possible to tune the ligand-centered transition by changing the substitution. The complexes also show the typical MLCT at around 470 nm.

In a later publication Tor et al. reported for the first time that coordination compounds, such as tris-chelate Ru(II) complexes, are good substrates for C–C

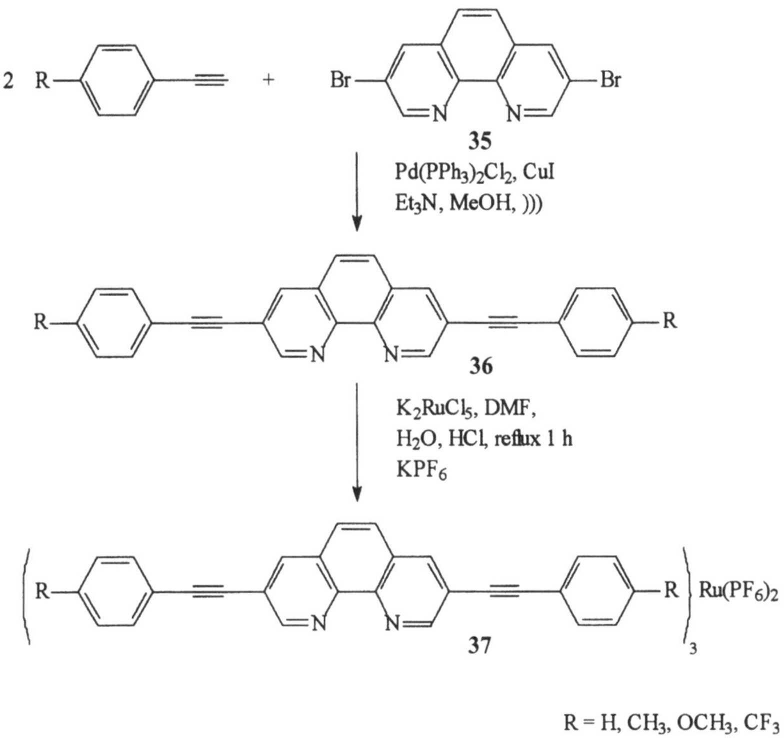

Scheme 6 Synthetic route to 3,8-bis(arylethynyl)-1,10-phenanthroline–Ru(II) complexes (Reproduced by permission of The Royal Society of Chemistry)

bond-forming reactions, such as Pd-catalyzed coupling reactions [58]. Therefore they prepared building blocks [$(bpy)_2$Ru(3-bromo- or ethynyl-1,10-phenanthroline)]$^{2+}$(PF_6^-)$_2$ **38/39** by refluxing 3-bromo- or ethynyl-1,10-phenanthroline and $(bpy)_2RuCl_2 \cdot 2H_2O$ in EtOH for 12 h. These blocks cross-coupled with mono-, di-, or triethynylated or -halogenated aromatic units, and $(PPh_3)_2PdCl_2$/CuI/Et_3N in DMF (the catalytic system) led to mono-, di-, and trinuclear Ru(II) complexes **40–43** in good yields, as shown in Fig. 3.

The longer degree of conjugation of the phenanthroline ligands in these complexes causes a bathochromic shift at the π–π^* band and the different nuclearity shows the 1:2:3 ratio of the extinction coefficient of the π–π^* as well as the MLCT bands. To avoid diastereomeric mixtures the authors established the first controlled synthesis of stereochemically defined multinuclear Ru(II) complexes [59].

Monofunctionalized and difunctionalized Ru(II) coordination complexes can be used as enantiomerically pure building blocks and can be cross-coupled to diastereochemically pure multi-Ru(II) complexes. The enantiomerically pure [$(bpy)_2$Ru(3,8-diethynyl-1,10-phenanthroline)]$^{2+}$(PF_6^-)$_2$ (**45**), [$(bpy)_2$Ru(3-ethynyl-1,10-phenanthroline)]$^{2+}$(PF_6^-)$_2$ (**44**), and [$(bpy)_2$Ru(3-bromo-1,10-phenanthroline)]$^{2+}$(PF_6^-)$_2$ (**46**) were isolated in their Δ and Λ forms. Chiral precursor

building blocks

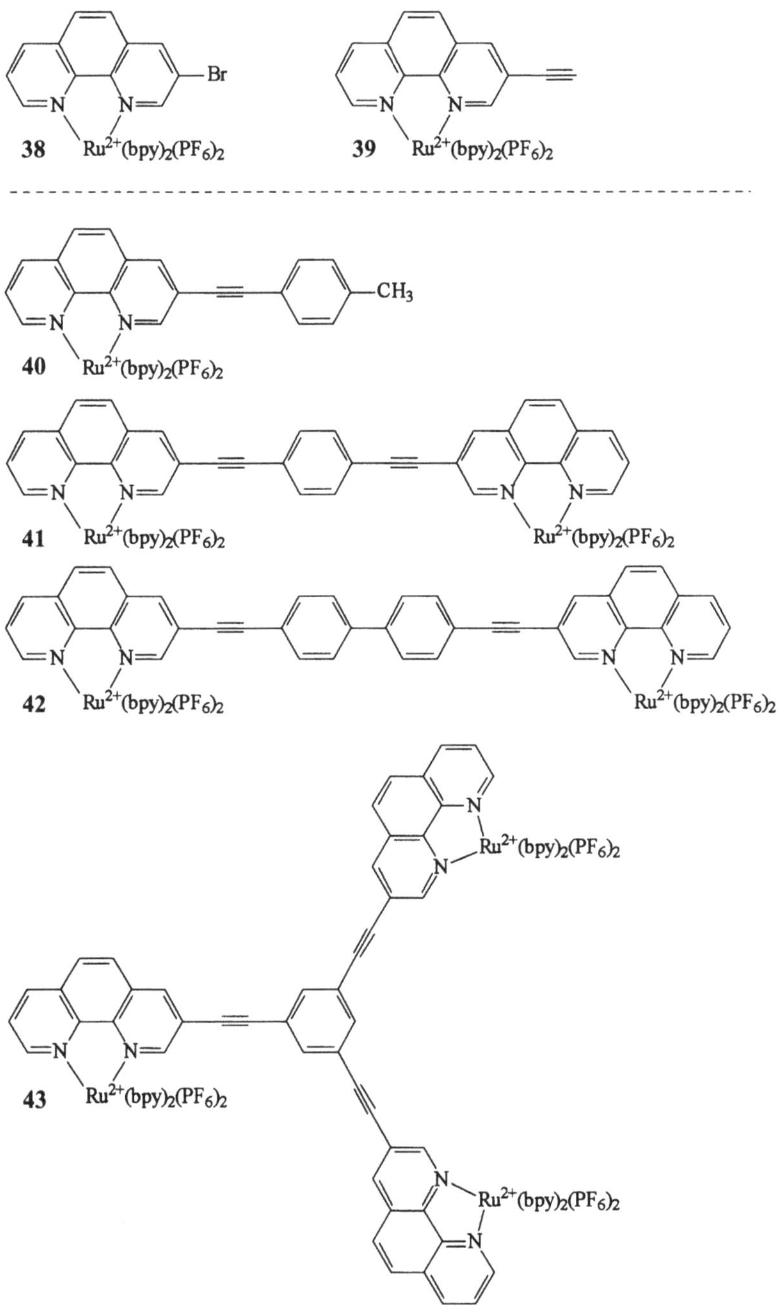

Fig. 3 Mono-, di-, and trinuclear phenanthroline–Ru(II) complexes

complexes containing (−)-*O,O′*-dibenzoyl-L-tartrate or (+)-*O,O′*-dibenzoyl-D-tartrate as ligands were treated with the phenanthroline derivatives, followed by exchanging the tartrate counterion with PF_6^-. Cross-coupling of these building blocks led to the well-defined and stereochemically pure di- and trinuclear Ru(II) complexes **47/48** (Scheme 7).

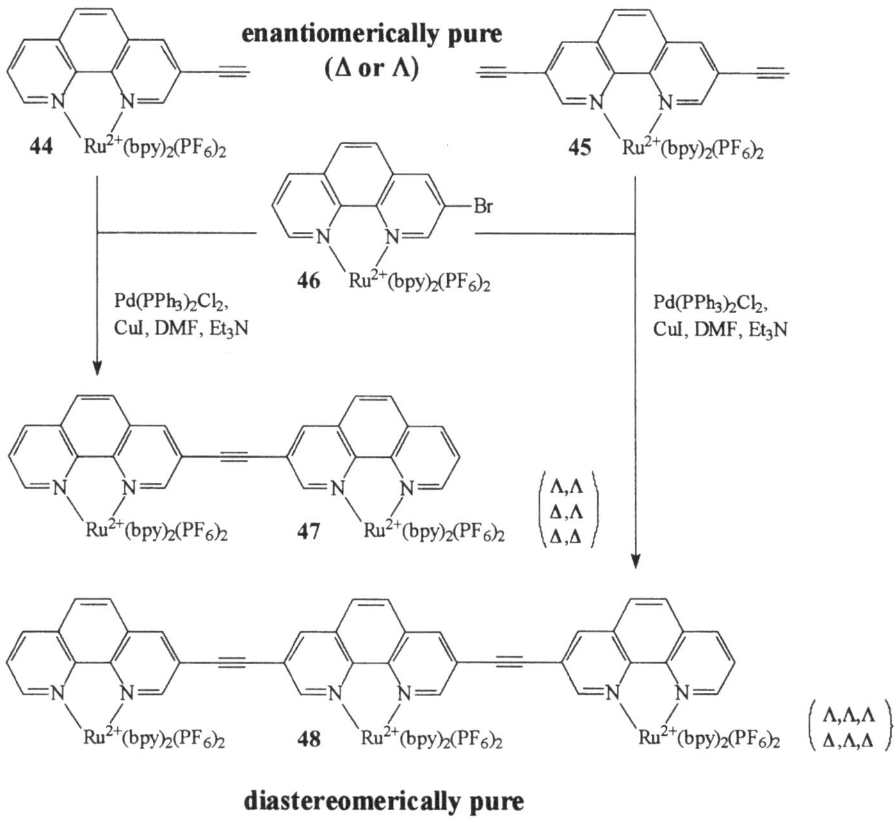

Scheme 7 Synthetic route to diastereomerically pure phenanthroline–Ru(II) complexes

The authors also go on to describe a convergent and powerful approach for the synthesis of heteronuclear complexes containing Ru(II) and Os(II) [60, 61]. Such heteronuclear complexes **49–53** (Fig. 4) are of great interest in order to study photoinduced energy and electron transfer [62]. Due to their outstanding redox and spectroscopic properties, Ru(II) and Os(II) have received a great deal of attention [19, 63–66]. Full control over structure and composition of the di- and trinuclear multimetallic complexes can be achieved by using bromo- or ethynyl-functionalized Ru/Os building blocks connected by Pd-catalyzed cross-coupling reactions.

The synthesis of chiral metal-containing polyaryleneethynylenes was achieved after extensive synthetic investigations, stereochemical investigations, hetero-

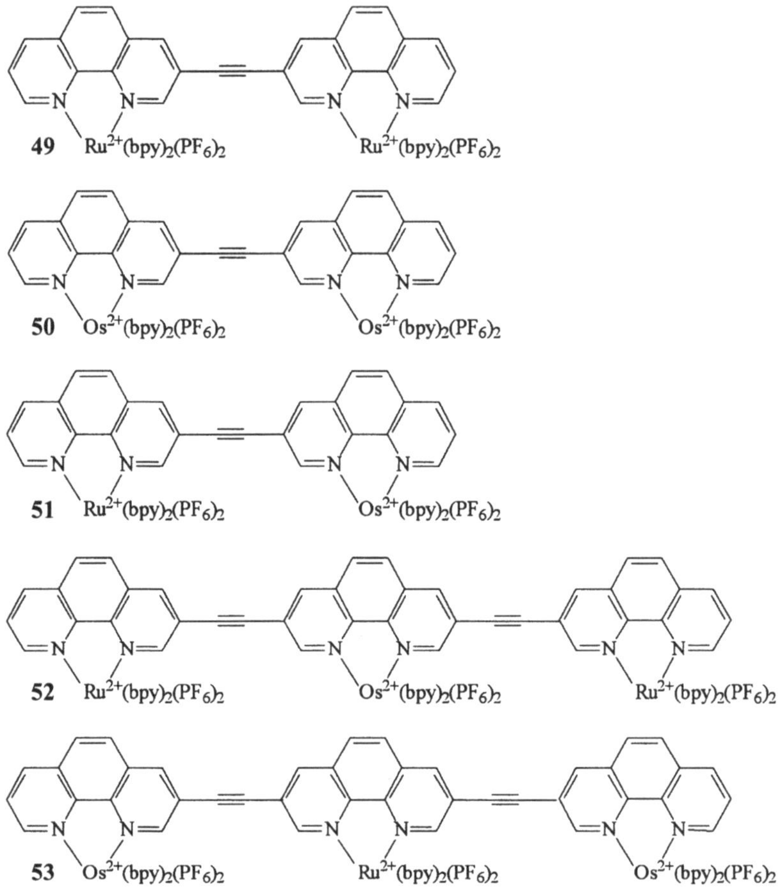

Fig. 4 Multimetallic phenanthroline complexes containing Ru(II) and Os(II)

nuclear studies, and studies on the absorption behavior of the model compounds and building blocks. Glazer and Tor describe a novel class of polymeric materials in which the metal ion is coordinated to the PAE backbone [67]. The predetermined absolute configuration at the metal centers in the chiral conjugated polymers requires enantiomerically pure difunctionalized complexes as starting monomer. Pd-catalyzed cross-coupling reaction between the enantiomerically pure Λ- or Δ-[$(bpy)_2$Ru(3,8-bromo-1,10-phenanthroline)]$^{2+}$$(PF_6^-)_2$ (**54**) and aromatic dialkynes (**55**) gave Ru-containing oligomers and polymers **56/57** (Scheme 8).

The reaction was carried out in DMF in the presence of [1,1-bis(diphenylphosphino)ferrocene]dichloropalladium(II), CuI, and Et_3N under sonication. 4-Nitroiodobenzene was used as a "capping" reagent. End-group analysis by ^{1}H NMR gave the average chain length by integration of the distinct aromatic signals of the terminal nitro-substituted benzene protons versus the characteristic bpy and aliphatic protons. After a short reaction time (1–3 h), oligomers

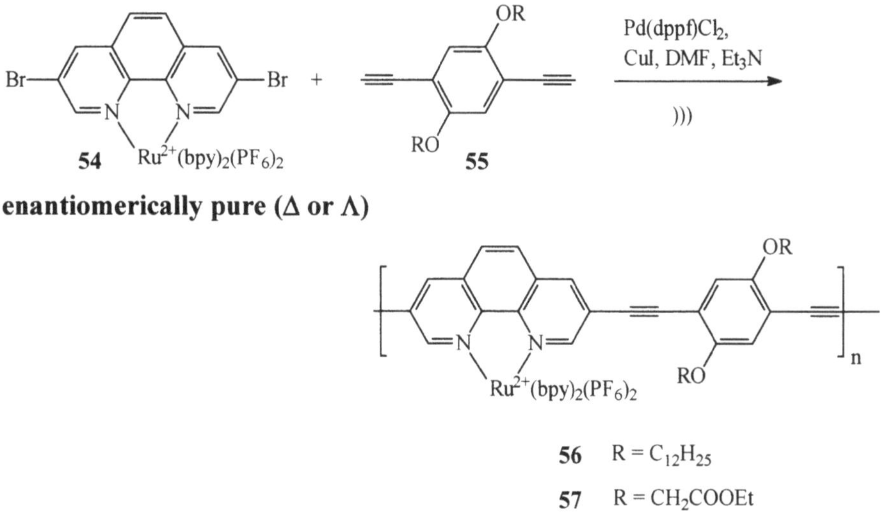

Scheme 8 Synthesis of the 1,10-phenanthroline–Ru(II) PAEs

with n=2–5 were obtained. Prolonged reaction times (12–16 h) yielded higher oligomers and polymers with n=9–12. The polymers show absorption at around 280 nm (bpy) and 438 nm (MLCT is overlapping with the absorption of the phenanthroline ligand with extended conjugation). Λ-polymer **56** (R=$C_{12}H_{25}$) for instance shows emission at 658 nm and a typical $Ru^{2+/3+}$ couple in cyclic and square-wave voltammetry with $E_{1/2}$=+1.10 V (vs. Ag/$AgNO_3$).

3.3 2,2′-Bipyridyl–Re(I) Complexes in the PAE Main Chain

Ley and Schanze [14, 68] have synthesized a series of PPE-type polymers that contain *fac*-(5,5′-diethynyl-2,2′-bipyridine)Re(I)$(CO)_3$Cl as part of the π-conjugated main chain by a Pd-mediated coupling reaction, as illustrated in Scheme 9. The polymers differ with respect to the mole ratio of 4,4′-diethynylbiphenyl (**59**) and (5,5′-diethynyl-2,2′-bipyridine)Re(I)$(CO)_3$Cl (**58**) used in the polymerization reaction mixture. Four different polymers have been prepared and contain 0 (P_0), 10 (P_{10}), 25 (P_{25}), and 50 (P_{50}) mol% of the Re(I) repeat units in the polymer chain, respectively (Scheme 9).

The molecular weight (M_n, GPC with polystyrene standard) is in the range of 8–14 kD (P_n~10–30). The M_n values for the metal-organic polymers are generally lower compared to that of the metal-free organic polymer (P_0). The authors therefore came to the conclusion that the Re–bipyridine monomer **58** is less reactive in the Pd-catalyzed cross-coupling reaction compared to the biphenyl monomer **59**. These polymers have also been characterized by ^{1}H and ^{13}C NMR, as well as by FTIR spectroscopy.

The metal-organic polymers (P_{10}, P_{25}, P_{50}) show two spectrally distinct absorption bands, one due to the π–π* absorption (λ_{max}=400 nm) of the polymer

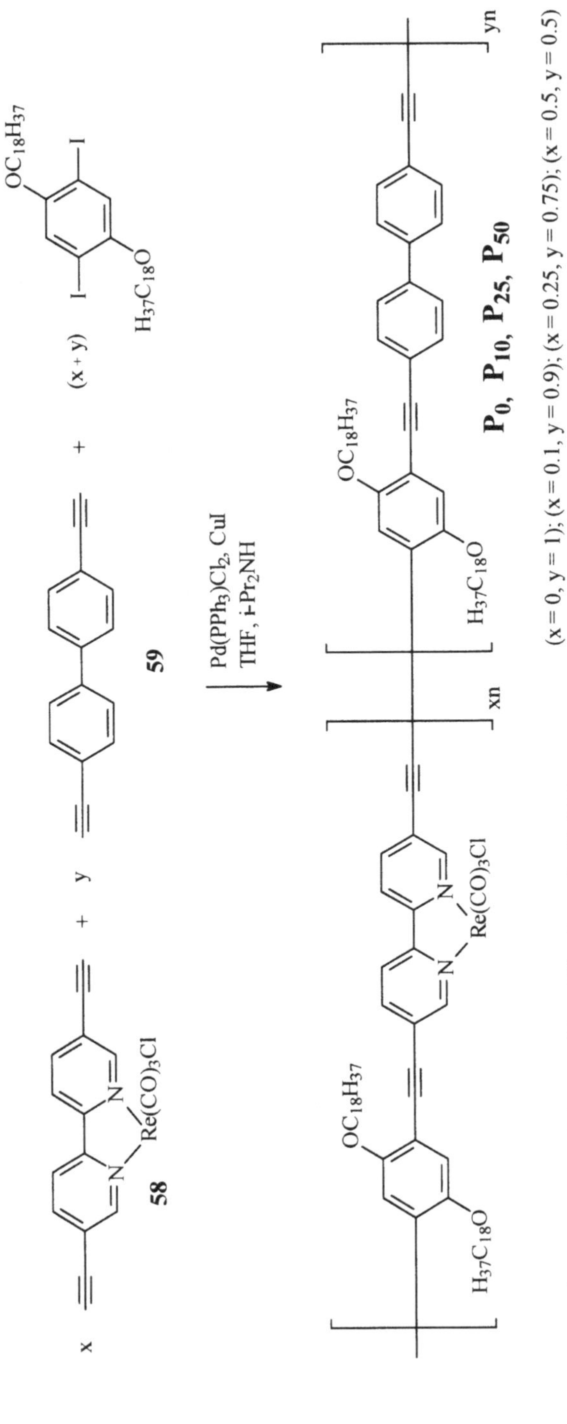

Scheme 9 Synthesis of the 2,2′-bipyridinylene-Re(I) PAEs

backbone and the other at lower energy which is attributed to the dπ(Re)→π^*_{poly} MLCT absorption (λ_{max}=440–540 nm) with an intensity that increases with the mole fraction of the (bpy)Re(CO)$_3$Cl (**58**) repeating unit in the polymer. The MLCT transition in the polymers is redshifted 40–50 nm from its position in model complex **60**, because the π^*_{poly} orbital (which is the acceptor for the MLCT transition) is significantly delocalized by the conjugated polymer backbone.

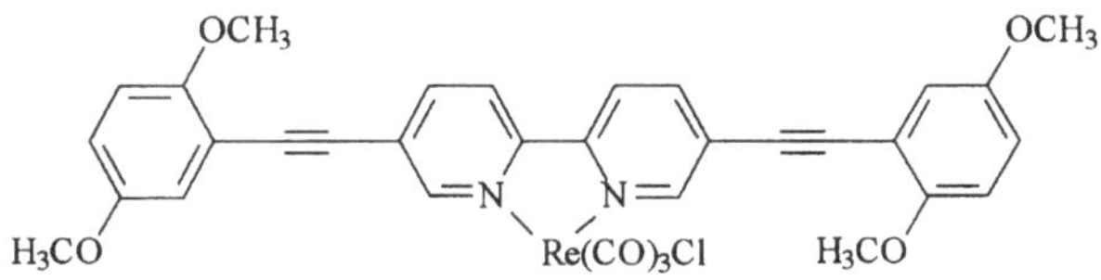

60

Comparison of the spectra of these Re(I)-containing polymers with those of the corresponding free oligomers demonstrates that metalation includes a substantial redshift of the lowest π,π* absorption. Metalation forces the bipyridyl unit into a planar conformation, thereby effectively increasing the conjugation length [13, 69]. In addition to this effect, the electrophilic metal center likely decreases the LUMO energy and consequently decreases the HOMO–LUMO gap.

Ley and Schanze have also examined the luminescence properties of the polymers P_0, P_{10}, P_{25}, and P_{50} in solution at 298 K, and in a 2-methyltetrahydrofuran solvent glass at 77 K. These spectroscopic studies reveal that fluorescence from the $^1\pi,\pi^*$ exciton state is observed at λ_{max}=443 nm, 2.80 eV in the polymers P_0–P_{50} at 298 and 77 K, but the intensity and lifetime of the fluorescence is quenched as the mole fraction of Re in the polymers is increased. This indicates that the metal chromophore quenches the $^1\pi,\pi^*$ state. The quenching is inefficient even when the mole fraction is large, suggesting that interchain diffusion of the $^1\pi,\pi^*$ exciton is slow compared to its lifetime [70]. Phosphorescence from the $^3\pi,\pi^*$ state of the conjugated polymer backbone is observed at λ_{max}=643 nm, 1.93 eV in P_{10}–P_{50} at 77 K, and emission at λ_{max}=690 nm, 1.8 eV is assigned to the dπ(Re)→π^*_{poly} MLCT transition.

3.4
Ferrocene as Part of PAEs

In 1997, Yamamoto et al. reported aryleneethynylene polymers with ferrocene (Fc) in the repeating unit [71]. Several aromatic diynes were coupled with 1,1′-diiodoferrocene **61** (Scheme 10, Table 3).

By using 1,6-diiodo-1′,6′-biferrocenylene (I-Fc-Fc-I) instead of I-Fc-I, the corresponding polymer (PAE-Fc5, Table 3) is given. Also attempted was the polycondensation with a "reverse-type" combination of monomer 1,1′-diethynylferrocene (≡Fc≡) and dihalo aromatic compound (X-Ar-X). This proved to

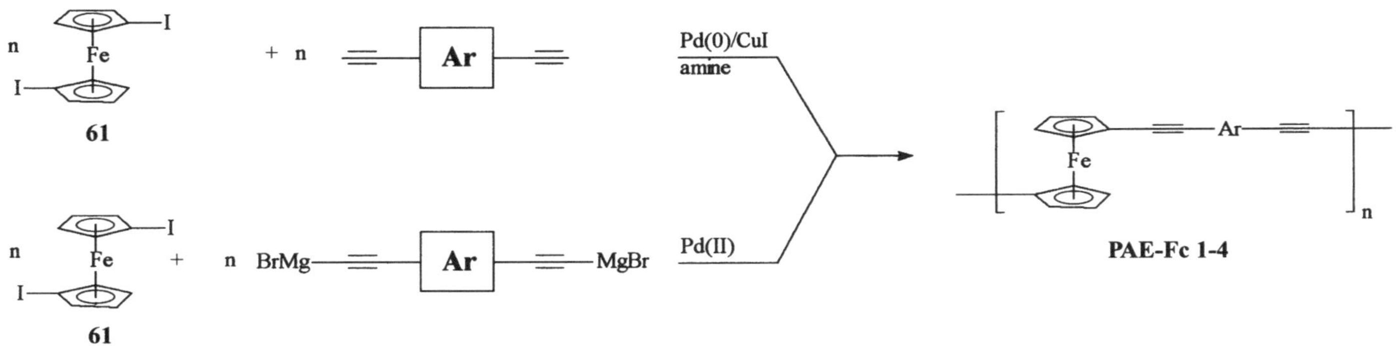

Scheme 10 Two different synthetic routes to the ferrocene PAEs

Table 3 Structures and properties of ferrocene PAEs

entry	polymer	Ar	10^{-4} $\overline{M}_w$ [a]	λ_{max} [nm]		E_{pa}	E_{pc}	E°
				π-π*	d-d	[V] [d]	[V] [d]	[V] [d, e]
PAE-Fc1	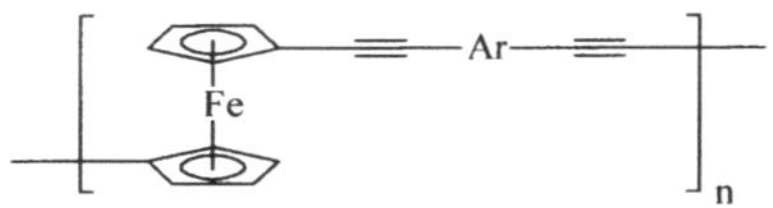	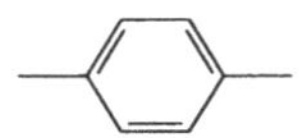	0.15[b]	325	415 ($CHCl_3$)	0.34[b]	0.17[b]	0.25[b]
PAE-Fc2		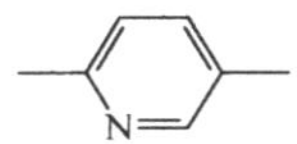	2.1	332 368	460 ($CHCl_3$) 526 (HCOOH)	0.35[b]	0.27[b]	0.31[b]

[a] Light scattering ($CHCl_3$).
[b] $CHCl_3$ soluble fraction.
[c] GPC (polystyrene standard).
[d] Ag/Ag^+, CH_3CN/CH_2Cl_2 (v/v=1:1).
[e] $E°=(E_{pa}+E_{pc})/2$.
Ref.: Yamamoto T, Morikita T, Maruyama T, Kubota K, Katada M (1997) Macromolecules 30:5390.

Table 3 (continued)

entry	polymer	Ar	10^{-4} $\overline{M}_w$ [a]	λ_{max} [nm]		E_{pa}	E_{pc}	E°
				π-π*	d-d	[V] [d]	[V] [d]	[V] [d, e]
PAE-Fc3		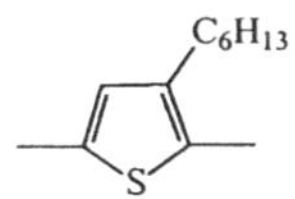	3.4	351	420 ($CHCl_3$)	0.35[b]	0.27[b]	0.31[b]
PAE-Fc4		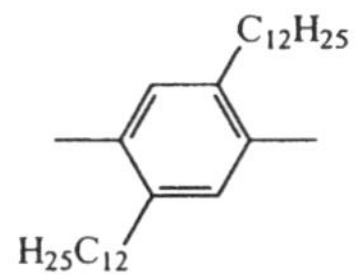	0.69[c]	335	442 ($CHCl_3$)	0.33	0.20	0.27
PAE-Fc5	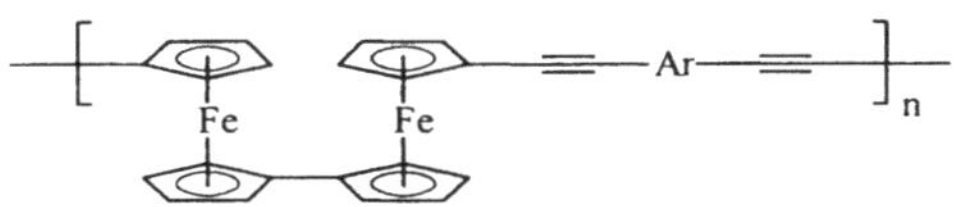	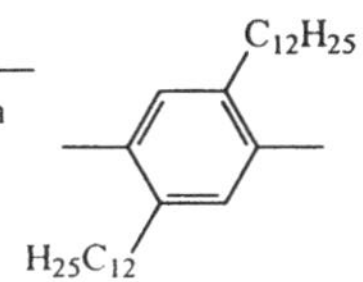	—	313	465 ($CHCl_3$)	0.23	0.03	0.13 (E_1)
						0.47	0.35	0.41 (E_2)

be unsuccessful. 1,1′-Diethynylferrocene, which was prepared in situ by the hydrolysis of $Me_3SiC{\equiv}C{-}Fc{-}C{\equiv}CSiMe_3$, is unstable and forms cyclized compounds, such as ferrocenophanes.

The polymers containing the arylene unit Ar with long alkyl spacers are soluble in organic solvents. These polymers are considered to take a random coil structure in the solution as judged from the relatively small degree of depolarization (p_v) measured in the light scattering analysis (PAE-Fc2: p_v=0.0065, PAE-Fc1: p_v<0.001). The polymers have high thermal stability in that thermal decomposition starts at about 400 °C (under N_2).

The polymers are redox active and show a higher oxidation potential compared with ferrocene [72], presumably due to the electron-accepting nature of the attached ethynyl groups [73]. The cyclic voltammogram peaks of PAE-Fc1–Fc5 are broadened compared to that of ferrocene. The broadening has been ascribed to an exchange of electrons between the metal centers through the main chain. If such electron exchange between Fc units takes place, the oxidation CV peak will include contributions from various oxidized states, and result in a broadened CV peak.

The Mössbauer spectrum of the iodine adduct of PAE-Fc1, containing p-C_6H_4 as the Ar group, reveals the partial oxidation of the Fe(II) unit to an $Fe(III)^+$ unit. In contrast, a similar iodine adduct of PAE-Fc2, containing pyridine-2,5-diyl as the Ar group, gives rise only to signals of the Fe(II) unit in its Mössbauer spectrum. The difference can be explained by the stronger π-accepting ability of the pyridine-2,5-diyl group.

3.5
Porphyrin Systems

Metalloporphyrin macromolecules often show low solubility in organic solvents. For their potential practical use in optical and electronic devices [74, 75] and solar-energy conversion technology [76, 77], obtaining polymers of metalloporphyrins with high molecular weights and good solubility is desired in order to ultimately prepare good quality films. Much effort has been devoted to the design of metalloporphyrin polymers with improved solubility. Yamamoto et al. [78] reported the synthesis of new porphyrin monomers. These monomers were useful in organometallic polycondensations and the resulting polymers were partially soluble. The monomers were synthesized by the MacDonald condensation of bispyrrole with aldehydes, followed by metalation with zinc. Two types of monomers were prepared and utilized in the polymer synthesis (Scheme 11). Polymer of type A is very soluble in organic solvents, while polymer of type B is only partially soluble in organic solvents. Polymer B with Ar= 2,5-thiophene unit and Ar′=1,4-$OC_{12}H_{25}C_6H_4$ has an M_n value of 12,200 g/mol, according to GPC analysis.

The UV–Vis spectra of the polymers show signals in the regions of the Soret band (417 nm) and the Q-band (547, 585 nm), a peak splitting. Zinc porphyrin itself, along with the monomeric compounds, does not show such splitting. The

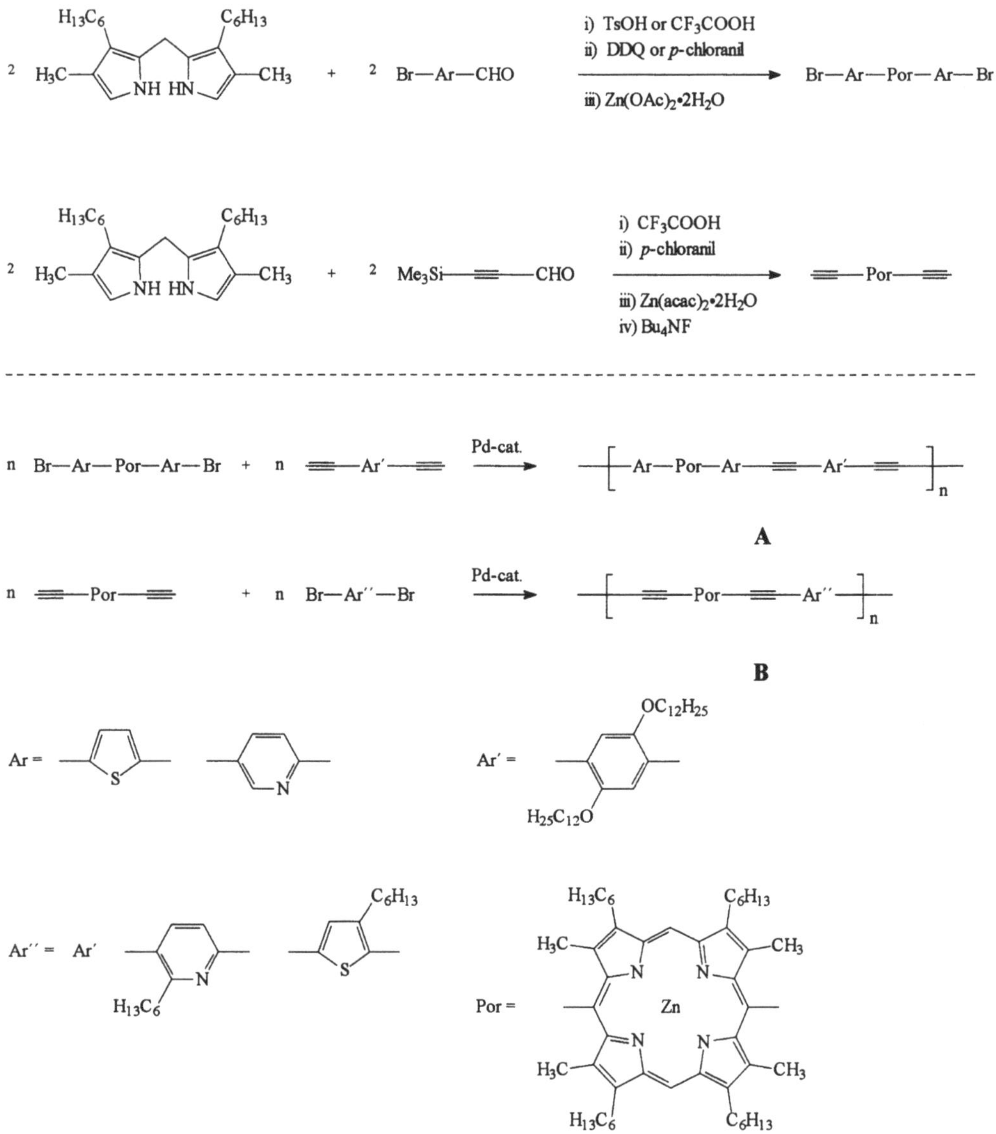

Scheme 11 Synthetic routes to porphyrinylene polymers **A** and **B**

splitting of the Soret band has been explained by the electronic coupling between the zinc porphyrin units.

When the –C≡C– unit is directly bonded to the zinc porphyrin unit (polymer of type B), the Soret and Q-bands are shifted to a longer wavelength, suggesting the formation of a highly π-conjugated system along the main chain, due to the lack of steric hindrance around the zinc porphyrin ring.

The Soret and Q-bands shift to a longer wavelength by about 10 nm upon the addition of the bidentate ligand, 1,4-diazabicyclo[2.2.2]octane (DABCO), to the chloroform solution of the polymer. A self-assembled ladder-type complex is observed. A similar self-assembled ladder-type complex of a porphyrin dimer with DABCO has been previously reported [79]. Casting of solutions of these polymers gave freestanding films, and the polymer films yielded UV–Vis data similar to those of polymers in chloroform.

The thin films are electrochemically active in both the oxidation and reduction regions, due to their having the electrochemically active zinc porphyrin units. The electrochemical processes are accompanied by color changes of the film, between black and yellowish-green in the oxidation region, and between red and yellowish-green in the reduction region.

3.6 Cyclobutadiene Complexes

Bunz et al. explored the possibility of doping PPE chains covalently with small amounts of fluorescence-quenching cyclobutadiene complexes, in order to endow their optical properties to the base polymer, PPE [80]. Due to their extensive experience of cyclobutadiene complexes in polymer synthesis [81], the authors prepared several polymers PAE-CoCp1–5 (Table 4) containing different contents of CoCp complexes. The quantum yields were determined by simple comparison of the intensities of the emitted light to that of a standard

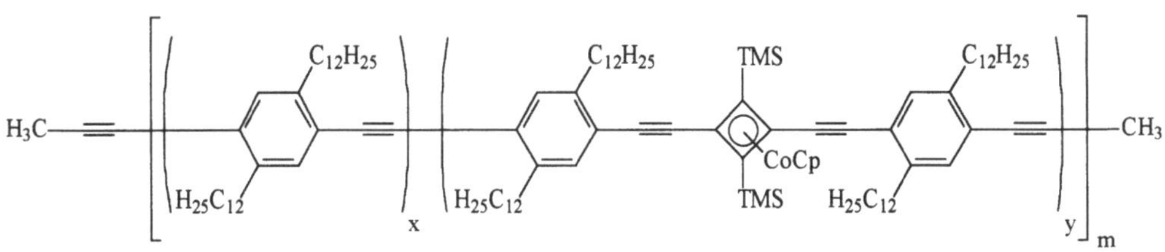

PAE-CoCp

Table 4 Structures and properties of cobalt–cyclopentadiene PAEs

Polymer	% organo-metallic	Ratio of all/organo-metallic units	P_n^a	Φ_{fl} [%]
PAE-CoCp1	0.07	1429	84	18
PAE-CoCp2	1	100	248	16
PAE-CoCp3	21	5	125	1.7
PAE-CoCp4	22	4.5	122	
PAE-CoCp5	33	3	43	

[a] GPC (polystyrene standard), $n=[(x+y)m]$.
Refs.: Altmann M, Bunz UHF (1995) Angew Chem Int Ed Engl 34:569; Harrison BC, Seminario JM, Bunz UHF, Myrick ML (2000) J Phys Chem A 104:5937; Steffen W, Kohler B, Altmann M, Scherf U, Stitzer K, Loye HC, Bunz UHF (2001) Chem-Eur J 7:117.

(Co-free PAE). In PAE-CoCp1, the fluorescence quantum yield is only 18% of that observed for Co-free PAE, even though the "quencher" substitutes less than 0.1% of the aryleneethynylene units. The fluorescence in solution disappeared in PAE-CoCp4, where every fifth unit is a cyclobutadiene complex. The mechanism by which this quenching occurs is via the cobalt-centered MLCT states [82, 83], conferred onto the polymer by the presence of cyclobutadiene complexes. Even in the solid state the polymers PAE-CoCp1–2 are nonemissive. It was therefore shown that incorporation of CpCo-stabilized cyclobutadiene complexes into PPEs even in small amounts leads to an efficient quenching of fluorescence in solution and in the solid state. Quenching occurs by inter- and intramolecular energy transfer [84].

3.7 Silole–Acetylene π-Conjugated Systems

While PAE-type polymers normally have relatively large bandgaps of about 2.1–2.6 eV [85–91], diethynylsilole-based polymers (described by Tamao et al.) show significantly narrower bandgaps (up to 1.8 eV [92]). A silacyclopentadiene ring is conspicuous by its low-lying LUMO level due to the σ^*–π^* conjugation in the ring [93]. The synthesis of polymers with silole–acetylene π-electronic systems was attempted using the Stille coupling reaction with several bis(stannylethynyl)arenes **63/64** and 2,5-dibromosiloles (**62**) [94], as shown in Scheme 12. While the use of distannyldiacetylene or bis(stannylethynyl)-2,5-pyridine led to insoluble polymeric materials, 1,4-phenylene and thienylene derivatives gave red and deep violet polymers PAE-Si1 and 2 (Table 5). Both polymers are air-stable and soluble in common organic solvents. Their absorption is bathochromically shifted at about 100–170 nm, and the bandgaps are rather small in comparison to those of poly(phenyleneethynylene) and poly(thienyleneethynylene) [88–89].

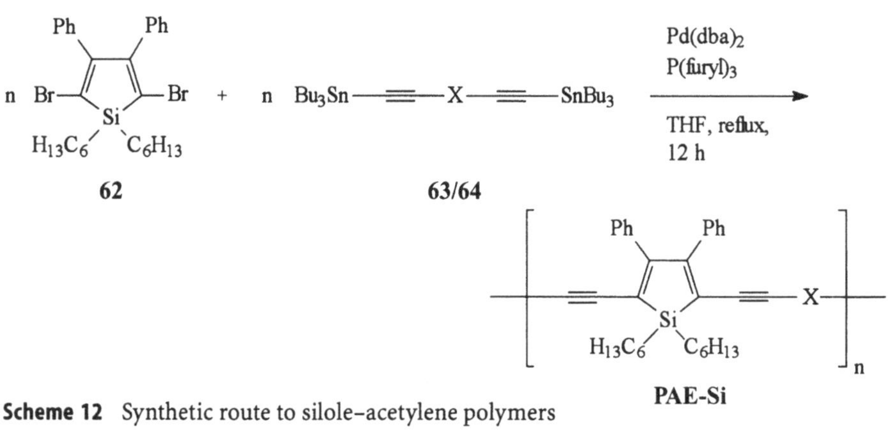

Scheme 12 Synthetic route to silole–acetylene polymers

Table 5 Structures and properties of silole–acetylene π-conjugated systems

Polymer	X	GPC[a]	yield [%]	λ_{max} ($CHCl_3$) [nm]	E_g^{opt} [eV][b]
PAE-Si1		$\overline{M}_n$=9000 g/mol $\overline{M}_w$=64000 g/mol	92	505,527	2.07
PAE-Si2	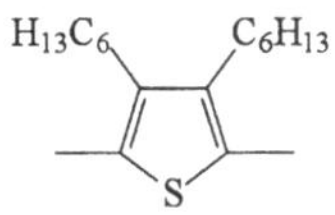	$\overline{M}_n$=13000 g/mol $\overline{M}_w$=63000 g/mol	68	576,605 (sh)	1.77

[a] Polystyrene standard.
[b] Estimated from the onsets of absorptions.
Ref.: Yamaguchi S, Iimura K, Tamao K (1998) Chem Lett 1:89.

4
Related Compounds

4.1
Phenyleneethynylene Oligomers (OPE) Possessing Metal Complexes at the Termini

Several hundred multinuclear metal polypyridine complexes have been prepared over the last few decades [19]. Of the available metals, rhenium(I), ruthenium(II), and osmium(II) cations have proved to be the most popular. In terms of constructing rigid polynuclear complexes, 2,2′:6′,2″-terpyridine, functionalized in the 4′-position [95], is a particularly attractive ligand. Its Ru(II) complexes have a long triplet-state lifetime, longer than that of the unsubstituted parent complex. This has led to the synthesis of a range of alkynylated complexes, e.g., **65**, that exhibit pronounced emission in the red region of the spectrum [96–99]. These polytopic ligands have also been found [100] to support long-range triplet energy transfers in mixed-metal Ru(II)/Os(II) dinuclear complexes, where the connecting organic framework promotes through-bond electron exchange.

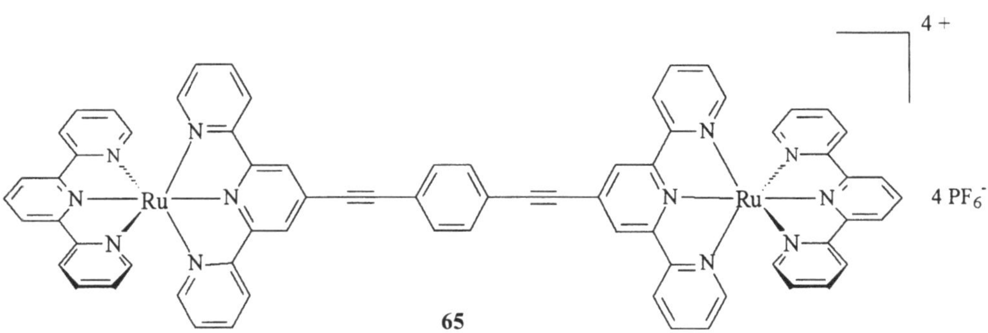

65

Other types of organic-based molecular wires have been proposed. Both the conductivity and its decay with length are crucial characteristics of such molecules. The goal is to address a single molecule with two metallic electrodes; thus, most of the experiments are based on the average response of a group of molecules, possibly arranged in a self-assembled nanoarchitecture. The groups of Yu [101] and Coudret [102] designed and synthesized a series of conjugated, rigid, rod-shaped wires. The molecules are asymmetrical with a redox site at one end and a connecting atom at the other end. Thiol groups are commonly used along with conjugated spacers such as oligo(phenyleneethynylene). Sita [103, 104] and coworkers reported the synthesis of ferrocene–$[C{\equiv}C{-}C_6H_4]_n$–SAc (n=1, 2, 3; Ac=acetyl) by the palladium-catalyzed Heck reactions using acetyl as a protecting group for arylthiols. Unfortunately, the solubility was found to be strongly dependent on the substituent carried by the sulfur atom. Indeed, the longest soluble molecule described was a methyl thioether-capped wire of 31 Å (n=3). Since gold–thioether interactions are weaker than thiol–gold ones, such capping groups are interesting for the formation of stable, mixed, self-assembled monolayers [105].

Classically, the solubility is improved by grafting alkyl or alkoxy spacer groups on the repetitive units of the conjugated backbone (Fig. 5a). Grafting the solubilizing groups at one or both ends of the OPE unit is another option (Fig. 5b). A single triisopropylsilyl group seems to be sufficient in order to solubilize an OPE consisting of five repeat phenyleneethynylene units [106]. Coudret and coworkers reported a third strategy in order to improve the solubility via the redox side (Fig. 5c). This group prepared OPEs with the $[Ru(bpy)_2(ppH)]^+$ unit (**66**) (Scheme 13) on the end. The cationic character of these complexes makes them soluble in various polar organic solvents. The

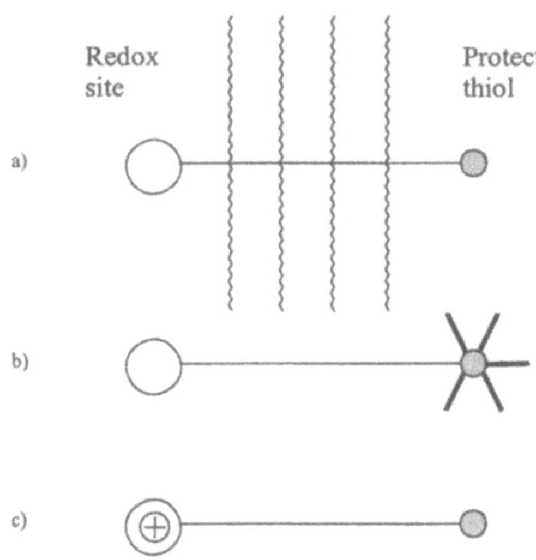

Fig. 5 Principles to reach solubility

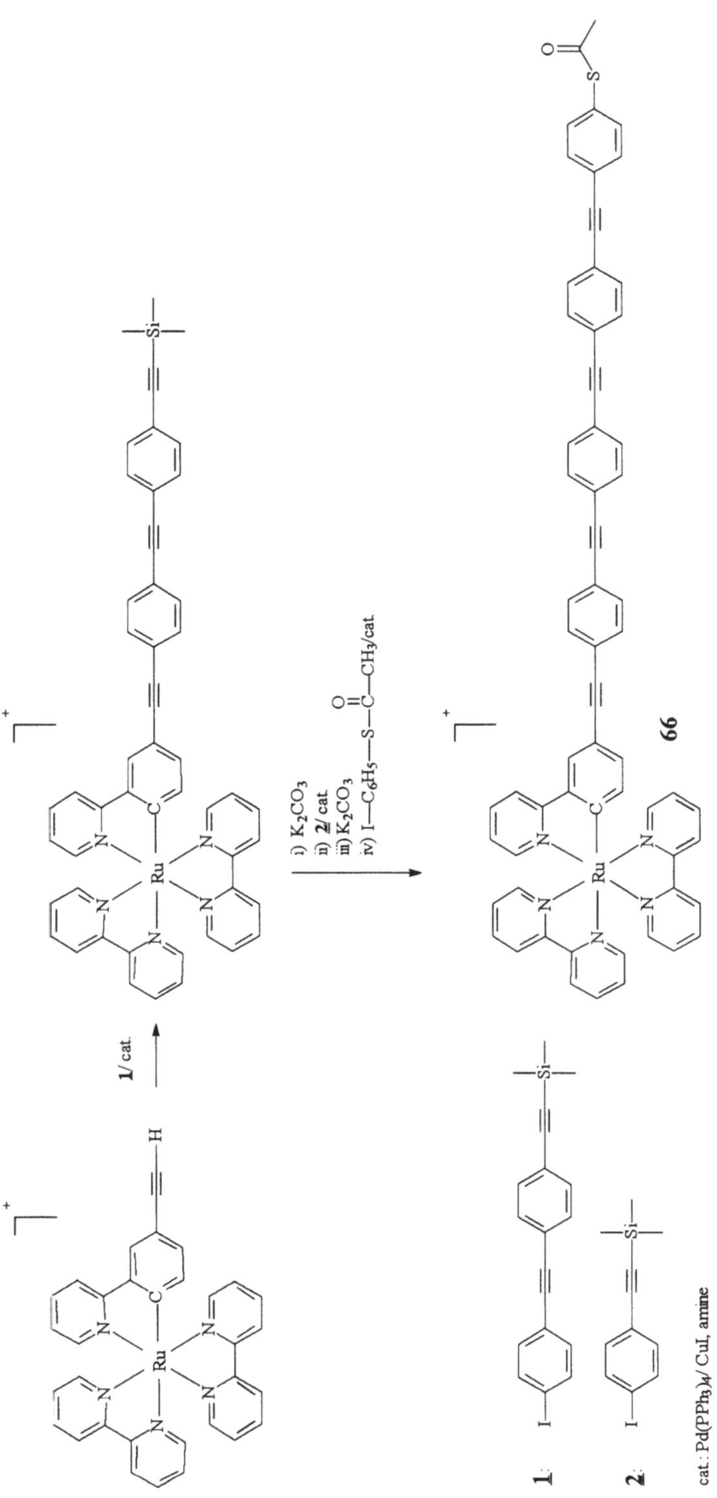

Scheme 13 Synthesis of the OPE having a protected thiol at the end

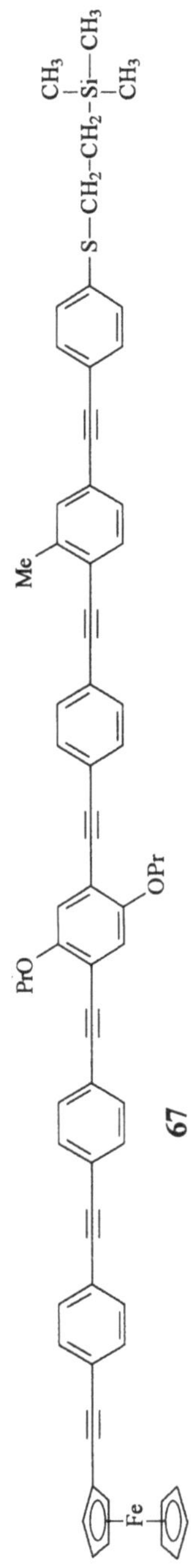

solubility allows a concise synthesis of conjugated redox-active thiols containing unsubstituted OPE from simple starting materials [102]. These authors believe that this strategy should provide a convenient access to long, conjugated, rigid molecules designed for various purposes, such as light-emitting devices.

Yu and coworkers reported the design and synthesis of a series of conductive OPEs possessing a ferrocene and thiol at each terminus [101]. The solubility is achieved by substitution at the phenyl groups with methyl and propoxy substituents (**67**). Several new reactions for preparing novel arenethiol-protected compounds (e.g., 2-(4-pyridinyl)ethyl-4′-(ethyl)phenyl sulfide) are reported.

4.2
PAEs as Chemosensors for Transition Metals

In order to quantify the transition metal ion concentration, Jones et al. [107] developed a highly sensitive fluorescent chemosensor in the form of dialkoxyphenyleneethynylene–thiophene copolymers **68/69**. The PAEs were functionalized on the thiophene unit with terpyridine (**68**), and included 2,2′-bipyridine (**69**) as a Lewis acid receptor. The terpyridine polymers [108] were found to respond quantitatively to transition metal ions at concentrations as low as 4×10^{-9} M (Ni^{2+}, Hg^{2+}, Cr^{6+}, and Co^{2+}). The additionally used bpy-PAE demonstrates that variation in the chelation at the receptor site is an important variable in tuning selectivity. The observed dynamic quenching mechanism, combined with the solubility of this material, provides the opportunity to extend these initial investigations to thin solid films for use in real-time monitoring applications.

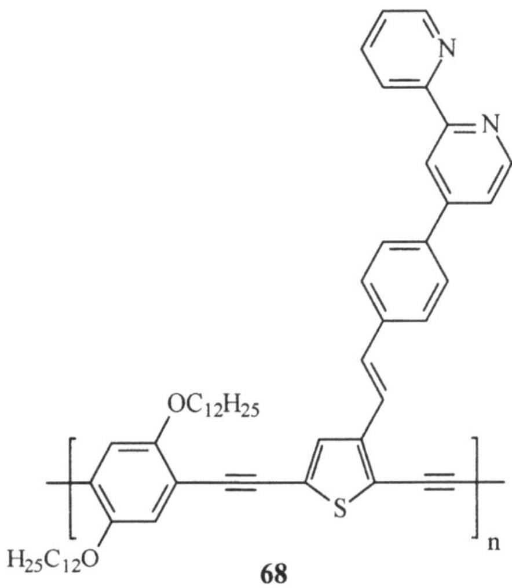

68

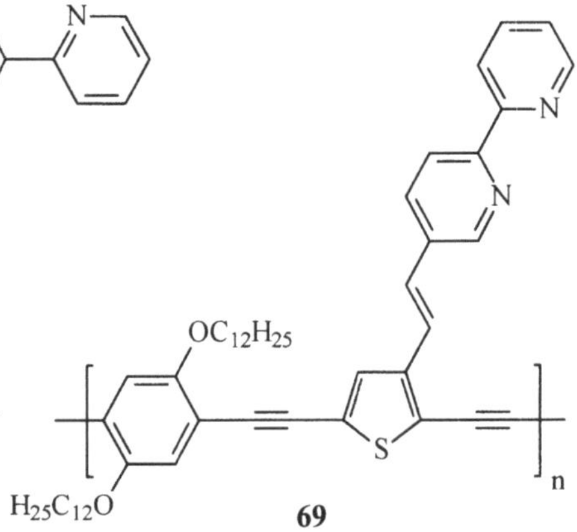

69

4.3
PAEs as Sensors for Potassium Ions

A new transduction mechanism based on the aggregation of conjugated sensory polymers induced K^+ ions as described by Swager et al. [109]. This new system displays enhanced sensitivity because of energy migration processes and has a high selectivity for K^+ over Na^+ ions.

They synthesized different poly(*p*-phenyleneethynylene)s substituted by 15-crown-5 (**70**) and designed a system in which potassium ions will induce polymer aggregation whereas sodium ions will not, so that a selective ion sensor would be produced. The 2:1 sandwich complex formed between K^+ ion and 15-crown-5 cause the polymers to aggregate, thus creating low-energy traps in the electronic structure of the polymer. Energy migration to these traps can result in a large response even at a dilute analyte concentration. The new poly(*p*-phenyleneethynylene) systems have a high sensitivity and offer the advantage of dual detection methods, as both UV/Vis and fluorescence spectroscopy can be employed. The absorbance and fluorescence spectra of several polymers in solution are essentially unchanged even after the addition of 1,500-fold excess of Li^+ or Na^+ ions. In contrast, addition of K^+ ions to a solution of the polymers produced a new redshifted peak (16-nm shift) in the absorption spectra and fluorescence quenching was also evident. The new absorbance and the quenching are the results of interpolymer π-stacking aggregation [110] which, in this case, is induced by K^+-ion bridges between two 15-crown-5 units on different polymer chains. The effectiveness of the interpolymer π-stacking aggregation can be governed by the bulk of the side groups (R) attached to the polymer [111].

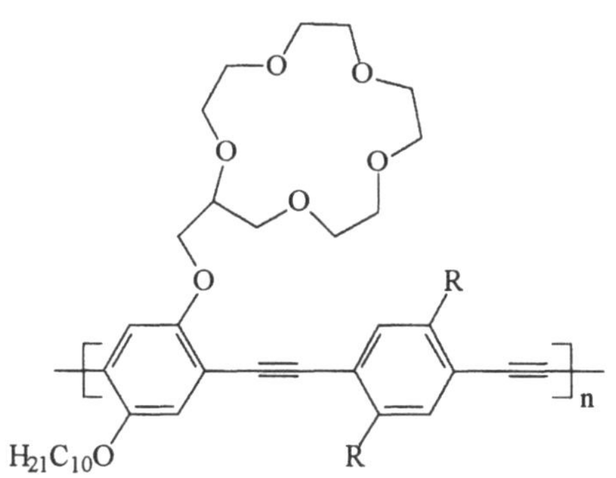

70

4.4
Rigid and Dendritic Nanosized Ruthenium Complexes

Osawa et al. [112] described the Pd-catalyzed one-step graft reaction of metallodendron units with a core coordination (metal center=Ni, Cu, Ru) compound. This afforded a polypyridine–metal-based stiff dendritic architecture which showed characteristics potentially useful for light-harvesting devices [113]. The Ru(II)–polypyridine-type complexes used are excellent candidates for chromophore components due to their outstanding excited-state characteristics in the visible region. The authors describe a new strategy to construct rigid and octahedral dendritic molecules by grafting, in one step, the rods of metallodendron units to that of a core metallodendron unit. With this method, the synthesis of a large-sized stiff metallodendrimer is feasible, and this is demonstrated by the formation of cationic complexes with 6, 12, and 18 peripheral Ru(II)–terpyridine units. The largest cationic compound **71** prepared in this fashion has an estimated diameter of 9 nm with 38 positive charges and 1,435 atoms. Preliminary photophysical data are also presented. The design comprises a tris(2,2′-bipyridine)metal unit placed at the center of a large molecule, from which (*p*-phenyleneethynylene) arms are extended from the 4,4′-positions of each bpy ligand. This structure is achieved by using a palladium-catalyzed cross-coupling reaction. Aside from its relevance to a light-harvesting device, this type of tris(bipyridine)metal-centered three-dimensional conjugated array has the possibility of being a good octupolar NLO device [114] and, moreover, the development of systematic syntheses is highly desirable.

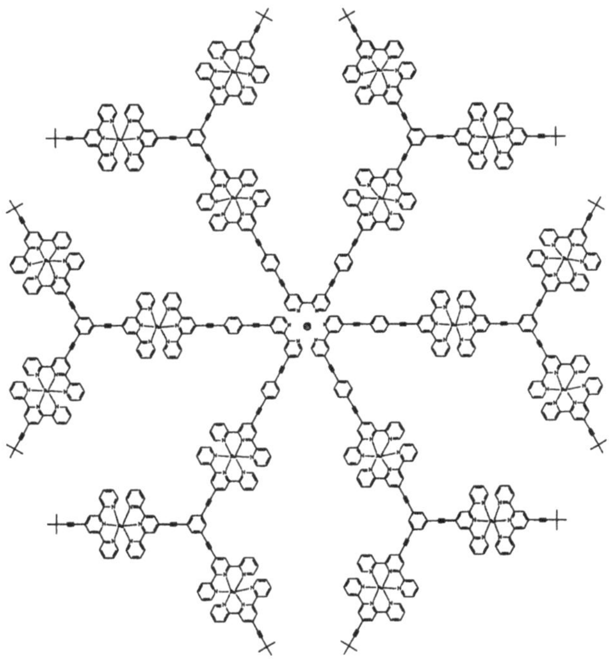

71

5 Conclusion

In recent years a great number of well-defined monodisperse low molecular organometallic oligoaryleneethynylenes have been synthesized. Less attention has been given to organometallic PAEs. In this review we attempted to give an overview of the field of PAE hybrid polymers with metals as inorganic components in the period from 1997 up to the present. Most of the building blocks used so far are based on Ru, Re, and Fe, and the ligands used are mainly of the polypyridine family or cyclopentadiene. The Sonogashira–Heck cross-coupling polycondensation of the diethynylarenes and dibromo-substituted metal chelates is the route used most in synthesis. The organometallic PAEs are electrochemically active, show a long-wavelength MLCT absorption, possess NLO properties, and show a strong photoemission under photoexcitation and electrochemical excitation. The potential practical use is in the field of optical and electronic devices and in solar energy conversion. In the future the research and practical use of organometallic PAEs will rapidly expand.

Acknowledgement Financial support given by the Deutsche Forschungsgemeinschaft is gratefully acknowledged.

References

1. Skotheim TA, Elsenbaumer RL, Reynolds JR (eds) (1998) Handbook of conducting polymers, 2nd edn. Marcel Dekker, New York
2. Shirakawa H (2001) Angew Chem 113:2642
3. MacDiarmid AG (2001) Angew Chem 113:2649
4. Heeger AJ (2001) Angew Chem 113:2660
5. Hadziioannou G, van Hutten PF (2000) Semiconducting polymers: chemistry, physics and engineering, 1st edn. Wiley-VCH, Weinheim
6. Kraft A, Grimsdale AC, Holmes AB (1998) Angew Chem Int Ed 37:402
7. Burroughes JH, Bradley DDC, Brown AR, Mackay NR, Friend RH, Burns LP, Holmes AB (1990) Nature 347:539
8. Giesa R (1996) J Macromol Sci Rev Macromol Chem Phys 36:361
9. Yamamoto T (1999) Bull Chem Soc Jpn 72:621
10. Bunz UHF (2000) Chem Rev 100:1605
11. Wang Q, Wang L, Yu L (1998) J Am Chem Soc 120:12860
12. Jiang B, Yang SW, Bailey SL, Hermans LG, Niver RA, Bolcar MA, Jones WE Jr (1998) Coord Chem Rev 171:365
13. Wang B, Wasielewski MR (1997) J Am Chem Soc 119:12
14. Ley KD, Schanze KS (1998) Coord Chem Rev 171:287
15. Long NJ (1995) Angew Chem 107:37
16. Paris JP, Brandt WW (1959) J Am Chem Soc 81:5001
17. Tokel NE, Bard AJ (1972) J Am Chem Soc 94:2862
18. Sauvage JP, Collin JP, Cambron JC, Guillerez S, Coudret C, Balzani V, Barigelletti F, De Cola C, Flamingni L (1994) Chem Rev 94:933
19. Balzani V, Juris A, Venturi M, Campagna S, Serroni S (1996) Chem Rev 96:759

20. Kalyanasundaram K, Grätzel M (1998) Coord Chem Rev 177:347
21. Kirch M, Lehn JM, Sauvage, JP (1979) Helv Chim Acta 62:1345
22. Benniston AC, Mackie PR, Harriman A (1998) Angew Chem Int Ed 37:354
23. Beer PD (1998) Acc Chem Res 31:71
24. Chin T, Gao Z, Lelouche I, Shin YK, Purandare A, Knapp S, Isied SS (1997) J Am Chem Soc 119:12849
25. Lee WY (1997) Microchim Acta 127:19
26. Knigh AW (1999) Trends Anal Chem 18:47
27. Wang Q, Wang L, Yu L (1998) J Am Chem Soc 120:12860
28. Peng Z, Gharavi AR, Yu L (1997) J Am Chem Soc 119:4622
29. Yu SZ, Hou S, Chan WK (2000) Macromolecules 33:3259
30. Yamamoto T, Yoneda Y, Maruyama T (1992) J Chem Soc Chem Commun 1652
31. Hotzel M, Urban S, Egbe DAM, Pautzsch T, Klemm E (2002) J Opt Soc Am 19:4645
32. Sun L, Hammarström L, Akermark B, Styring S (2001) Chem Soc Rev 30:36
33. Ciardelli F, Tsuchida E, Woerhle D (eds) (1996) Macromolecular metal complexes. Springer, Berlin Heidelberg New York
34. Pautzsch T, Klemm E (2002) Macromolecules 35:1569
35. Pautzsch T, Blankenburg L, Hager M, Egbe DAM, Klemm E (2003) Proc SPIE 5215
36. Pautzsch T, Blankenburg L, Klemm (2003) Polym Sci A Polym Chem 42:722
37. Romero FM, Ziessel R (1995) Tetrahedron Lett 36:6471
38. Maerker G, Case HF (1958) J Am Chem Soc 80:2745
39. Sullivan BP, Salmon DJ, Meyer T (1978) J Inorg Chem 17:3334
40. Hage R, Dijkhuis AHJ, Haasnoot JG, Prins R, Reedijk J, Buchanan BE, Vos JG (1988) Inorg Chem 27:2185
41. Rau S, Ruben M, Büttner T, Temme C, Dautz S, Görls H, Rudolph M, Walther D, Brodkorb A, Duati M, O'Connor C, Vos GJ (2000) J Chem Soc Dalton Trans 3649
42. Hartwig JF, Kawatsura M, Hauck SJ, Shaughnessy KH, Alcazar-Roman LM (1999) J Org Chem 64:5575
43. Wolfe JP, Buchwald SL (1996) J Org Chem 61:1133
44. Harris MC, Buchwald SL (2000) J Org Chem 65:5327
45. Yamamoto T, Nishiyama M, Koie Y (1998) Tetrahedron Lett 39:2367
46. Kim SW, Shim SC, Kim DY, Kim CY (2001) Synth Met 122:363
47. Pautzsch T, Rode C, Klemm E (1999) J Prakt Chem 341:548
48. Birckner E, Grummt UW, Göller AH, Pautzsch T, Egbe DAM, Al-Higari M, Klemm E (2001) J Phys Chem A 105:10307
49. Grummt UW, Pautzsch T, Birckner E, Sauerbrey H, Utterodt A, Neugebauer U, Klemm E (2004) J Phys Org Chem 17:199
50. Hotzel M, Rentsch S, Egbe DAM, Pautzsch T, Klemm E (2001) Synth Met 119:545
51. Sammes PG, Yahioglu G (1994) Chem Soc Rev 23:327
52. Dietrich-Buchecker CO, Guilhem J, Khemiss AK, Kintzinger JP, Pascard C, Sauvage JP (1987) Angew Chem Int Ed Engl 26:661
53. Chambron JC, Dietrich-Buchecker CO, Heitz V, Nierengarten JF, Sauvage JP, Pascard C, Guilhem J (1995) Pure Appl Chem 67:233
54. Vögtle F, Lüer I, Balzani V, Armaroli N (1991) Angew Chem Int Ed Engl 30:1333
55. Goodman MS, Weiss J, Hamilton AD (1994) Tetrahedron Lett 35:8943
56. Tzalis D, Tor Y (1995) Tetrahedron Lett 36:6017
57. Bosnich B (1969) Acc Chem Res 2:266
58. Tzalis D, Tor Y (1996) Chem Commun 1043
59. Tzalis D, Tor Y (1997) J Am Chem Soc 119:852
60. Connors PJ Jr, Tzalis D, Dunnick AL, Tor Y (1998) Inorg Chem 37:1121

61. Kalyanasundaram K, Nazeeruddin MK (1994) Inorg Chim Acta 226:213
62. Ward MD (1995) Chem Soc Rev 121
63. Harriman A, Sauvage JP (1996) Chem Soc Rev 41
64. Sauvage JP, Collin JP, Chambron JC, Gillerez S, Coudret C, Balzani V, Barigelletti F, De Cola L, Flamigni L (1994) Chem Rev 94:993
65. Barigelletti F, Flamigni L, Collin JP, Sauvage JP (1997) Chem Commun 333
66. Harriman A, Ziessel R (1996) Chem Commun 1707
67. Glazer EC, Tor Y (1999) Polym Reprints 40:513
68. Walters KA, Ley KD, Cavalaheiro CSP, Miller SE, Gosztola D, Wasielewski MR, Bussandri AP, van Willingen H, Schanze KS (2001) J Am Chem Soc 123:8329
69. Ley KD, Li Y, Johnson JV, Powell DH, Schanze KS (2000) Chem Commun 1749
70. Ley KD, Whittle CE, Bartberger MD, Schanze KS (1997) J Am Chem Soc 119:3423
71. Yamamoto T, Morikita T, Maruyama T, Kubota K, Katada M (1997) Macromolecules 30:5390
72. Kaufman F, Cowand DO (1970) J Am Chem Soc 92:6198
73. Yamamoto T, Yamada W, Takagi M, Kizu K, Maruyama T, Ooba N, Tomaru S, Kurihara T, Kaino T, Kubota K (1994) Macromolecules 27:6620
74. Anderson HL, Martin SJ, Bradley DDC (1994) Angew Chem Int Ed Engl 33:655
75. Wagner RW, Lindsey JS, Seth J, Palaniappan V, Brocian DF (1996) J Am Chem Soc 118:3996
76. Wennerstoem O, Ericsson H, Raston J, Svensson S, Pimlott W (1998) Tetrahedron Lett 30:1129
77. Kuciauskas D, Liddell PA, Lin S, Johnson TE, Weghorn SJ, Lindsey JS, Moore AL, Moore TA, Gust D (1999) J Am Chem Soc 121:8604
78. Yamamoto T, Fukushima N, Nakajiama H, Maruyama T, Yamaguchi J (2000) Macromolecules 33:5899
79. Taylor PN, Anderson HL (1999) J Am Chem Soc 121:11538
80. Steffen W, Bunz UHF (2000) Macromolecules 33:9518
81. Altmann M, Bunz UHF (1995) Angew Chem Int Ed Engl 34:569
82. Rengel H, Altmann M, Neher D, Harrison BC, Myrick ML, Bunz UHF (1999) J Phys Chem B 103:10335
83. Harrison BC, Seminario JM, Bunz UHF, Myrick ML (2000) J Phys Chem A 104:5937
84. Steffen W, Kohler B, Altmann M, Scherf U, Stitzer K, Loye HC, Bunz UHF (2001) Chem Eur J 7:117
85. Sanechika K, Yamamoto T, Yamamoto A (1984) Bull Chem Soc Jpn 57:752
86. Yamamoto T, Takagi M, Kizu K, Maruyama T, Kubota K, Kanbara H, Kurihara T, Kaino T (1993) J Chem Soc Chem Commun 797
87. Takagi M, Kizu K, Miyazaki Y, Maruyama T, Kubota K, Yamamoto T (1993) Chem Lett 913
88. Yamamoto T, Yamada W, Takagi M, Kizu K, Maruyama T, Ooba N, Tomaru S, Kurihara T, Kaino T, Kubota K (1994) Macromolecules 27:6620
89. Moroni M, Moigne JL, Luzzati S (1994) Macromolecules 27:562
90. Moroni M, Moigne JL, Pham TA, Bigot J-Y (1997) Macromolecules 30:1964
91. Ma L, Hu Q-S, Musick KY, Vitharana D, Wu C, Kwan CMS, Pu L (1996) Macromolecules 29:5083
92. Yamaguchi S, Iimura K, Tamao K (1998) Chem Lett 1:89
93. Tamao K, Yamaguchi S, Ito Y, Matsuzaki Y, Yamabe T, Fukushima M, Mori S (1995) Macromolecules 28:8668
94. Stille JK (1986) Angew Chem Int Ed Engl 25:508
95. Grosshenny V, Romero FM, Ziessel R (1997) J Org Chem 62:1491

96. Benniston AC, Grosshenny V, Harriman A, Ziessel R (1994) Angew Chem Int Ed Engl 33:1884
97. Grosshenny V, Harryman A, Ziessel R (1995) Angew Chem Int Ed Engl 34:1100
98. Harriman A, Mayeux A, De Nicola A, Ziessel R (2002) Phys Chem Chem Phys 4:2229
99. Harriman A, Hissler M, Khatyr A, Ziessel R (2003) Eur J Inorg Chem 955
100. Ziessel R, Hissler M, El-Ghayoury A, Harriman A (1998) Coord Chem Rev 178:1251
101. Yu CJ, Chong Y, Kayyem JF, Gozin M (1999) J Org Chem 64:2070
102. Hortholary C, Coudret C (2003) J Org Chem 68:2167
103. Hsung RP, Chidsey CED, Sita LR (1995) Organometallics 14:4808
104. Hsung RP, Babcock JR, Chidsey CED, Sita LR (1995) Tetrahedron Lett 36:4525
105. Li XM, de Jong MR, Inoue K, Shinkai S, Huskens J, Reinhould ND (2001) J Mater Chem 11:1919
106. Anderson S (2001) Chem Eur J 7:4706
107. Zhang Y, Murphy CB, Chatterjee S, Jones WE Jr (2000) Polym Mater Sci Eng 83:527
108. Jiang B, Zhang Y, Sahay S, Chatterjee S, Jones WE Jr (2000) Proc SPIE 3856:212
109. Kim J, McQuade DT, McHugh SK, Swager TM (2000) Angew Chem 112:4026
110. Kim J, McQuade DT, McHugh SK, Swager TM (1999) Polym Prepr Am Chem Soc Div Polym Chem 40:748
111. McQuade DT, Kim J, Swager TM (2000) J Am Chem Soc 122:5885
112. Osawa M, Hoshino M, Horiuchi S, Wakatsuki Y (1999) Organometallics 18:112
113. Balzani V, Campagna S, Denti G, Juris A, Serroni S, Venturi M, (1998) Acc Chem Res 31:26
114. Dhenaut C, Ledoux I, Samuel DW, Zyss J, Bourgault M, Bozec HL (1995) Nature 374:339

Adv Polym Sci (2005) 177: 91–149
DOI 10.1007/b101380

Supramolecular Organization of Foldable Phenylene Ethynylene Oligomers

Christian R. Ray · Jeffrey S. Moore (✉)

Departments of Chemistry and Materials Science and Engineering and
the Beckman Institute for Advanced Science and Technology, University of Illinois,
Urbana, IL 61801, USA
moore@scs.uiuc.edu

Abstract This review summarizes the supramolecular organization of foldable phenylene ethynylene (PE) oligomers and polymers. *m*-Phenylene ethynylenes (*m*PEs) have demonstrated the ability to undergo a cooperative, solvophobic collapse from a random conformation into a compact helical structure. The stability of this compact structure can be altered through modification of various nonspecific and specific interactions. A result of the solvophobic collapse of the *m*PE backbone is the formation of a solvophobic cavity where small-molecule guests may bind with some measure of specificity. Use of chiral guests or chiral moieties in the backbone or side chains can lead to a bias of the handedness of the helical form. Further supramolecular organization was observed in solution in the form of higher order aggregates and at the air–water interface. In the solid state it was found that *m*PE packing was dependent on the stability of the folded state. Application of these folding principles to imine-containing *m*PEs allowed for the biasing of combinatorial libraries toward the most stable folded state or host–guest complex, as well as the formation of high molecular weight polymers. Strategies for creating water-soluble *m*PEs are also explored. The theoretical groundwork for the emerging field of *o*-phenylene ethynylenes (*o*PEs) will also be discussed.

Keywords Phenylene ethynylene oligomers · Supramolecular organization · Guest binding · Foldamers · Arylene ethynylene

Abbreviations and Symbols

PE	Phenylene ethynylene
*m*PEs	*Meta*-phenylene ethynylenes
*o*PEs	*Ortho*-phenylene ethynylenes
*p*PEs	*Para*-phenylene ethynylenes
Tg	Triethylene glycol monomethyl ether
$\Delta G(CH_3CN)$	ΔG value of fully folded structure
ΔG_{prop}	Energy involved in adding monomer units to an oligomer chain
p_{fold}	Probability that a given conformation will reach the folded state before unfolding
ΔG_{nuc}	Free energy of nucleation event
F_u	Fraction of oligomers that are in the unfolded state
TMS	Trimethylsilyl
CD	Circular dichroism
SAXD	Small-angle X-ray diffraction
WAXD	Wide-angle X-ray diffraction
MW	Molecular weight
M_n	Number average molecular weight
kDa	Kilodaltons
Q	Heat generated
UV	Ultraviolet

1 Introduction

Biopolymers, specifically proteins and polynucleotides, display the ultimate supramolecular organization as they spontaneously and reversibly fold into well-defined conformations through noncovalent interactions [1]. The study of the processes that bring about this supramolecular organization can prove difficult because of the sheer number of repeat units involved. This has led researchers to use synthetic analogs to aid in the study of the specific interactions that bring about such complex structures [2–6]. The phenylene ethynylene (PE) backbone has been used in this fashion to gain insight into how aromatic stacking and solvophobic effects bring about the formation of a compact supramolecular structure in solution. The advantages of the PE backbone are the relative ease of synthesis and the modular fashion in which components of the system can be changed. This gives great control over the molecular architecture based on the backbone and side-chain components chosen.

The diversity of the PE family is largely due to the backbone linkage. Systems with *para* linkages typically act as "rigid rods" or "hairy rods" depending on the side-chain substitution. These compounds have been extensively studied for use as molecular wires, in optical systems, and in chemosensor systems among many other applications [7]. The *ortho* and *meta* linkages, on the other hand, offer multiple conformations about the phenylene ethynylene unit which can give rise to many different ordered states in solution. Studying the driving forces that bring about these different states can then grant insight into how biological systems are able to form such complex and varied structures.

This review will focus on systems that are able to adopt supramolecular order in both intra- and intermolecular fashion. Because of this, *para*-phenylene ethynylenes (*p*PE)s will not be covered in this review. However, there are several excellent reviews of *p*PE synthesis, structure, and function in the recent literature that discuss the intermolecular order of those systems [7–15]. Similarly, phenylene ethynylene macrocycles, which form a wide range of supramolecular aggregates with interesting stacking behavior, will not be covered but have been extensively reviewed [16–22].

2
Meta-Phenylene Ethynylenes

In the following sections the supramolecular properties of *m*-phenylene ethynylenes (*m*PE)s will be discussed. The research has been broken into sections in the hope of grouping different supramolecular concepts such as guest binding, higher order aggregate formation, and solid-state organization. Sects. 2.1–2.3 could be loosely termed the "fundamental research" section where the basic physical organic chemistry is studied. Sects. 2.4–2.10 begin to lay the foundation for applications and extension to other fields.

2.1
Determination of the Folded-State Structure

A series of oligo *m*PEs (**1–9**) were substituted with triethylene glycol monomethyl ether (Tg) side chains, which impart solubility in a wide range of organic solvents [23]. The side chains were attached through an ester linkage making the aromatic groups electron deficient, enhancing the propensity of the backbone to undergo π–π stacking [24]. The *meta* linkage of the backbone allows the oligomer chains to exist in a random coil conformation when effectively solvated. However, computational studies indicated that oligomers of sufficient chain lengths, seven or eight repeat units, could self-organize into

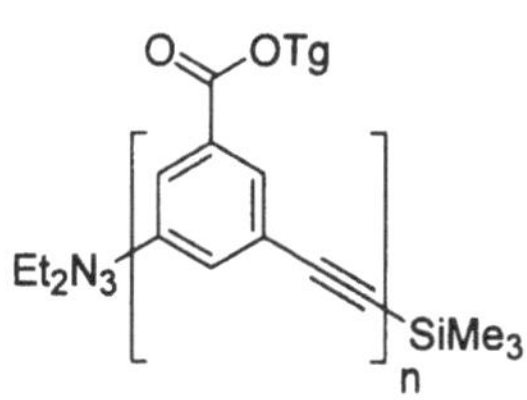

n = 2, 4, 6, 8, 10,
12, 14, 16, 18 (**1-9**)

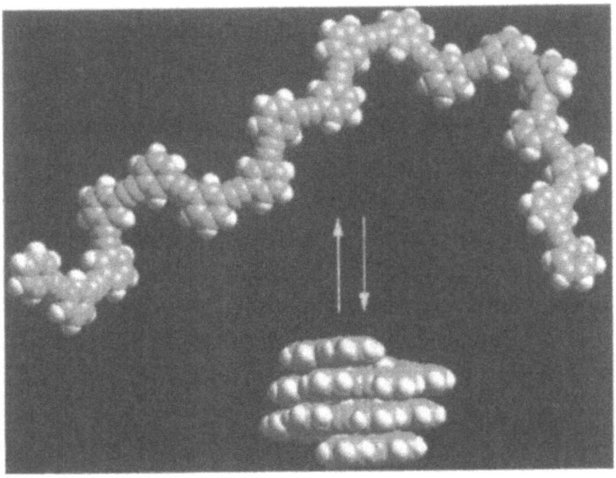

Fig. 1 A space-filling model showing the conformational equilibrium between helical and random coil states for a *m*PE oligomer (n=18). Side chains have been omitted for clarity

a helical conformation when the solvent quality was poor (Fig. 1) [23]. This solvophobic collapse of the aromatic backbone maximizes π–π interactions as well as positive side-chain–solvent contacts while minimizing the solvophobic effects between the oligomer and bulk solvent [23].

The presence of a folded state was evaluated by studying the spectroscopic properties of **1–9** in both chloroform and acetonitrile, which are "good" and "bad" solvents, respectively [23, 25, 26]. The molar absorptivity of **1–9** in chloroform increased in a linear fashion with respect to oligomer length (Fig. 2). However, when UV spectra were taken in acetonitrile two linear trends were identified. Oligomers **1–5** displayed a slope similar to that observed in chloroform, while **4–9** displayed a much smaller slope. The decrease in slope is

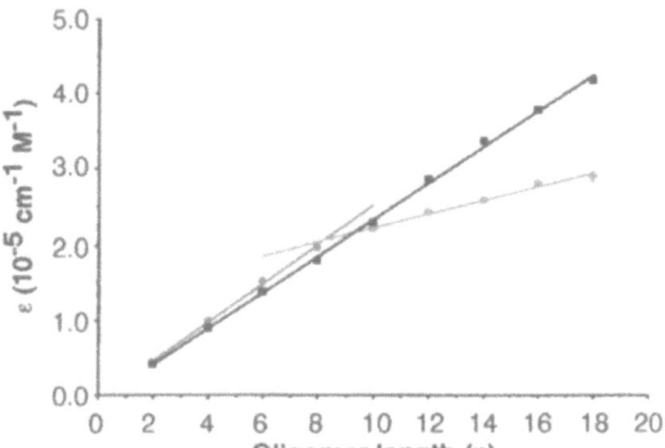

Fig. 2 The molar extinction coefficient ε (303 nm) for oligomers **1–9** in chloroform (*straight line*) and acetonitrile (*broken line*). The lines are linear fit to the data

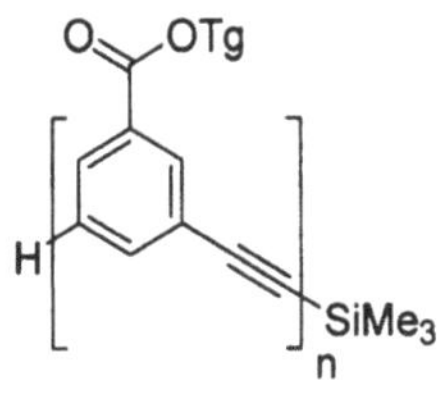

n = 2, 4, 6, 8, 10,
12, 14, 16, 18 (**10-18**)

attributed to the hypochromic effect, which is sensitive to the distance between chromophores and their relative orientation [27]. These effects are common in DNA and RNA and are suggestive of a helical conformation of the *m*PE backbone [23]. Solvent-dependent effects were also observed in ^{1}H NMR experiments [23]. In acetonitrile, upfield shifting of aromatic resonances was observed for the longer length oligomers (**6–9**), which is attributed to the π-stacking of aromatic units. However, negligible shifting of aromatic resonances was observed in chloroform [23, 24]. Finally, the fluorescence spectra of **10–18** (where the triazene unit has been replaced by a hydrogen atom, as the triazene group causes fluorescence quenching) displayed both solvent and chain length dependence (Fig. 3) [25]. The fluorescence spectra of **10–18** exhibited a peak at

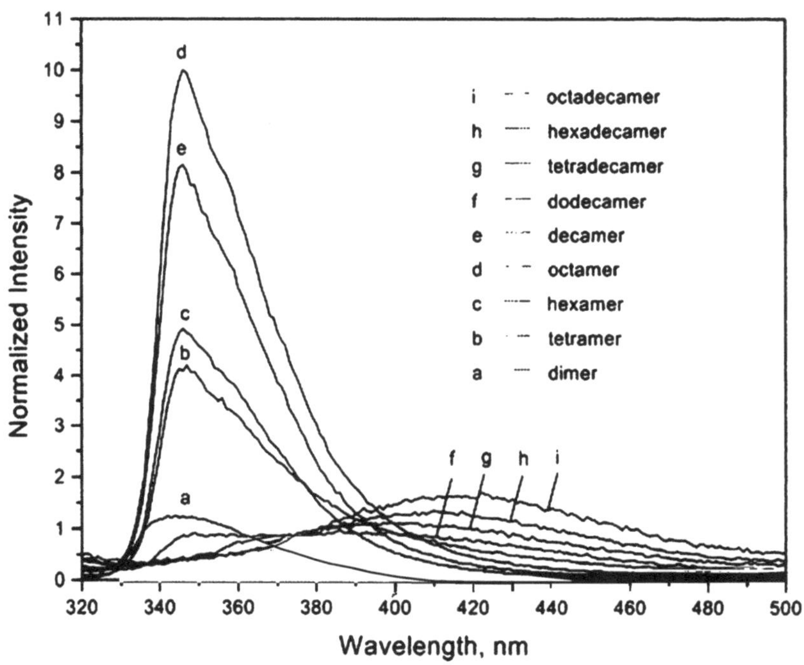

Fig. 3 Fluorescence spectra of **10–18** demonstrating characteristic peaks for unfolded (350 nm) and folded (420 nm) structures and the redshifting of emission to an excimer-like state

350 nm that increased in intensity with chain length in chloroform. In acetonitrile, almost identical spectra were obtained for **10–13**. However, when the chain length reached ten repeat units and above (**14–18**), quenching of the fluorescence intensity was observed (Fig. 3). This redshifted peak is described as an excimer-like state because the π-stacking that brings about quenching is present in the ground state, as demonstrated by examination of excitation spectra. These three independent spectroscopic techniques all suggest a compact, folded conformation in acetonitrile solution at room temperature once the chain exceeds nine units in length [25].

The consistent solvent- and chain-length-dependent phenomena suggest that a helical structure is being formed; however, double-helical, "knotlike" or alternatively folded structures could not be ruled out (Fig. 4). Double-helical structures are not believed to be present since there is no concentration dependence below 10 μmol for the 12-mer in acetonitrile at room temperature. To rule out misfolded knotlike structures, a series of oligomers were designed with a methyl group in the internal position of the helix (**19–27**) [28]. The

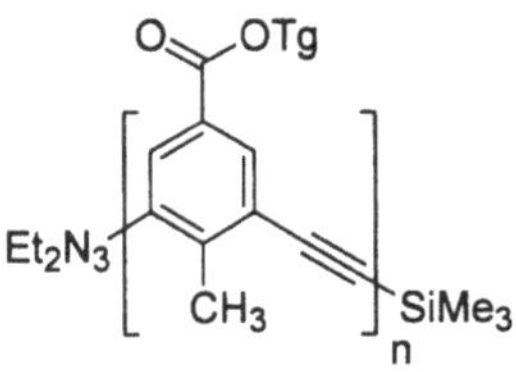

n = 2, 4, 6, 8, 10, 12, 14, 16, 18 (**19-27**)

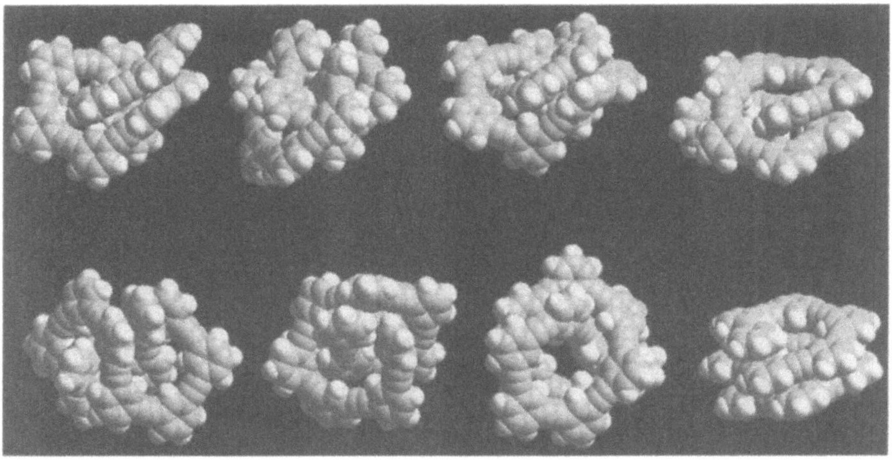

Fig. 4 Lowest energy conformations resulting from a Monte Carlo conformational search on an unsubstituted phenylene ethynylene dodecamer. The ideal helical conformation is shown in the lower right-hand corner. All of the other conformations are destabilized by the addition of bulky side chains

addition of a methyl group both stabilizes the helical form by increasing the solvophobicity of the backbone and decreases the size of the helical cavity, thus limiting the "threading" of the oligomer chain through the interior of the helix. Fluorescence and UV studies of **19–27** were similar to those of **10–18**, but showed an increased propensity to adopt a folded conformation. The ability to form a folded state when alternative structures had been eliminated by oligomer design once again suggests that the *m*PE backbone is able to form a helix in solution [28].

A structural model of the *m*PE helix would ideally come from X-ray crystal structure analysis; however, the Tg side chains make it difficult to obtain high-quality X-ray crystals. An alternate method to gain information about the helical state is double spin labeling. Spin labeling has been used in both proteins and *p*PEs to determine distances between structures [29–32]. In order to determine the helical pitch of *m*PEs, spin labels were installed such that they were four, five, and six repeat units away from each other (**28–30**) [33]. To determine if the spin labels affected the *m*PE folding the UV spectra of **28–30** were obtained. Characteristic random coil absorbances were observed in chloroform, while peaks indicative of a folded state were found in acetonitrile. Also, a crystal structure of the monomer showed that the spin label did not interfere with π-stacking. The high dielectric constant of acetonitrile makes it a poor solvent for ESR studies, so the folding properties of **28–30** were determined in ethyl acetate, a solvent better suited for ESR [33]. In ethyl acetate an intermediate state was observed consisting of predominantly folded oligomers, allowing ESR experiments to be performed.

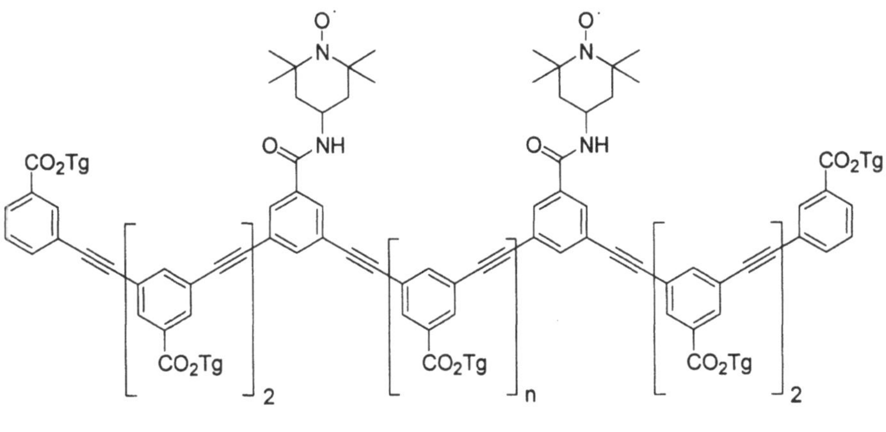

28 (n = 4), **29** (n = 5), **30** (n = 6)

When in the folded state, the spin labels are placed directly above, or above and adjacent to each other (Fig. 5). Once the labels are in close proximity to each other they begin to interact, causing line broadening in the ESR spectrum which increases as the distance between spin labels decreases. To show that the

Fig. 5 Schematic representation of doubly spin-labeled phenylene ethynylene oligomer in the unfolded and folded states, showing the increased proximity of the spin labels in the folded state vs the unfolded state

line broadening is folding dependent, the ESR spectra of **28–30** were taken in chloroform, and no line broadening was observed [33]. A monolabeled oligomer showed no line broadening in ethyl acetate or chloroform, ruling out intermolecular spin-label interaction. The ESR spectra of **28–30** in ethyl acetate showed line broadening for all oligomers, with the greatest broadening observed for **29** suggesting a helical pitch of six. Attempts to determine the exact distances between the spin labels were unsuccessful even at low temperature because of the dynamic nature of the helix [33].

2.2 Helix Coil Transition

Many biological polymers display a cooperative transition from an ordered to disordered state [34] including the helix coil transitions observed for both peptides [35] and nucleic acids [36]. Synthetic systems that are able to undergo a cooperative helix coil transition can complement biopolymer studies and are of potential interest for density-responsive materials.

The folding reaction of *m*PE oligomers can be described by the equilibrium between a random coil conformation and a helical conformation (Eq. 1) [25].

$$\text{random conformation} \rightleftharpoons \text{helix} \quad (1)$$

When a helix is formed, the initial n_0 residues must be fixed into place without any gain in stability through monomer–monomer interactions. This process of helix nucleation is related to a free energy change, ΔG_{nuc}, which is described by the relationship $-RT\ln(\sigma)$ where σ accounts for the entropic cost of fixing the free rotation between monomer units and is related to the cooperativity of the transition. Addition of monomer units to an already formed helix has the benefit of gaining enthalpic stabilization through the formation of favorable monomer–monomer interactions. This stabilization overrides the entropic cost of fixing the backbone conformation and organizing side chains. The overall process of adding monomer units to a growing helical chain is described as helical propagation (ΔG_{prop}). The free energy change for helical propagation is given by $-RT(n-n_0)\ln(s)$; where n is the number of residues in the chain, n_0 is equal to the number of monomers required to form one turn of the helix, and s is the helix propagation constant, which accounts for the enthalpic gain of monomer–monomer interactions. The overall equilibrium of the helix coil transition (K_{eq}) can then be related to these parameters σ and s through the helix coil model as shown in Eq. 2 [37].

$$K_{eq} = \sigma^{n - n_o} \tag{2}$$

Applying the helix coil theory to computational studies of the *m*PE backbone suggests that above a critical chain length of seven or eight repeat units the backbone can adopt a helical structure. The attachment of additional monomer units would further stabilize the helical structure and increase the cooperativity of the folding reaction [23].

To determine if the helix coil model accurately describes the folding reaction of the *m*PE oligomers, a series of solvent denaturation titrations were performed on **13–18** [25,38]. The fluorescence intensity of the oligomers at 350 nm (I_{350}, shown to correspond to the unfolded state) was recorded in a solvent composition ranging from 100% acetonitrile (fully folded) to 100% chloroform (fully unfolded). The plot of I_{350} vs solvent composition gave a sigmoidal-shaped curve for **14–18** and a relatively straight line for **13** (Fig. 6a). The sigmoidal shape of the curves is indicative of a cooperative process where the longer length oligomers demonstrated a more cooperative transition. To determine the folding stability of the reaction, the raw data were treated using the protein denaturation analysis [39], which allows the fraction of chain molecules that are folded or unfolded to be determined and, therefore, the $\Delta G_{folding}$ is obtained for a given solvent composition (Fig. 6b). Plotting the $\Delta G_{folding}$ values of the transition region vs solvent composition gives a straight-line relationship that can be extrapolated to 0% chloroform (Fig. 6c). This value at 0% chloroform represents the value of ΔG for the fully folded structure ($\Delta G(CH_3CN)$). The $\Delta G(CH_3CN)$ of **14–18** was determined (Table 1) and plotted vs chain length (Fig. 6d), giving a linear fit which showed that above 12 repeat units the stability of the *m*PE folded state increased by 0.7 kcal mol^{-1} per monomer, in agreement with the predictions of the helix coil model [25].

Determination of the timescale of the folding reaction was accomplished through nanosecond kinetics studies on **15**, which showed that the folding reaction occurs on a submicrosecond timescale, similar to small helical peptides

Table 1 Solvent-induced unfolding of phenylene ethynylene oligomers

Chain length	$[CHCl_3]_{1/2}$ (vol%)	*m* (cal mol^{-1})	$\Delta G(CH_3CN)$ (kcal mol^{-1})
12-mer[a] (**15**)	56	54±2	−3.0±0.1
14-mer[a] (**16**)	64	68±2	−4.3±0.1
16-mer[a] (**17**)	69	84±2	−5.8±0.1
18-mer[a] (**18**)	72	99±4	−7.1±0.3
18-mer[b] (**18**)	65	101±4	−6.5±0.2

[a] Determined by fluorescence quenching; [b] Determined by UV absorption values.

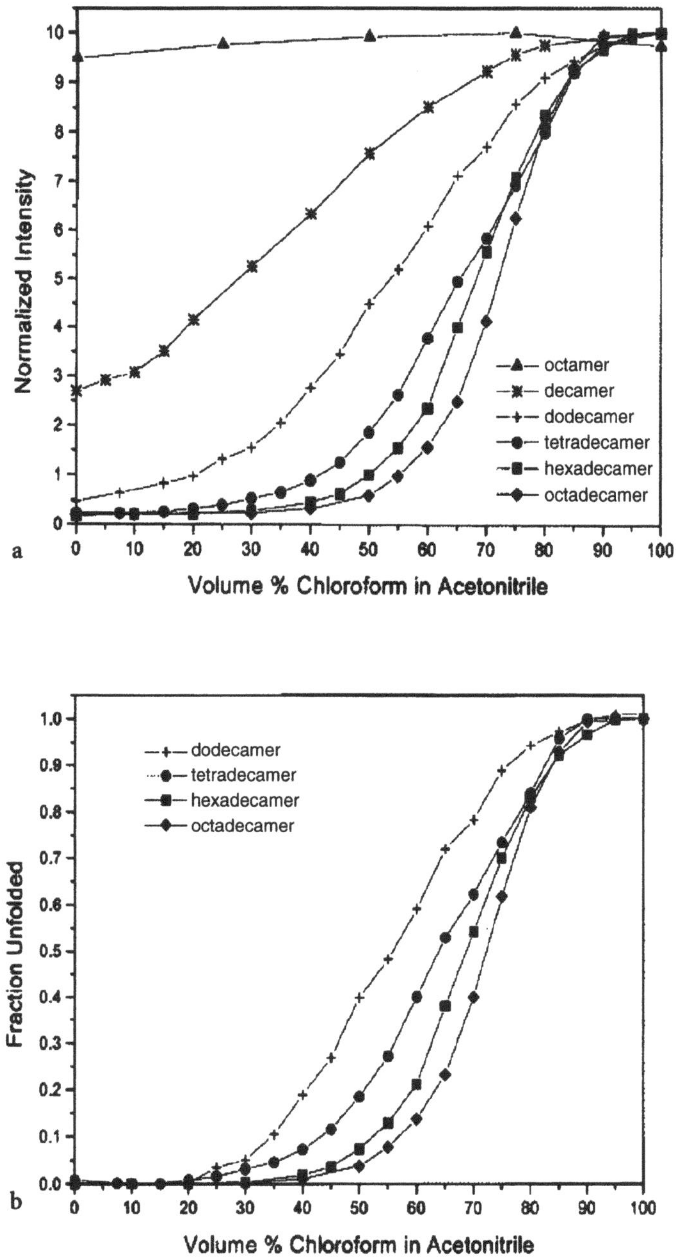

Fig. 6 **a** Solvent denaturation studies for oligomers **13–18**. **b** Fraction of unfolded oligomers for a given solvent composition. **c** Extrapolation of ΔG to give $\Delta G(CH_3CN)$. **d** $\Delta G(CH_3CN)$ vs chain length showing the strong correlation between oligomer length and helix stability

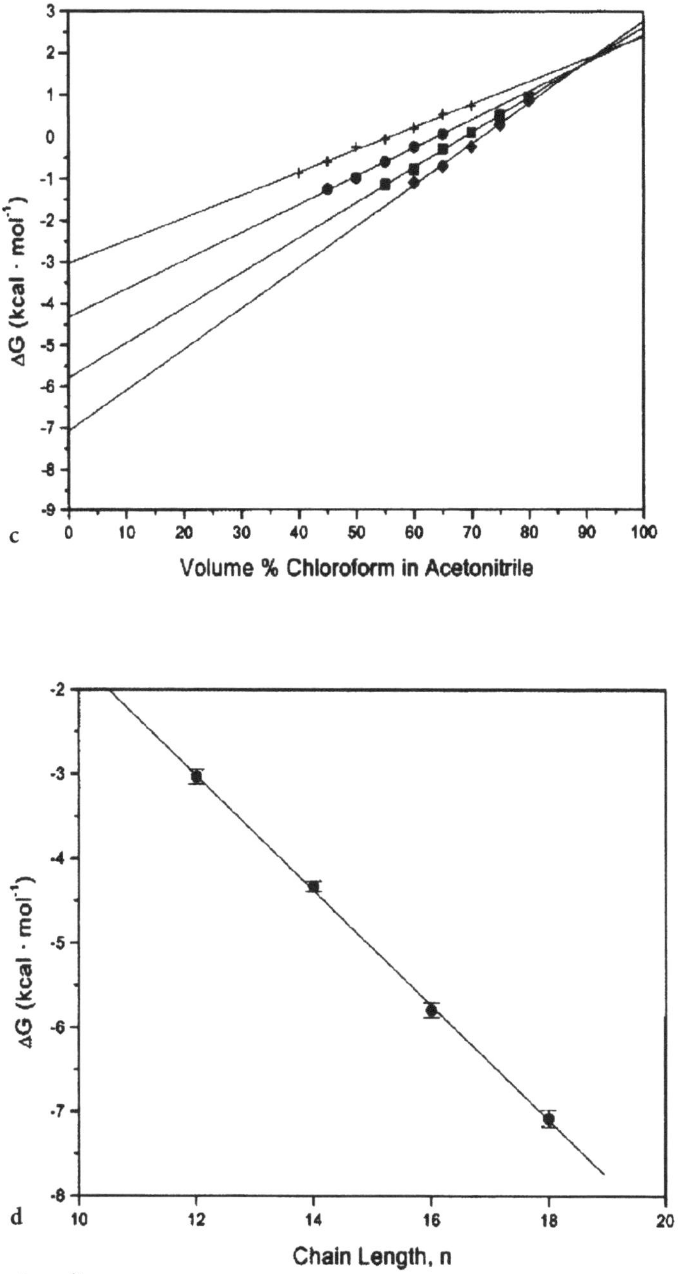

Fig. 6 (continued)

[40]. Helix coil theory dictates that once the helix nucleus is formed, folding should be rapid, leading to single-exponential kinetics [41]. To determine if *m*PE follows single-exponential kinetics, temperature-jump (*T*-jump) relaxation studies were performed [42]. Previous results indicated that thermal denaturation of the *m*PE backbone was possible [23], allowing the temperature-dependent kinetics of folding to be studied. Samples of **15** were prepared at 0.5 μM in a 1:1 THF:MeOH mixture, conditions that limit temperature-induced aggregation while placing the folding transition near room temperature [43]. Steady-state fluorescence of **15** at different temperatures showed two distinct peaks that correspond to the folded and unfolded states. Singular value decomposition of these spectra as a function of temperature displayed a sigmoidal transition from a folded to an unfolded state, allowing the determination of an equilibrium constant *K*(*T*). Denaturation of the samples by laser *T*-jump [44] allows the relaxation of **15** back to the folded state to be monitored by fluorescence spectroscopy every 14 ns with a 0.5-ns time resolution. This relaxation profile was then fitted to an exponential, biexponential, and stretched exponential function to give the rate of folding. At temperatures below 315 K, the data are better fit by the latter two functions, indicating folding was not adhering to single-exponential kinetics. Combining the exponential growth rates from a single-exponential fit ($k_{obs}=k_f+k_b$) with the approximate equilibrium constant determined from thermal denaturation studies ($K(T)=k_f/k_b$), appropriate forward and backward rate constants were determined [43]. An Arrhenius plot shows that the rate of folding decreases slightly as the temperature increases, indicating a greater energy barrier for folding at higher temperature (Fig. 7). This is consistent with a π-stacked transition state, which

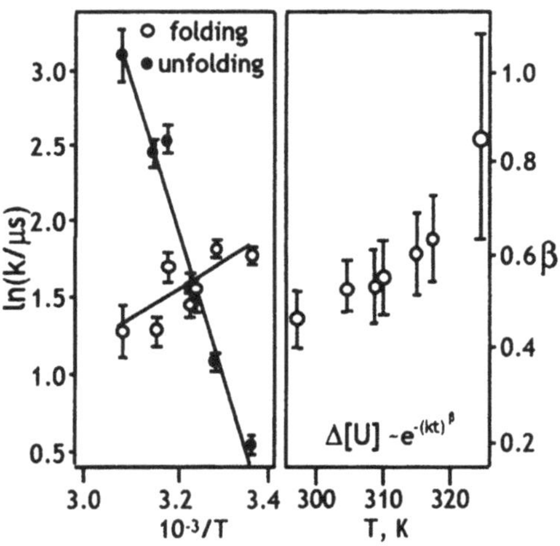

Fig. 7 *Left*: Arrhenius plot of the approximate folding and unfolding rates of **15**. *Right*: At low temperatures, the relaxation is better fitted by a stretched exponential or biexponential; the stretching exponent is shown here as a convenient measure of the deviation from single exponentiality

would require at least five torsions to be fixed resulting in a higher entropic cost at elevated temperature.

The lack of single-exponential kinetics was a strong indication that the folding reaction was not strictly adhering to a two-state model [43]. To better describe the folding process a lattice model was suggested where the reaction coordinate was chosen to be the number of π-stacked interactions (Fig. 8). This model indicates that folding becomes favorable when one π-stacking interaction is present. As additional contacts are made, the free energy decreases further. The deviance from ideal exponential kinetics is explained by the equilibrium between folding and unfolding in partially folded states. The folding reaction can then be described by helical ⇆ partially folded ⇆ unfolded. Application of this model to **15** allowed the nearly quantitative description of folding transitions and equilibrium constants at multiple temperatures.

The validity of the nonideal helix coil folding behavior was upheld by a theoretical calculation which looked at a series of molecular dynamics (MD) folding simulations [41]. In these simulations the reaction coordinate was chosen to be p_{fold}, or the probability that a given configuration will reach the folded state before unfolding [45]. On this coordinate unfolded and folded states will have p_{fold} values of 0 and 1, respectively, and intermediate transition states will have values around 0.5. The MD simulations showed that the folding reaction exhibits a series of sharp transitions from the unfolded to intermediate and finally folded state, which implies that each one of these states is separated by distinct energy barriers (Fig. 9). These simulations demonstrate that many different pathways for folding are possible and that folding may proceed through many intermediates before the most stable conformation is adopted. When these simulations were fit to a double exponential, the characteristic folding times were similar to those obtained experimentally [43].

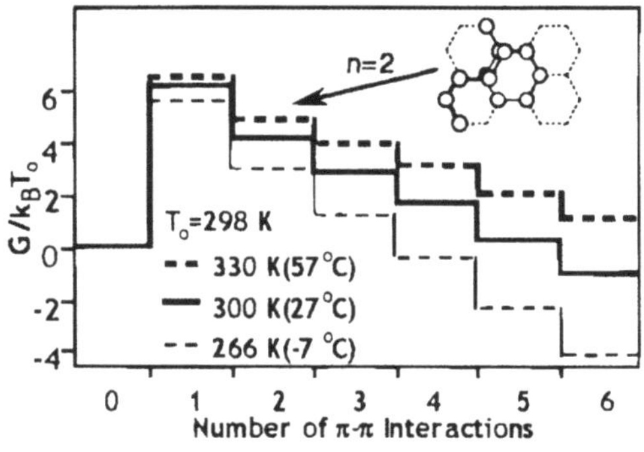

Fig. 8 Free energy profiles derived from the lattice model. The folded side of the barrier slopes to the folded state very gently. Numerous small barriers (k_BT, not shown) connect planar states with different n. A representative planar–planar lattice conformer is shown

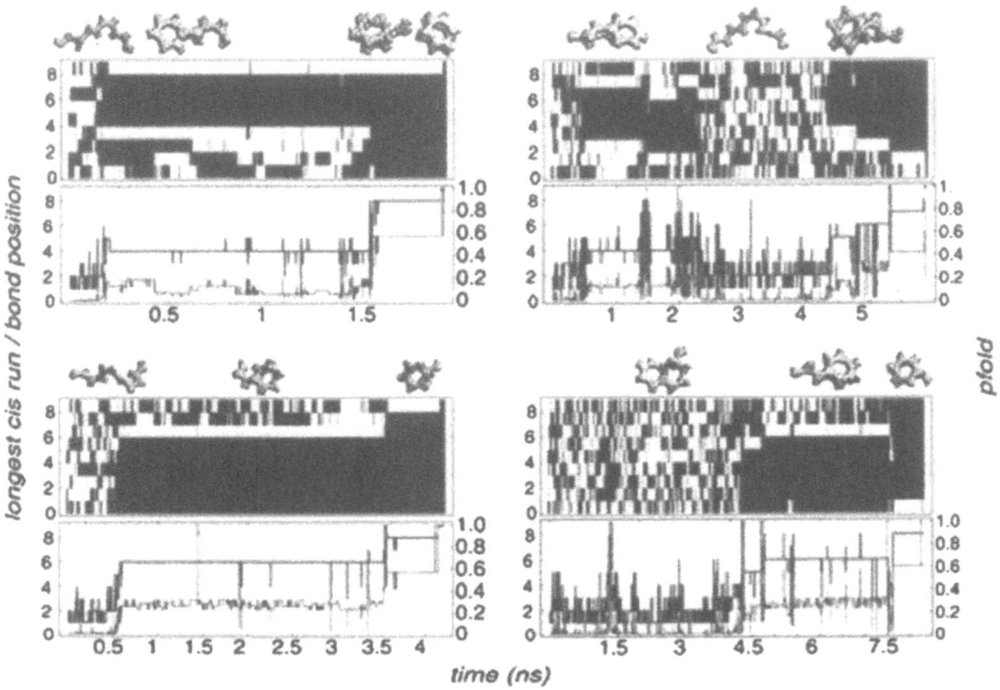

Fig. 9 Comparison of folding trajectories. On the *top portion* of each trajectory the nine *m*PE dihedral angles (nine minus the two terminal dihedral angles) between phenyl rings are shown versus time (*white* as *trans*, *cis* as *black*). The *bottom portion* represents the longest stretch of folded structure versus time. We see that there is a great deal of variety in the folding pathways. In the *upper right* frame, we see folding to and unfolding from a trapped state

2.3 Modification of the Folding Reaction

The *m*PE folded state is largely determined by nonspecific interactions between the solvophobic backbone and solvent, and the π–π interactions between backbone monomer units. Modifying the strength or nature of these interactions or adding additional types of interactions results in a change in the equilibrium of the folding reaction. One such modification is altering the solvophobicity of the oligomer backbone through the relatively simple process of changing solvent properties [26]. The effects of different solvents on the folding reaction were determined by monitoring UV absorbances that have been identified to correspond to both the folded and random states [23, 25]. The ratio of these absorbances has been determined in acetonitrile (0.672) and chloroform (0.938), which are the benchmarks to describe the folded nature of an oligomer. Folding properties were then determined in a wide range of solvents. Surprisingly, almost all solvents promoted helix formation. The exceptions to this trend are the chlorohydrocarbons, which act as strong denaturants. This is most likely

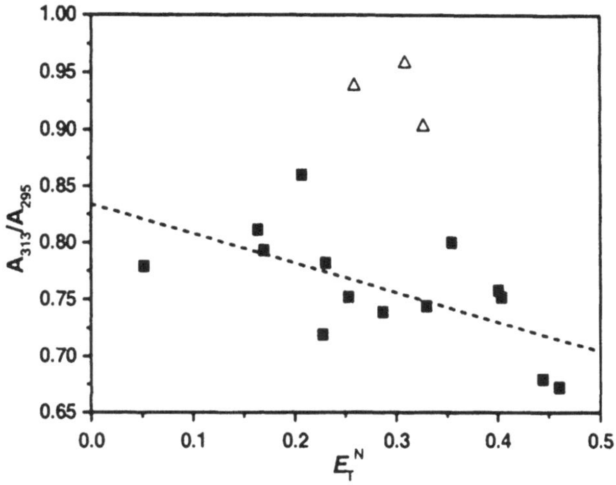

Fig. 10 Ratio of UV absorbances vs solvent polarity. *Squares* represent solvents which bring about a folded or partially folded conformation. *Triangles* represent denaturating solvents ($CHCl_3$, CH_2Cl_2, 1,2-dichloroethane). The overall trend of increased folding nature in more polar solvents is evident

due to their ability to form CH–π bonds with the aromatic backbone [46], diminishing the intrabackbone π–π interactions and thereby destabilizing the folded state [47]. A weak correlation between increased solvent polarity and helical stability was also observed (Fig. 10) [26]. This is consistent with an increased driving force for the *m*PE backbone to undergo a solvophobically driven collapse to limit solvent backbone contacts in the more polar environment.

The attractive monomer–monomer interactions of *m*PEs described in the helix coil theory are dominated by π–π stacking [23–25, 43]; therefore, modifying the strength of these interactions will have a large effect on helix stability. The sensitivity of the folding reaction to π-stacking was determined by varying the electronic nature of the *m*PE aromatic backbone, as it is known that π-stacking is influenced by the electronic nature of the system [48, 49]. A range of electron-rich and -poor aromatic backbones were accessed through ester (**31**), phenolic ether (**32**), and benzylic ether (**33**) side chains [47]. The propensity of π–π stacking (and therefore folding stability) of backbones **31–33** was determined through the ^{1}H NMR aggregation studies of hexameric macrocycles [24]. Upfield shifting of diagnostic ^{1}H NMR resonances indicative of

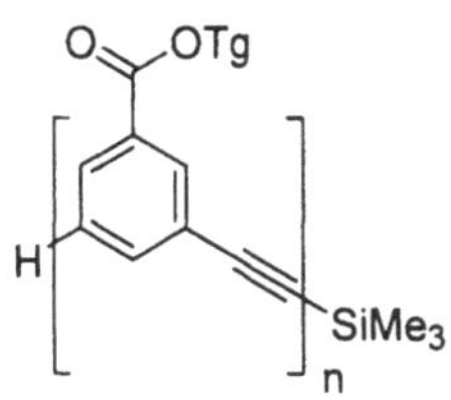

31

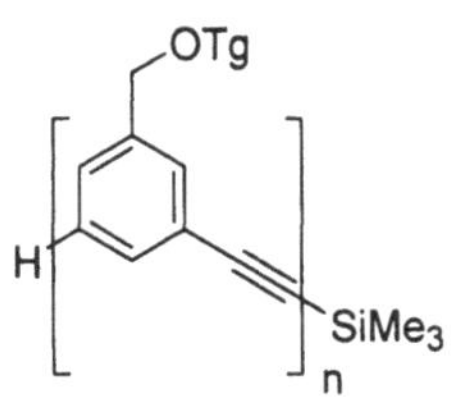

32

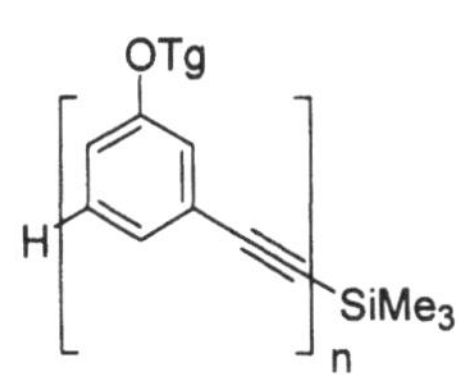

33

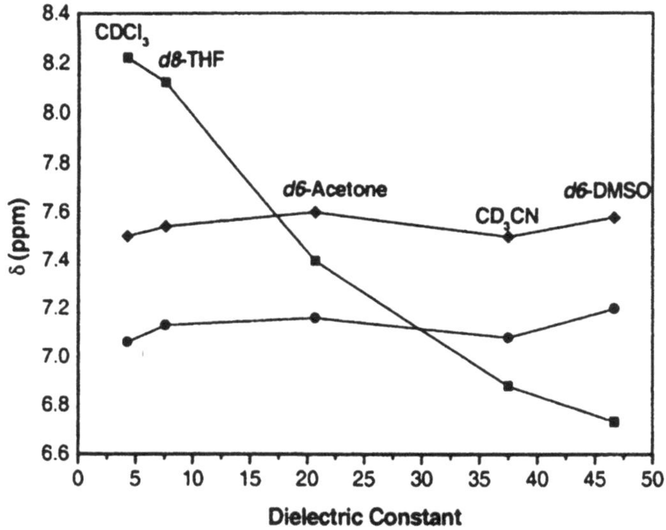

Fig. 11 ^{1}H NMR shifts of **31** (*filled square*), **32** (*filled diamond*), and **33** (*filled circle*) showing the presence of aggregation in increasing solvent polarity for **32**

aggregation were observed for **31** as the polarity of the solvent was increased (Fig. 11). No such shifting was observed for **32** and **33**, indicating a lack of aggregation most likely due to disfavored π-stacking or side-chain steric effects. The folding stability of linear analogs of **31–33** (n=18) was determined by solvent denaturation studies (UV spectroscopy). Typical folding behavior was observed for the native **31**; however, denaturation of **32** indicated the oligomer existed in a partially folded state in pure acetonitrile, while **33** showed no evidence of folding whatsoever. Fluorescence spectroscopy of **31–33** confirmed these results; the spectra of **32** and **33** showed no folded character in organic solvents (Fig. 12). Upon changing the solvent mixture of **32** and **33** to 1:1 acetonitrile:water, a redshifting of the emission was observed, but this may be attributed to intermolecular aggregation of oligomer chains instead of intramolecular folding. The increase in folding stability from **33** (not folded, electron rich) to **32** (partially folded) and to **31** (fully folded, electron poor) clearly shows the strong effect the electronic nature of the backbone has on the folded state.

Helix formation in the helix coil model describes a nucleation event where the first turn of the helix is formed at an energetic loss. The free energy for this nucleation event (ΔG_{nuc}) has been experimentally determined to be 3.6 kcal mol^{-1} for the native mPE system, corresponding to 0.9 kcal mol^{-1} for each torsion in the first helical turn [50]. By fixing a phenylacetylene bond torsion, the ΔG_{nuc}, and therefore the overall stability of the folded state, could be lowered by approximately 0.9 kcal mol^{-1}. To determine if this was possible, the folding properties of two constitutional isomers containing hydrogen bond (H bond) donor–acceptor systems, where **34** can form intramolecular H bonds

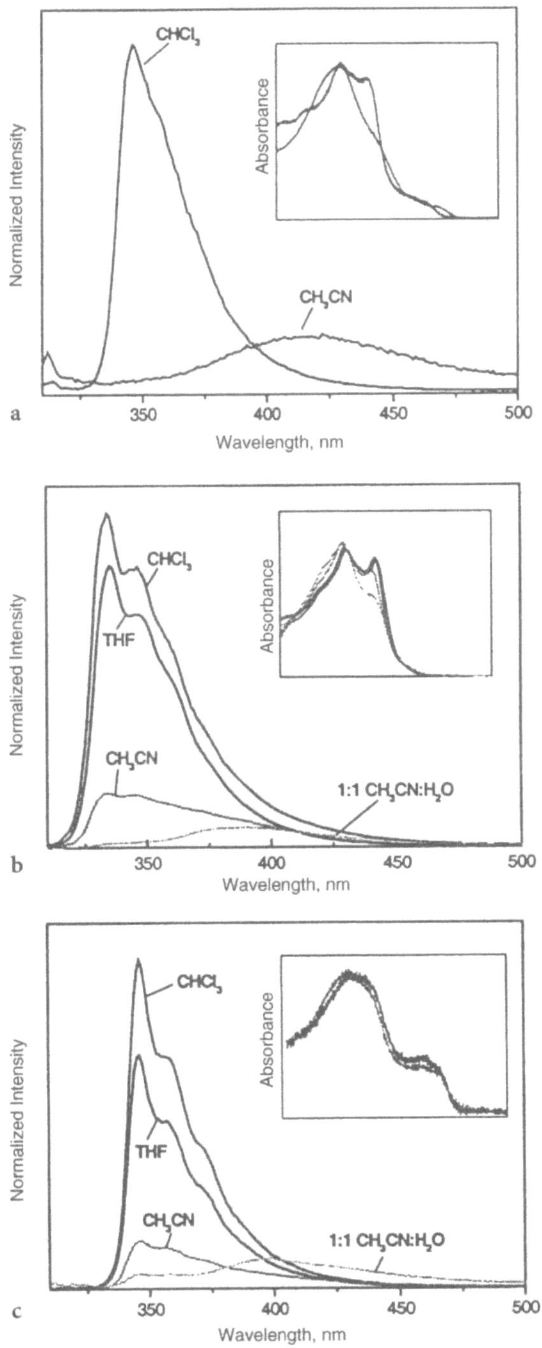

Fig. 12 Fluorescence spectra of **31** (**a**), **32** (**b**), and **33** (**c**) showing the inability of **32** and **33** to fold in organic solvents. The *insets* show the corresponding absorption spectra, where the presence of a second UV band indicative of folding can be observed for **31** and **32**, but not **33**

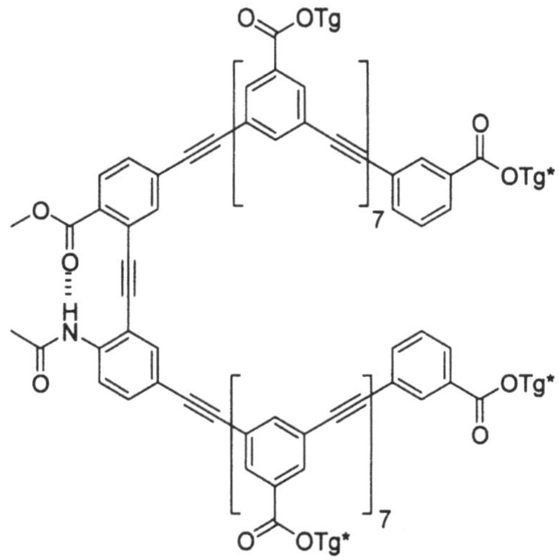

34

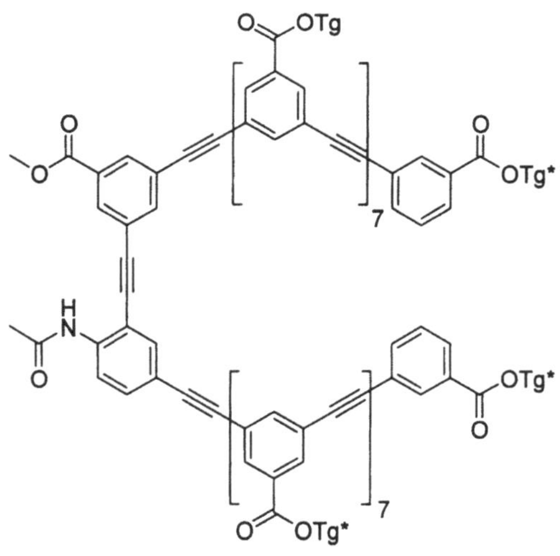

35

and **35** cannot, were studied through solvent denaturation experiments (UV spectroscopy) (Fig. 13). Both oligomers displayed sigmoidal curves with **34** shifted to the right, indicating the oligomer was resisting denaturation compared to **35**. Solvent denaturation analysis [39] of the curves revealed that the $\Delta G(CH_3CN)$ for **35** was -5.8 ± 0.6 kcal mol^{-1}, almost identical to that of the native oligomer **17** (-5.8 ± 0.1 kcal mol^{-1}, fluorescence), indicating that the side-chain modification had no effect on the stability of the folded state. Hydrogen bond-forming **34**, however, required 5.3% more denaturant to reach the mid-point of the transition (F_u 0.5) and has a $\Delta G(CH_3CN)$ of -7.0 ± 0.7 kcal mol^{-1}.

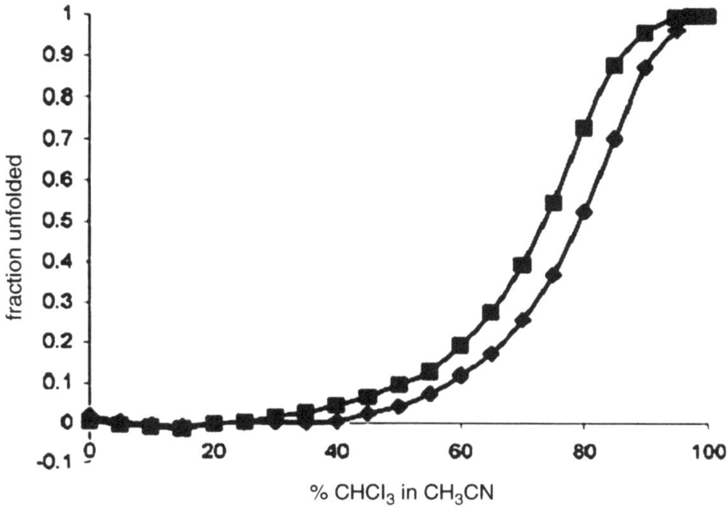

Fig. 13 Solvent denaturation titrations of **34** (*filled diamond*) and **35** (*filled square*) showing the full range of folded and unfolded character

This indicates that fixing of the torsion about the phenylacetylene bond leads to a $\Delta\Delta G(CH_3CN)$ of 1.2 kcal mol^{-1}, very close to the expected stabilization for ΔG_{nuc} upon fixing of one torsion. This demonstrates that stabilization of the folding nucleus through supramolecular interactions leads to the stabilization of the folded-state structure.

The formation of secondary structure in *m*PEs is typically driven by nonspecific interactions: solvophobic forces and π–π interactions. Protein folding, on the other hand, often utilizes a combination of nonspecific and specific interactions, where hydrophobic interactions drive the folding while directional forces define the final structure [51, 52]. To mimic this concept, metal-binding moieties were installed into the cavity of the *m*PE helix. Cyano groups were placed at the *ortho* position of every other repeat (**36**), forming two roughly trigonal planar coordination sites [53]. The distance from the CN nitrogen to

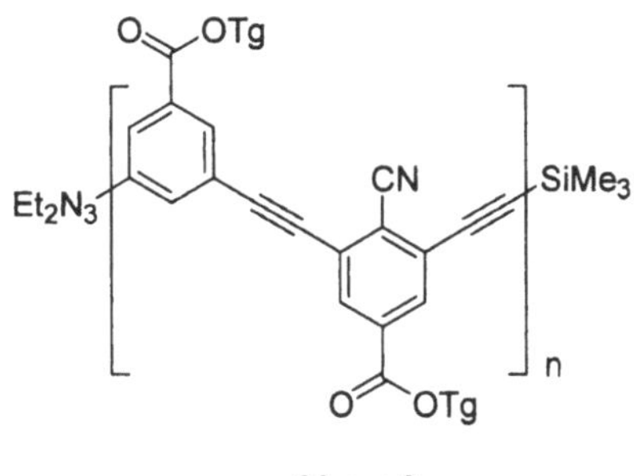

36 n = 6
37 n = 3

the helical axis is about 2.1 Å, consistent with metal–nitrile ligation [54, 55]. The metal chosen for these studies needed to match the coordination spheres of **36**; therefore Ag^+, which typically adopts a trigonal planar geometry, was chosen in the form of AgO_3SCF_3. Solvent conditions were required that would not form a helical structure through solvophobic effects but would still allow folding upon addition of metal ions. The absorbance of **36** in THF showed peaks characteristic of a mostly unfolded state, making THF an acceptable solvent. To ensure that the addition of AgO_3SCF_3 did not drastically alter the solvent properties, thereby inducing folding through solvophobic effects, or alter the ran-

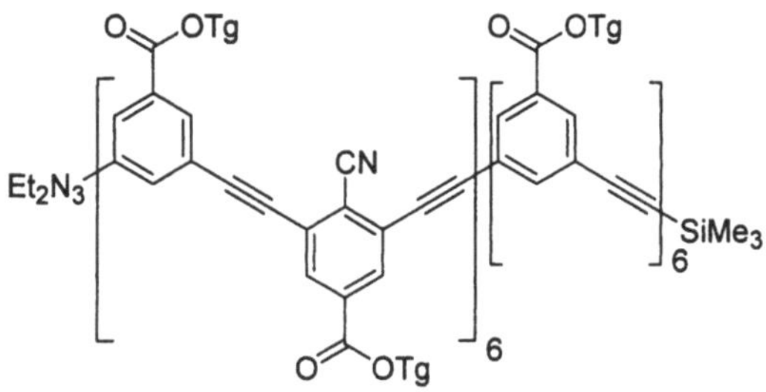

38

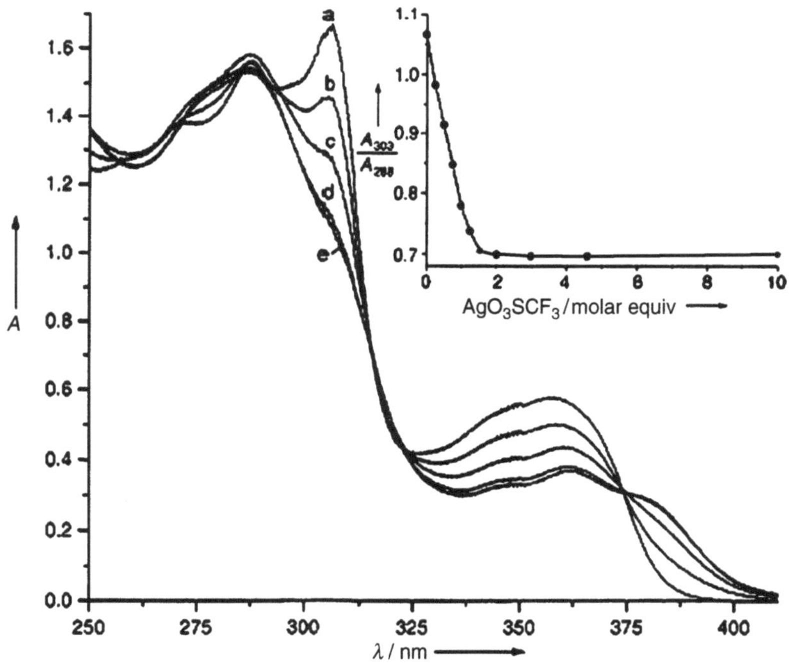

Fig. 14 UV absorbance spectra of **36** in THF with increasing amounts of AgO_3SCF_3: (a) 0, (b) 0.5, (c) 1.0, (d) 1.5, and (e) 2.0 equiv. *Inset*: Plot of A_{303}/A_{288} versus the number of molar equivalents of AgO_3SCF_3. A=absorbance (c(**36**)=6.7×10^{-6} M)

dom conformation through side-chain interactions, AgO_3SCF_3 was added to a solution of **15** in THF. No changes were observed in the UV spectra, indicating that metal binding was not occurring in the control study.

Addition of silver triflate to a solution of **36** in THF resulted in a decrease in the absorbance at 306 nm while the absorbance at 288 nm did not change (Fig. 14), indicating that metal binding was driving helix formation [23, 25]. Evaluation of the Ag^+ titration curve (Fig. 14, inset) shows that the ratio of the two absorbances decreased until two equivalents of AgO_3SCF_3 were added. This suggests the formation of $[\mathbf{36}^{\cdot}Ag_2]^{2+}$ through a two-step process (Fig. 15) where the product of the association constants (K_1K_2) was estimated to be greater than 10^{12} M^{-2}. To determine the effect of solvophobic forces on the metal binding, oligomers with three alternating cyano *m*PE units were studied. The addition of

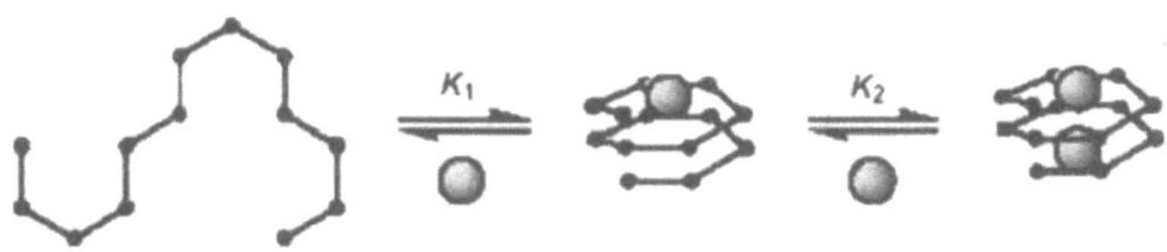

Fig. 15 Representation of the metal-induced formation of helical structures. The metal ions (Ag^+) are shown as spheres

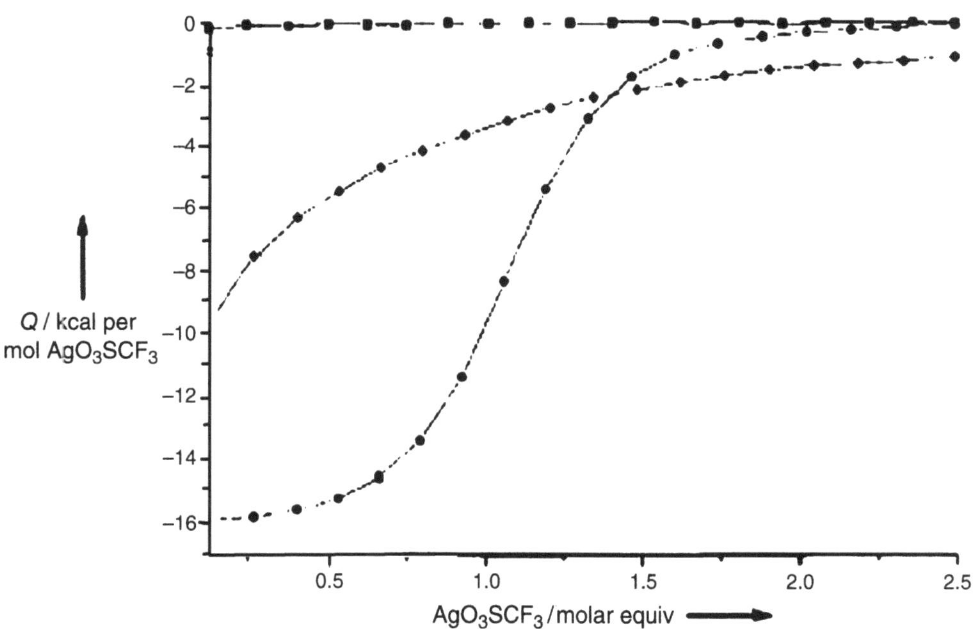

Fig. 16 Heat generated (Q) as a function of the amount of AgO_3SCF_3 added to solutions of **36** (*filled circle*), **38** (*filled diamond*), and **37** (*filled square*) in THF, as determined by isothermal titration microcalorimetry (T=28 °C, c=0.2 mm)

AgO_3SCF_3 to **36** was followed by isothermal titration microcalorimetry and showed a sigmoidal curve that reached a maximum at 2 equiv Ag^+, indicative of cooperative binding of two metal ions and confirming the UV studies (Fig. 16). The same experiment with hexamer **37** showed no heat evolution, indicating that binding did not occur. Addition of AgO_3SCF_3 to **38**, which is **37** with six additional *m*PE units, showed the evolution of heat due to binding, but there was no defined end point of the titration. The titration was then monitored by UV spectroscopy and an association constant of $2\times10^4\ M^{-1}$ was determined. This lower association constant suggests that in the formation of $[\mathbf{36}\cdot Ag_2]^{2+}$ $K_2 \gg K_1$, indicative of a cooperative process of the two metal ions. The inability of **37** to bind silver ions shows the importance of the nonspecific solvophobic interactions in bringing about a somewhat folded structure that encourages metal binding.

2.4 Binding of Guest Molecules

A by-product of the solvophobic collapse of the *m*PE backbone is the formation of a hydrophobic cavity. Similar cavities have been explored for use as hosts in molecular recognition; however, these cavities are typically quite rigid and synthetically difficult to access [4, 6]. The formation of a hydrophobic cavity through supramolecular folding of the *m*PE backbone offers a modular platform for the study of small molecule binding with supramolecular hosts.

A convenient method to demonstrate guest binding in *m*PE systems is circular dichroism spectroscopy (CD). CD is a method which allows the determination of chiral excess in a system [56]. In the absence of a chiral influence, the *m*PE helix exists as a racemic mixture of both M and P helices and displays no CD signal. Upon the addition of a chiral guest, diastereomeric complexes are

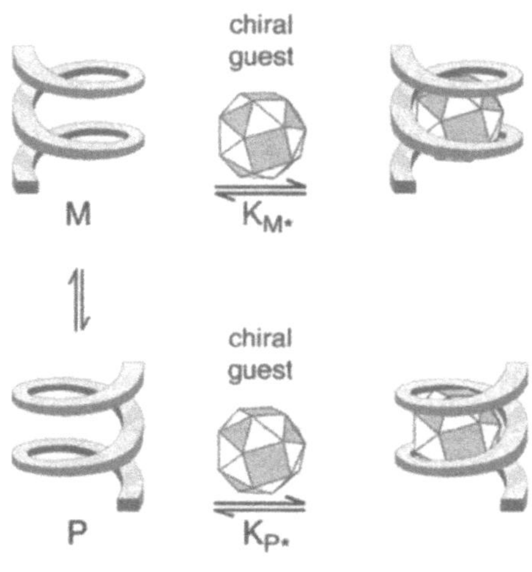

Fig. 17 Pathway for formation of a diastereomeric excess of one twist sense of helix that could be monitored by CD spectroscopy

formed with association constants K_p and K_m. If these association constants are not equal, an excess of one complex will be formed and a CD signal will be observed (Fig. 17). In both the folded and unfolded state **15** shows no CD signal; however, upon addition of the chiral monoterpene (–)-α-pinene (**39**), a large Cotton effect was observed in the wavelength range where the oligomer absorbs (Fig. 18) [57]. Binding of (+)-α-pinene gave the mirror-image CD signal to the one observed for (–)-α-pinene, indicating that the Cotton effect is a result of a complex between **15** and α-pinene (Fig. 18).

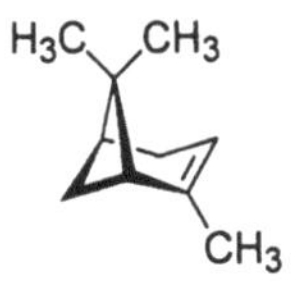

39
(-)-α-pinene

The properties of the oligomer–pinene complex were determined through a series of CD experiments [57]. The strength of the host–guest complex was obtained by adding **39** until saturation of the CD signal. An isodichroic point was observed in the spectrum, indicating one type of complex is formed the stoichiometry of which was determined to be 1:1 through a Benisi–Hildebrandt plot ($(1\ \Delta\varepsilon^{-1})$ vs $(1\ [\text{guest}]^{-1})$). This allowed the CD data to be fit to a 1:1 binding isotherm giving a K_{11} of 6,380 M^{-1}. The nature of the binding was determined to be a solvophobic driving force as the K_{11} of binding is linearly related (over the range investigated) to the water percentage in acetonitrile (Fig. 18). Oligomer **15** was found to be a somewhat general host, as various other terpenes were able to bind with relatively similar values of K_{11}; however, **39** formed the strongest complex. Evidence that this solvophobically driven binding was occurring inside the cavity instead of on the exterior was shown by limiting the cavity size with methyl groups (**24**). When binding experiments were performed with **24** and (–)-α-pinene, the K_{11} was found to be 40 M^{-1}, over 100-fold smaller than was observed for **15** indicating that guest binding occurs inside the helix.

As *m*PE oligomers become longer, the aspect ratio of the helical cavity also increases, such that a rodlike shape complements the cavity better than a spherical shape. To take advantage of this change in aspect ratio a chiral, rodlike piperazine guest (**40**) was synthesized [58]. The affinity of a host–guest com-

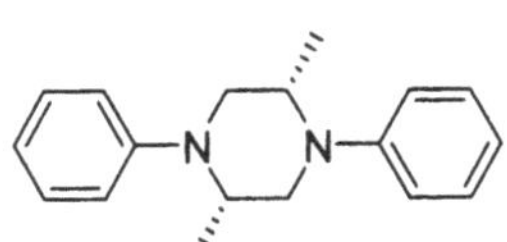

40

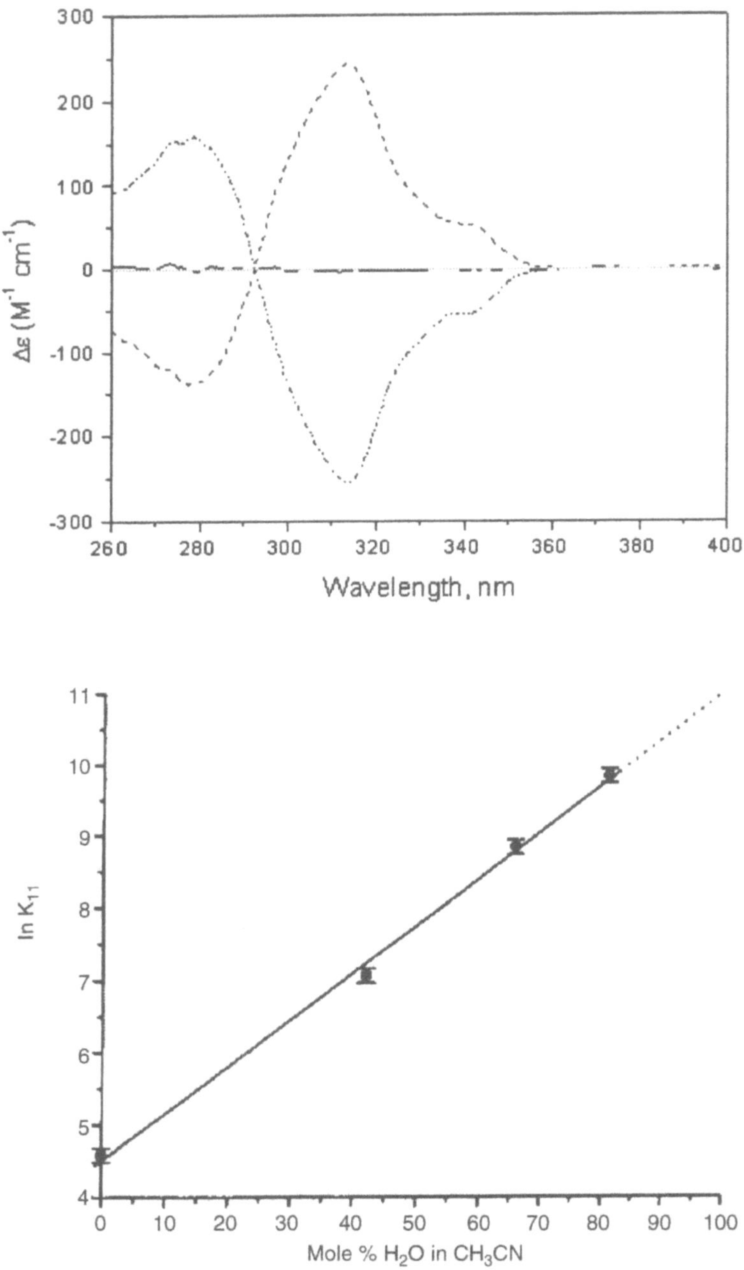

Fig. 18 *Top*: CD spectra. Oligomer **15** (*solid line*); oligomer **15** in the presence of 100 equiv of (–)-α-pinene (*dotted line*) and 100 equiv of (+)-α-pinene (*dashed line*) in a mixed solvent of 40% water in acetonitrile (by volume) at 295 K. [**15**]=4.2 μmol. *Bottom*: Plot of $\ln K_{11}$ for **15** against the solvent composition. The *solid line* is the least-squares linear fit (correlation coefficient=0.9987), and the *dotted line* is the extrapolation to 100% water. Error bars are from the nonlinear fitting of the data to K_{11}

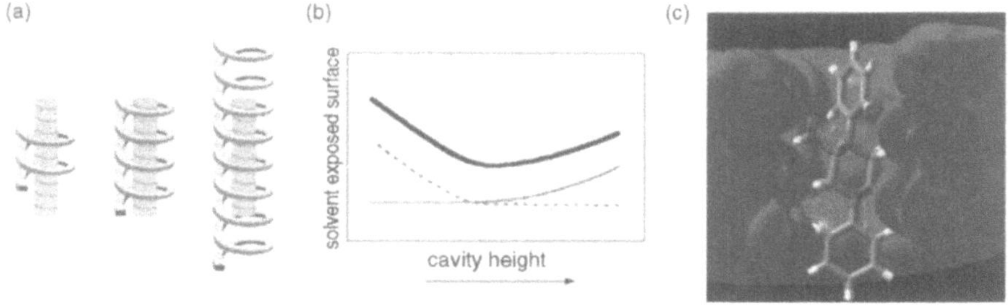

Fig. 19 **a** Schematic diagram illustrating the binding of a rodlike guest to helical oligomers of different lengths. The cavity height is determined by the length of the oligomer. **b** Solvent-exposed surface of the oligomer cavity (—) and the rodlike guest (---) in a complexed state as a function of cavity height. The total amount of solvent-exposed surface (—) shows a minimum that predicts a cavity length with the highest affinity for the rodlike guest. **c** Minimized structure of **18** with **40** determined by a Monte Carlo docking algorithm

plex is believed to be based on the area of contact between the interacting molecular surfaces [59, 60]. Therefore, a given guest should have an optimum-sized *m*PE host that creates the maximum contact between the two surfaces while limiting the total exposed surface (Fig. 19). To determine this optimum-sized *m*PE for guest **40**, binding studies were performed with **13–18** and longer oligomers **41–43**. Addition of multiple equivalents of **40** brought about saturation of the induced CD signal for all oligomers studied and allowed the determination of a 1:1 binding stoichiometry. The association constants of these complexes spanned a large range with the largest K_{11} values for **41** and **42** (n=20, 22), indicating that these lengths offered the most surface-area interaction (Fig. 20). The decrease in binding affinity of **43** is most likely due to the increase in amount of solvent-exposed surface now present in the helix. This important example shows that a given guest displays higher binding affinities for relatively specific lengths of *m*PEs.

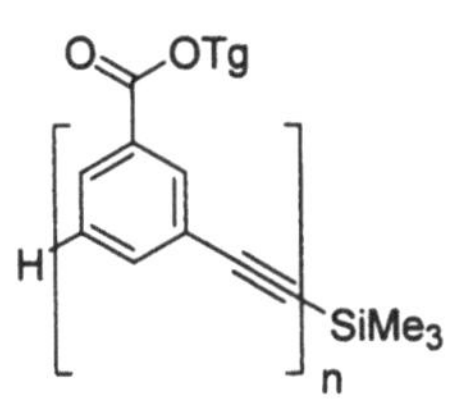

41 n = 20
42 n = 22
43 n = 24

The mechanism of the guest binding offers an interesting question. During the formation of a host–guest complex, does the guest thread itself into the interior of the helical cavity (pathway 1), or does the *m*PE oligomer wrap around

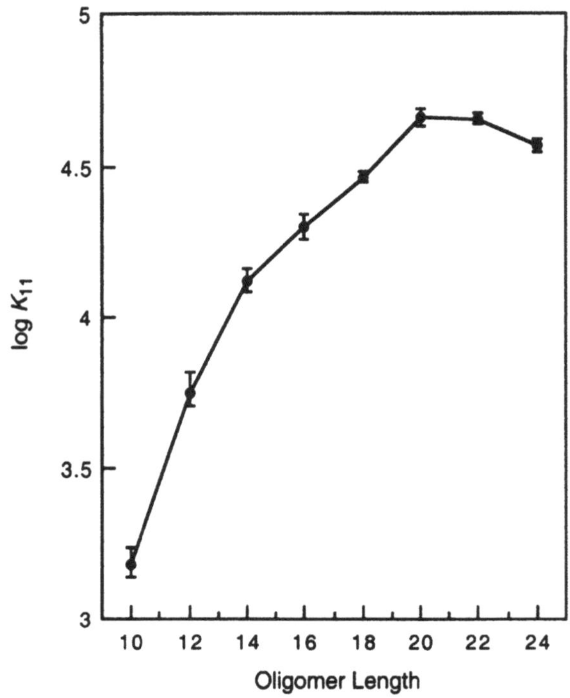

Fig. 20 Plot of log K_{11} against oligomer length. The binding affinity of **40** reaches a maximum value with the 20-mer (**41**) and 22-mer (**42**)

the guest (pathway 2) (Fig. 21)? To determine which pathway is more likely, **40** was modified with bulky end-group substituents to give a dumbbell-shaped guest (**44**) [61]. Modeling studies showed that the bulky substituents prevent **44** from entering into the helical cavity without some type of helical rearrangement. Therefore, the inability of **44** to bind to *m*PE oligomers would suggest that guest binding occurs through a threading mechanism (pathway 1). Binding studies of oligomers **17**, **18**, and **41–43** with the dumbbell-shaped guest showed

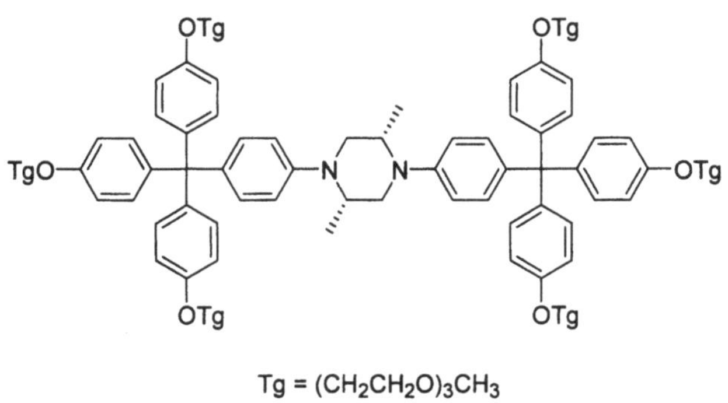

Tg = $(CH_2CH_2O)_3CH_3$

44

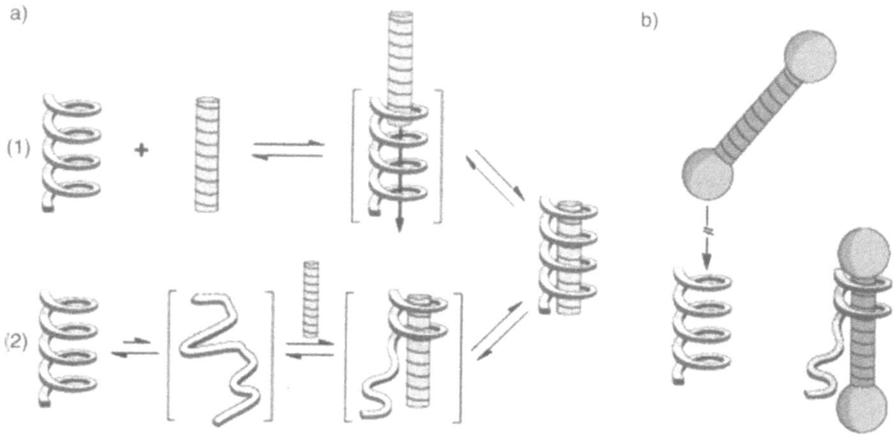

Fig. 21 Foldamer association with rodlike ligands. **a** Two possible limiting mechanisms are the threading of a guest into an oligomer cavity without disruption of oligomer structure (1), and the wrapping of an unfolded or partially unfolded oligomer around the chiral ligand (2). **b** Mechanistic probe with bulky end groups that will make guest binding through pathway 1 unfavorable

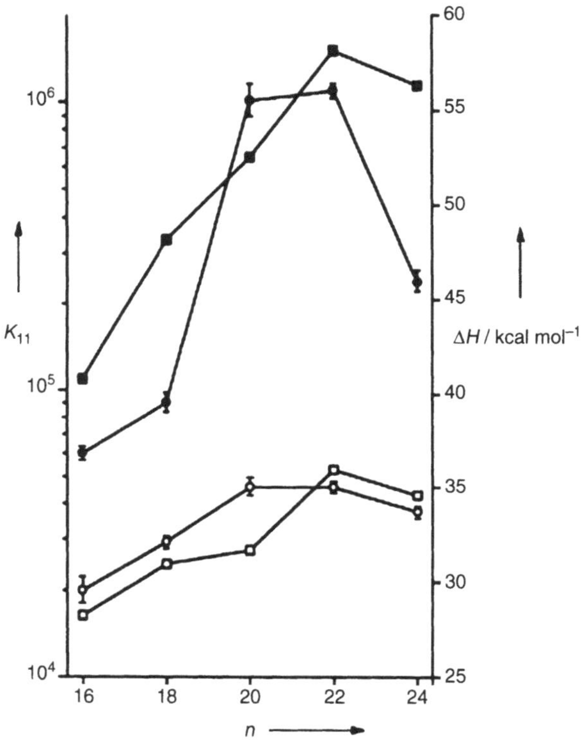

Fig. 22 Plot of log K_{11} values (*solid squares and circles*) and modeled heat of association (*open squares and circles*) against oligomer length for ligands **40** (*circles*) and **44** (*squares*). Experimental K_{11} values for **40** are taken from ref. [58] for direct comparison to **44**

an induced CD signal in all cases, indicating the *m*PE oligomer was able to reorganize in order to wrap around the dumbbell-shaped guest (pathway 2). However, evaluation of the kinetics of this process showed that binding of **44** was occurring much slower than what was observed for the binding of **40**. This suggests that when binding a bulky guest the *m*PE backbone is able to wrap around the guest, but when possible pathway 1 is the preferred mode of binding. The bulky end groups of **44** result in an increase in the positive contacts between the host and guest, leading to an increase in binding affinities when compared to **40** (Fig. 22). Because of the rigid nature of the guest, oligomers **42** and **43** (n=20, 22) were able to form contacts with both ends of the guest, giving an even larger increase in the affinity. For this reason, the binding of the dumbbell-shaped guest with oligomers **42** and **43** was quite specific.

2.5 Introduction of a Twist Sense Bias to the Folded State

The ability of the *m*PE helix to bind guests showed the potential of the system toward molecular recognition and guest binding. The biasing of the twist sense through methods other than guest binding would offer a chiral environment for binding. New methods for transferring chiral information to the oligomer backbone were desired.

Installation of chiral side chains allows the introduction of a chiral element without drastic changes to the system. Chiral Tg (Tg*) (**45**) [62] or apolar (*S*)-3,7-dimethyl-1-octanoxy (Ap*) (**46**) [63] side chains both place the asymmetric unit close to the oligomer backbone, thereby imparting the largest bias possible for the side chain. The UV spectral analysis and solvent denaturation studies of **45** were almost identical to those of native oligomers, indicating that the chiral side chains do not disrupt the stability of the folded state significantly [62]. In chloroform, **45** showed no twist sense bias as evidenced by the lack of CD signal, while in acetonitrile, a Cotton effect was observed indicating the transfer of chiral information and the formation of a twist sense bias in the folded state. The acetonitrile concentration needed to observe a Cotton effect was much higher than is needed to bring about folding, suggesting that the backbone adopts "helix-like" order well before the side chains impart bias upon the twist sense (Fig. 23). Only after the side chains have efficiently ordered

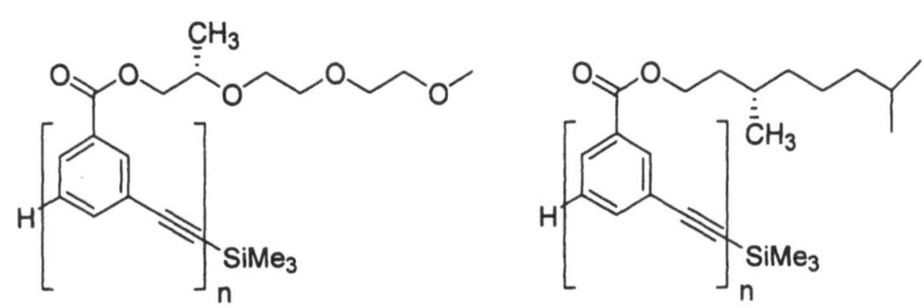

45 **46**

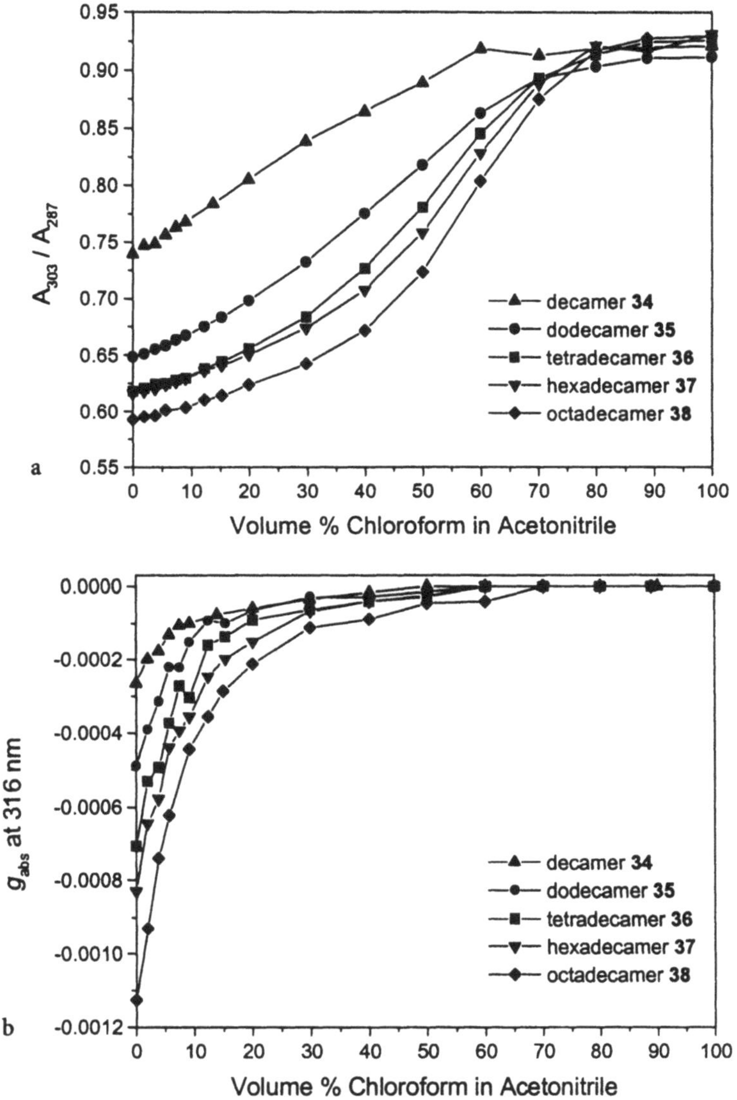

Fig. 23 Solvent denaturation curves of **45** measured by **a** the UV absorbance ratio A_{303}/A_{287} and **b** the circular dichroism (g_{abs} at 316 nm) showing the need for higher vol% acetonitrile to bring about twist sense bias than that needed for folding

around the collapsed backbone structure is the chiral information transferred to a helix [62]. Similar results were observed for **46**, where "it is apparent that the transfer of chirality is a highly cooperative process that requires a progression of order beyond the initially formed helical state [63]." Another possible explanation for the lag in chiral response is a congregation of loosely "helix-like" backbone conformations which may rapidly interconvert between relatively compact, but not yet helical states. This would lead to UV spectra that suggest a folded state while no ordered diastereomeric excess has been formed. At high concentration of acetonitrile, the chiral side chains are able to impart chirality to the helix; however, the induced twist sense bias is relatively weak. Although this approach leads to a rather small twist sense bias, the folding characteristics of the oligomers are not modified.

In an effort to impart a stronger twist sense bias on the *m*PE helix, a chiral binaphthol derivative was placed directly into the *m*PE backbone [64]. The chiral binaphthol unit was placed both at the center (**47**) and terminus (**48**) of the oligomer in order to determine how the position of the chiral unit would affect the twist sense bias and folded state stability. Solvent denaturation studies (UV

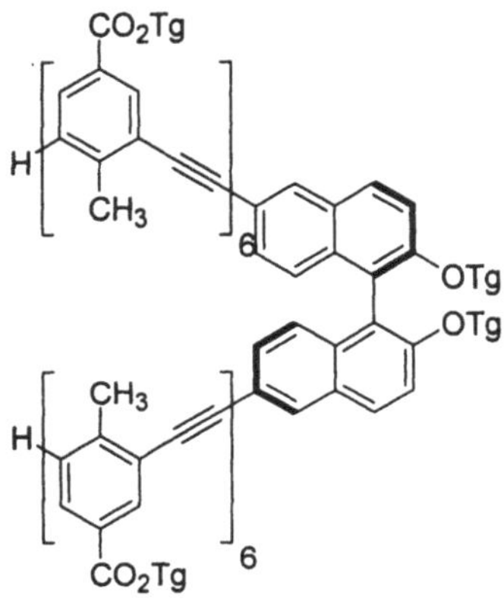

47

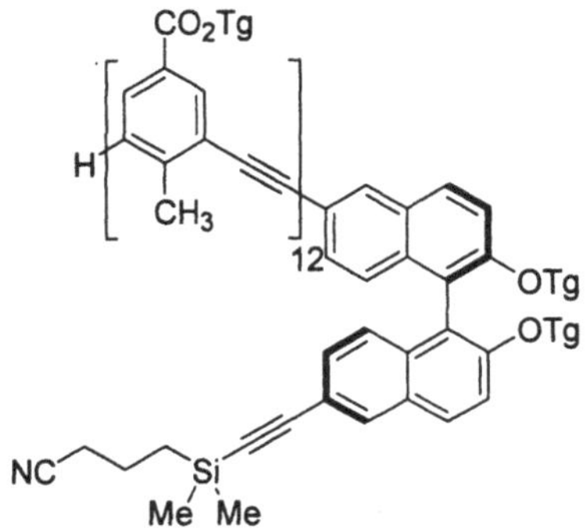

48

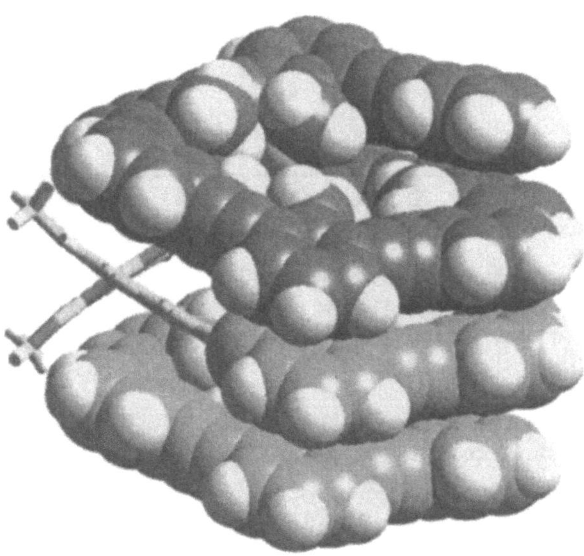

Fig. 24 Minimized structure of **47** showing the kink in the *m*PE backbone brought about by the binap moiety

and fluorescence) of **47** and **48** indicated that both oligomers adopt a folded state in acetonitrile, with the folded state of **48** between 3 and 5 kcal mol^{-1} more stable (depending on the method of determination). This is attributed to the binaphthol unit forming a kink in the backbone that disrupts π–π interactions (Fig. 24), which is more destabilizing in the center of the helix versus the terminus. The position of the binaphthol unit also affects the ability to impart a twist sense bias on the oligomers. The effect was found to be almost twice as large for **47**, where the chiral unit is in the center of the helix, than **48**. The introduction of a chiral element directly into the *m*PE backbone, regardless of the position, creates a large twist sense bias but at the price of decreased folded state stability.

The twist sense bias of the *m*PE helix can also be modified through the use of a chiral tether [65]. This tether links two *m*PE segments that are not able to fold independently, giving an oligomer of combined length that favors the formation of a helical structure in poor solvents. The presence of a chiral moiety in the tether creates an asymmetric environment that may impart a twist sense bias on the forming helix. The chiral center of the tether is a (+)-tartaric acid derivative with either a hydrophobic isopropylidene protecting group (**49**), a trimethylsilyl (TMS) protecting group (**50**), or the free diol (**51**). Analysis of the UV spectra showed that **49–51** exist in a random coil conformation in chloroform and a folded state in acetonitrile. The presence of a folded state

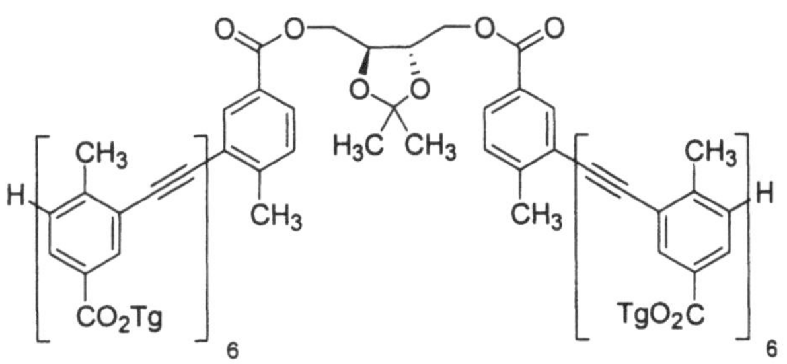

49

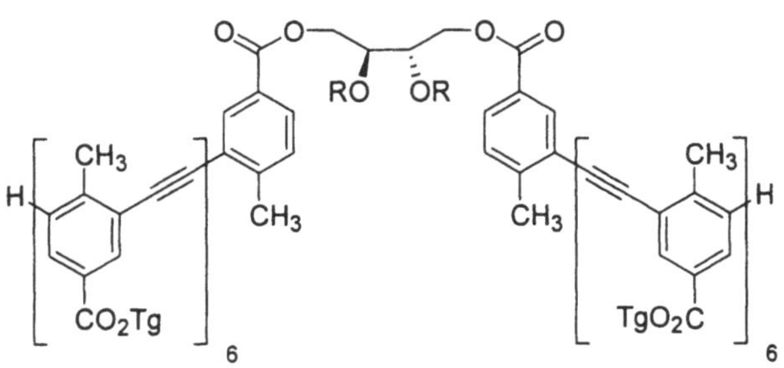

50 R = $SiMe_3$
51 R = H

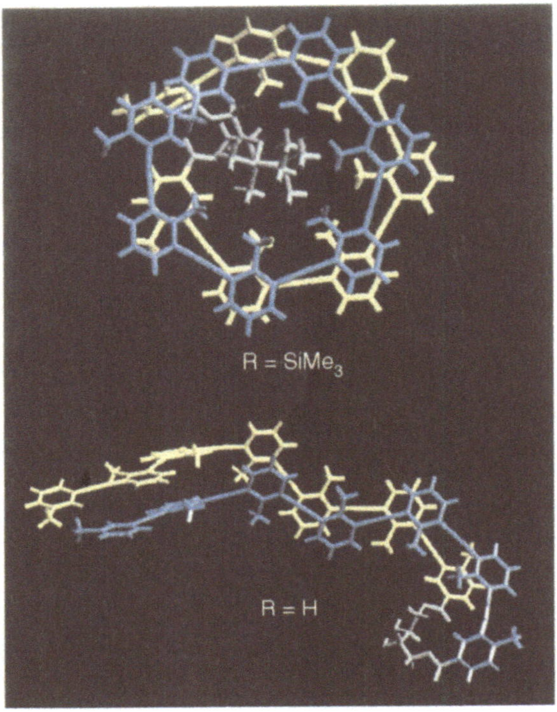

Fig. 25 Minimized structures of an analog of **50** (*top*) and **51** (*bottom*) in which the ester side chains were omitted

indicates that the two nonfolding sections were able to adopt a favorable folding conformation in the presence of the tether. This suggests that the tether is able to bring the nonfolding sections in close proximity without disrupting positive monomer–monomer interactions. The largest twist sense bias was observed for **50**, while **49** showed only a weak CD signal. Given the similar hydrophobic nature of the protecting groups this was unexpected. Evaluation of computer modeling shows the hydrophobic TMS ethers project into the helical cavity while the flexible backbone of the tether allows the *m*PE backbone to π-stack (Fig. 25). Alternatively, the relatively inflexible ketal (**49**) prohibits the adoption of conformations that allow π–π stacking to occur. The CD signal of **51** was nearly identical in both acetonitrile and chloroform, suggesting that some other π-stacked folded state other than a helix was formed (Fig. 25). Although the tether strategy was able to create a large twist sense bias for **50**, the TMS groups effectively block the inside of the cavity. This makes the tether method of twist sense bias less attractive for use in asymmetric reactions or molecular recognition applications.

2.6 Higher-Order Aggregates

The formation of higher-order aggregates is an early step in the effort to assemble functional, nonbiological, multimolecular species. To determine if higher aggregates are formed in *m*PEs, CD was used to monitor the thermal denaturation of **45** over the temperature range –10 to +80 °C in 20 vol% water in acetonitrile [66]. A plot of the maximum CD intensity of **45** (*n*=8–18) vs temperature shows that, depending on the oligomer chain length, there are two distinct processes taking place (Fig. 26). Oligomers **45** (*n*=8–12) show a typical sigmoidal transition from a folded state to an unfolded state through thermal denaturation, as evidenced by the decrease in the CD signal at higher temperature. However, **45** (*n*=14–18) demonstrate an increase in the CD signal as the temperature increases until a maximum value is reached, at which point the intensity of the signal decreases. The isodichroic point which these three oligomers display indicates that a single type of species is being formed for the three largest oligomers. Because the nonsigmoidal denaturations were only observed for oligomers that strongly favor the folded state, it was suggested that the change in the shape of denaturation curves is caused by the association of the helix into column-like aggregates (Fig. 27).

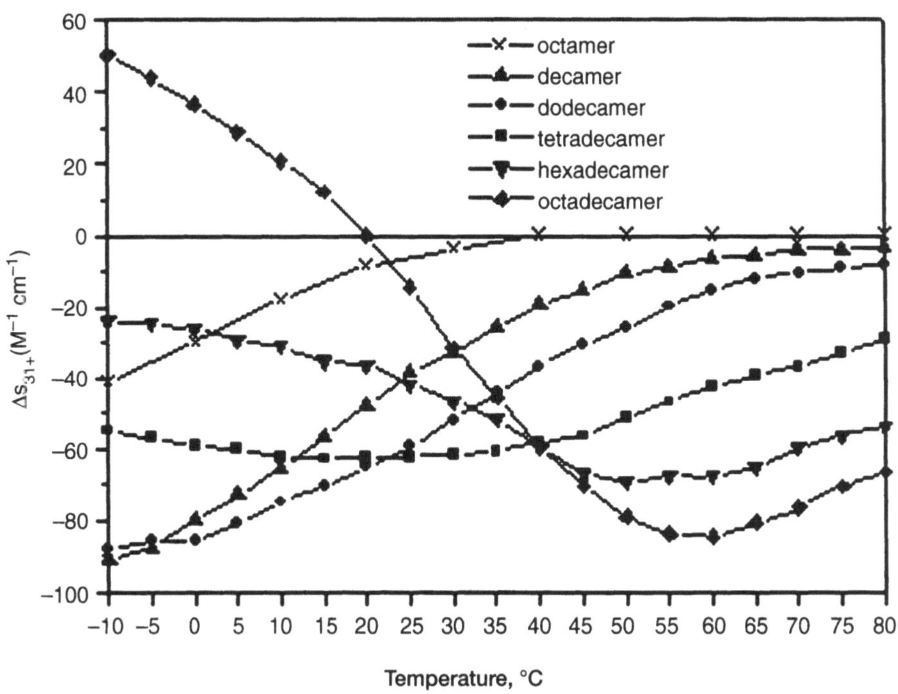

Fig. 26 Plots of $\Delta\varepsilon_{314}$ vs temperature for octamer **45** (*n*=8) through octadecamer (*n*=18) in 20 vol% water in acetonitrile. All solutions were cooled from 80 to –10 °C

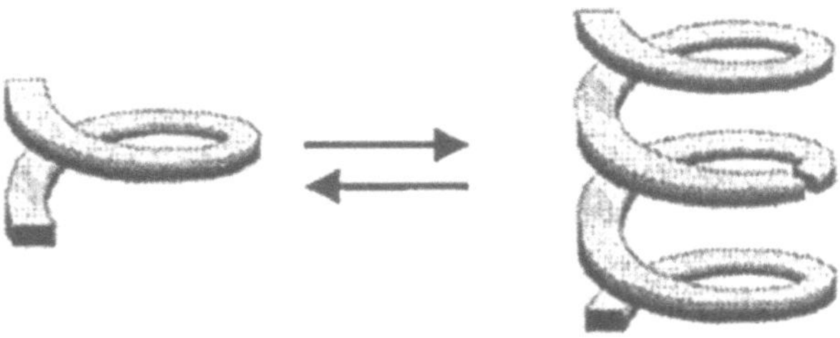

Fig. 27 Proposed aggregation mode of the oligomers

Stacking of a chiral helix on top of an achiral helix could potentially impart a twist sense bias on the aggregate as a whole. To determine if chirality transfer was taking place in column-like aggregates, "sergeant and soldiers" experiments were performed where achiral oligomers were doped with different percentages of chiral oligomers and the resulting CD was observed [67]. The CD maxima of mixtures of chiral (**45**, n=18) and achiral oligomers (**18**) were studied in different solvent compositions to determine if chirality was transferred through aggregation. In 100% acetonitrile the CD signal was linear with respect to the percentage of chiral oligomer, indicating that the signals observed in 100% acetonitrile can be attributed to intramolecular effects (Fig. 28). However, as the vol% water in acetonitrile was increased, the CD signal deviated positively from linearity, showing that the chirality is transferred from the chiral to achiral oligomers through aggregation. The amplification of chirality in the "sergeant and soldiers" experiments is not very large, most likely due to the lack of directionality in the solvophobic driving of folding and stacking and the absence of diastereomeric purity of aggregates [66]. Nonetheless, these experiments show that higher-order *m*PE aggregates are being formed in mixed aqueous/organic solvents. The presence of a chiral element can induce a twist sense bias for the entire aggregate.

Amphiphilic *p*PEs have been shown to adopt interesting supramolecular order at the air–water interface [68–72]. To investigate if similar properties would be observed in *m*PEs, the side-chain arrangement of the *m*PE backbone was altered to give amphiphilic polymers (**52–54**). The arrangement of these polymers at the air–water interface was investigated using Langmuir films. The

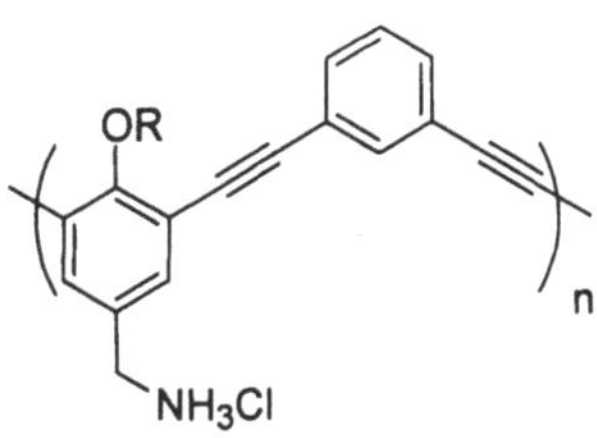

52 R = n-C_5H_{11}
53 R = n-C_8H_{17}
54 R = n-$C_{12}H_{25}$

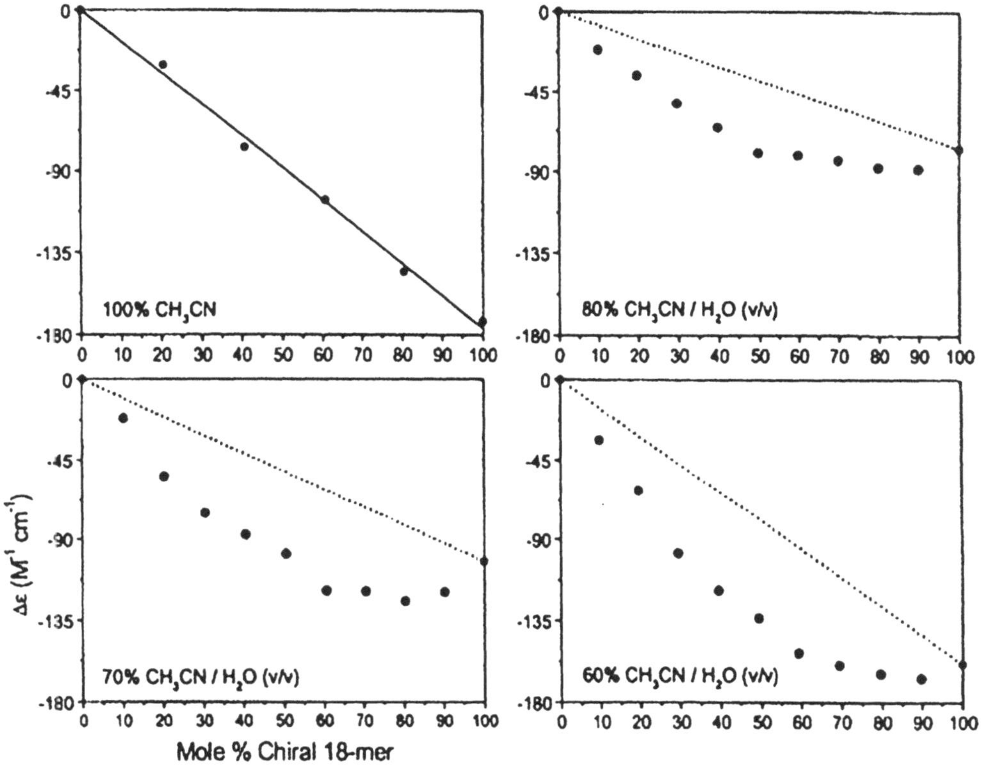

Fig. 28 Plot of $\Delta\varepsilon_{314}$ vs mol % chiral octadecamer **45** (n=18) for solutions of varying amounts of chiral and achiral **18** (n=18) octadecamers in different concentrations of water/acetonitrile (v/v). All spectra were recorded in solutions with a total oligomer concentration of 3.3 μM. The *dotted lines* are the expected signals that should arise upon dilution of a sample containing only chiral octadecamer. The *solid line* is the least-squares linear fit of the chiral octadecamer dilution data (*top left*, correlation coefficient=0.998

surface area of each repeat was calculated by extrapolating the steepest slope of the pressure–area isotherm (Fig. 29) to zero pressure giving surface areas of 41, 42, and 45 Å^2 for **52–54**, respectively, which corresponds to an edge-on orientation. This arrangement places the hydrophobic backbone and alkyl side chains in air while the hydrophilic cationic amine is immersed in the aqueous phase (Fig. 30) [73, 74]. In this arrangement the aromatic faces are separated by approximately 4 Å, with the adjacent polymer chains slightly offset in order to achieve maximum π-stacking [24, 75, 76]. The degree of association of the polymers in the monolayer was determined to be close to unity for **52**, which corresponds to a gaslike state. Meanwhile, the degree of association for **53** and **54** is closer to two, consistent with the more hydrophobic nature of polymers with large alkyl side chains. This explains the sharp change in slope of the pressure–area isotherm at 38 mN/m for **52**, likely due to monolayer collapse. Polymers **53** and **54**, with longer more hydrophobic alkyl side chains and a greater

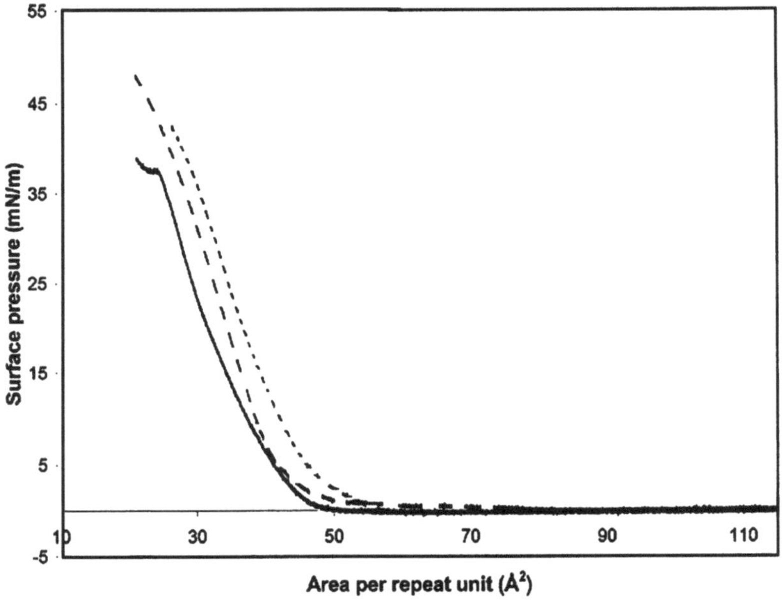

Fig. 29 Pressure–area isotherms for **52** (*solid line*), **53** (*dashed line*), and **54** (*dotted line*) showing monolayer formation

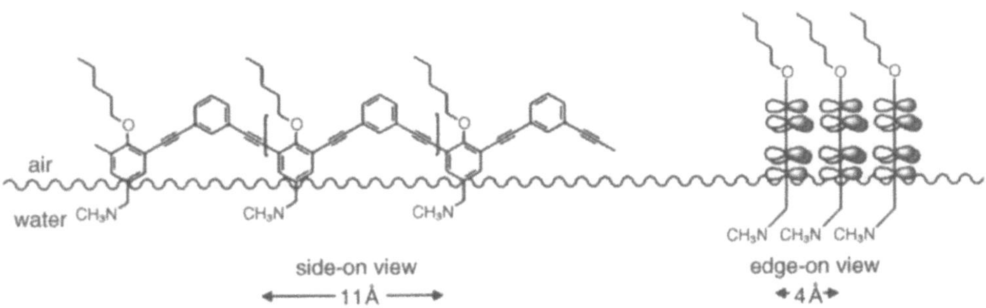

Fig. 30 *Left*: Side view of a single chain of amphiphilic polymer **52** at the air–water interface. *Right*: End-on view of three adjacent π-stacking polymer chains

degree of association, do not collapse up to 48 mN/m. An interesting application of this supramolecular order is the ability of these amphiphilic polymers to bring about lysis in phospholipid vesicles in a similar fashion to antimicrobial peptides and biomimetic polymers [74].

2.7 Solid-State Properties

The ability of *m*PEs to form a compact, folded conformation in solution [23, 25] led researchers to look at the solid-state properties. Unfortunately, the Tg side chains needed to impart solubility in organic solvents make the oligomers amorphous, and therefore difficult to characterize in the solid state. X-ray powder diffraction and thin-film UV spectroscopy have been shown to be very useful in the characterization of *p*PEs, and have been used extensively in these studies [7].

The solid-state structures of **10–18** were determined through small-angle X-ray powder diffraction (SAXD), wide-angle X-ray powder diffraction (WAXD), and thin-film UV experiments [77]. Samples of the oligomers were prepared by drawing the melted oligomers into capillary tubes and annealing them, leading to an increase in the order of the X-ray diffraction as well as redshifting of absorbance in UV spectroscopy. This suggested that the oligomers must undergo significant reorganization to reach the thermodynamically most stable solid-state conformation. The X-ray diffraction of the oligomers gave *d*-spacings characteristic of a lamellar, interdigitated conformation where the size of the lamellae is dependent on the chain length of the oligomers (Fig. 31). UV spectra of annealed films confirmed that the oligomers are in an extended conformation, showing spectra corresponding to a "transoid" backbone. This lamellar arrangement takes advantage of the aromatic π–π stacking while eliminating the exposed surface of the helical cavity.

Oligomers substituted with a methyl group at the *ortho* position (**19–27**) display more stable folded states in solution than **10–18** because of a more solvophobic backbone and increased solvent exclusion in the helical cavity [23, 25]. In trying to determine the solid-state properties of the methyl-substituted

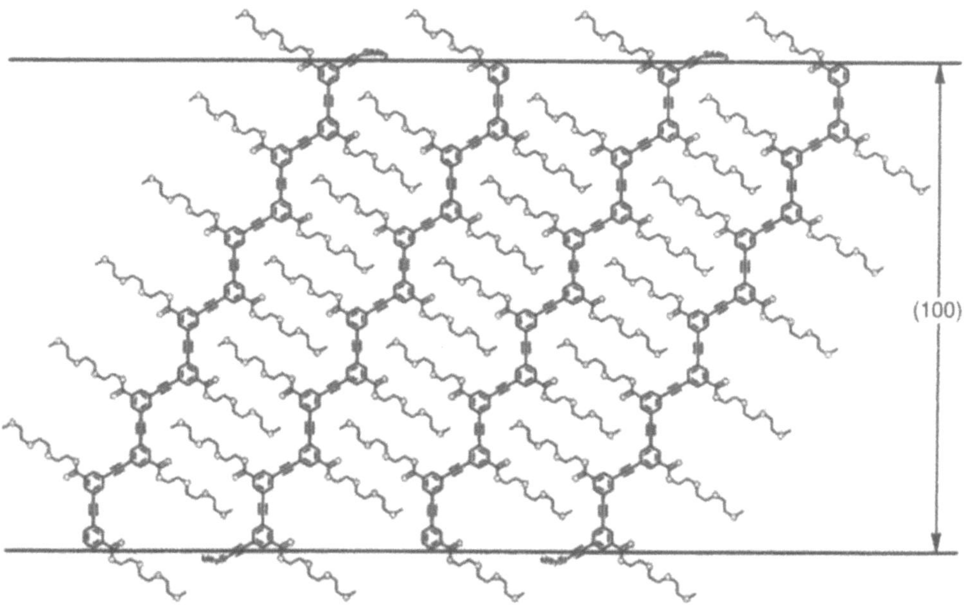

Fig. 31 Schematic diagram of the packing model for hydrogen-series oligomers in the solid state (oligomer **15** is shown)

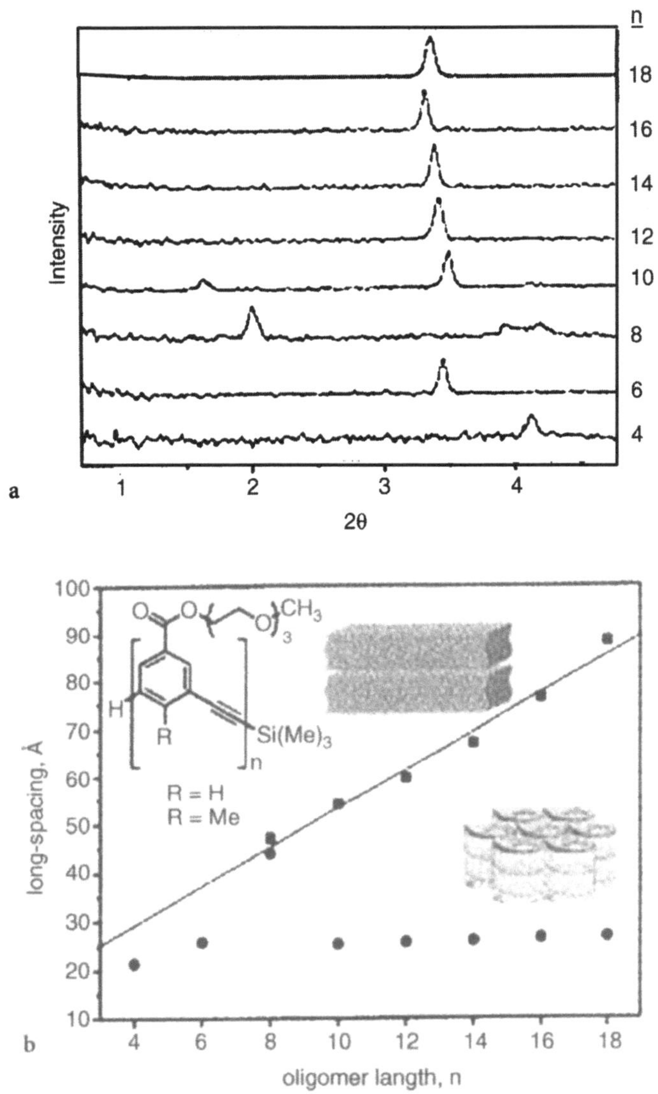

Fig. 32 **a** SAXD profiles for methyl-series oligomers (**20–27**, n=4–18) at room temperature. **b** Plot of long spacings versus oligomer length for two oligo(m-phenylene ethynylene) series as observed in SAXD measurements. Hydrogen-series oligomers (**11–18**) (prepared from the melt) pack in a lamellar arrangement (*filled square*, pictured *top*), and thus the long spacings depend linearly on chain length [77]. In contrast, **20–27** (prepared from solvent evaporation) exhibit long spacings that are independent of chain length ($n \geq 10$), suggesting the possibility of helical columns (*filled circle*, pictured *bottom*)

oligomers it was found that **24–27** do not undergo melting transitions, but begin to decompose at elevated temperatures making it necessary to develop a different method of sample preparation. Concentrated solutions of **19–27** in methylene chloride were drawn into capillary tubes, and the solvent was allowed to evaporate at room temperature over a period of days [78]. The WAXD and SAXD patterns of these samples show length dependence characteristic of their solid-state structure. Oligomers that favored folding in acetonitrile solution adopt hexagonal packing (**21**, **23–27**), while **19**, **20**, and **22** show *d*-spacings characteristic of a lamellar phase (Fig. 32). To confirm the presence of a hexagonal conformation in the solid state, thin films of **21** and **23–27** were cast. The UV spectra of the films showed an all-*cisoid* conformation, indicating the formation of helical nanotubes. Addition of the methyl group at the *ortho* position allows the solid-state organization to be changed from lamellar to hexagonal nanotubes, demonstrating control over the solid-state conformation [78, 79].

Modification of the solid-state structure of *m*PE oligomers can also be accomplished through guest templation. It had previously been shown that addition of silver triflate to **36** brings about helix formation in solution [23, 25]. To determine if the same phenomenon occurred in the solid state, SAXD and WAXD profiles of **36** with and without silver triflate (prepared through evaporation) were obtained [80]. The solid-state profiles of **36** with silver triflate were characteristic of a hexagonal phase, while in the absence of silver triflate WAXD and SAXD spectra indicative of lamellar conformation were observed. To determine if the solid-state structure of *m*PE oligomers could be templated through the use of less-specific interactions, the solid-state organization of **10–18** and **41–43** was determined in the presence of rodlike guest **40** [80]. The WAXD and SAXD profiles of **10–18** combined with **40** were consistent with a lamellar conformation, similar to samples prepared in the absence of guest. However, in the presence of **40**, oligomers **41–43** displayed *d*-spacings characteristic of a hexagonal packing motif. The high binding affinities of **40** for **41–43** in solution [58] suggest that the guest is able to bind to the host and template the formation of a hexagonal conformation in the solid state. Solid-state UV spectroscopy of **42** and **40** confirmed that the oligomer exists in a *cisoid* conformation in the solid state, while the presence of an induced CD signal suggested that **40** had templated hexagonal ordering through incorporation into the cavity of the oligomer.

2.8 Imine-Containing Oligomers

A reversible ligation technique was desired that would allow the use of the *m*PE system in dynamic combinatorial libraries (DCL) in an effort to identify "masterpiece" sequences [81]. The imine bond metathesis is known to have an equilibrium constant close to unity, undergoes reactions at reasonable rates at room temperature, and has a geometry that is compatible with the phenylacetylene unit [82]; therefore it was chosen as a component for *m*PEs.

The quantification of macrocycle self-association is a useful tool in determining the likelihood a given backbone will adopt a helical conformation through solvophobic driving forces [24, 47]. To determine the effect of the imine bond on the native *m*PE system, the aggregation of a hexameric imine-containing macrocycle (**55**) was studied [83]. Increasing either the concentration or solvent polarity of solutions of **55** led to upfield shifting of the ^{1}H NMR resonances, indicative of aggregation. The concentration-dependent aromatic shifts were fit to either an isodesmic (THF) or modified isodesmic (acetone) model [84], depending on the solvent, to give association constants (K_E) of 4,800±1,800 M^{-1} and 290±50 M^{-1}, respectively. In both cases the K_E of **55** was lower than that of the native *m*PE macrocycle (13,000 M^{-1}) [85]. This is due in part to the extra degree of freedom available to the imine bond that is not present in the native structure. Additionally, the ^{1}H NMR shifting in acetone did not fit the isodesmic model very well, requiring a modified isodesmic model to accurately describe the macrocycle aggregation in acetone.

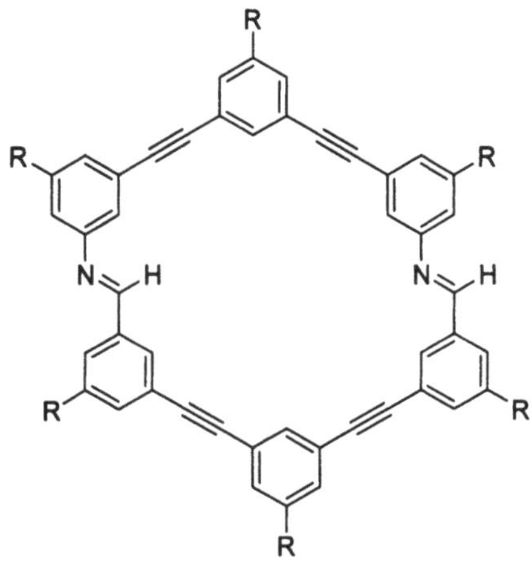

R = $CO_2(CH_2CH_2O)_3CH_3$

55

The modified isodesmic model may imply the need to take the dipole of the imine bond into account in the formation of aggregates. Two different orientations are possible for the formation of stacked macrocycles. The antiparallel fashion has the advantage of attractive dipole interactions that result in a charge-neutral, "tight" dimer (Fig. 33). Stacking in a parallel sense has no attractive dipole interactions and leads to an overall dipole. Dimerization (K_2) proceeds in an antiparallel fashion such that addition of subsequent monomer unit(s) to the dimer no longer has attractive dipole interactions, resulting in a lower K_E value. A model based on this behavior was used to describe the ^{1}H NMR aggregation data in acetone and gave a much better fit with a K_2 over four times as large as K_E (12,700 M^{-1} vs 2,570 M^{-1}) (Fig. 34). This value of K_2 is on the same order as

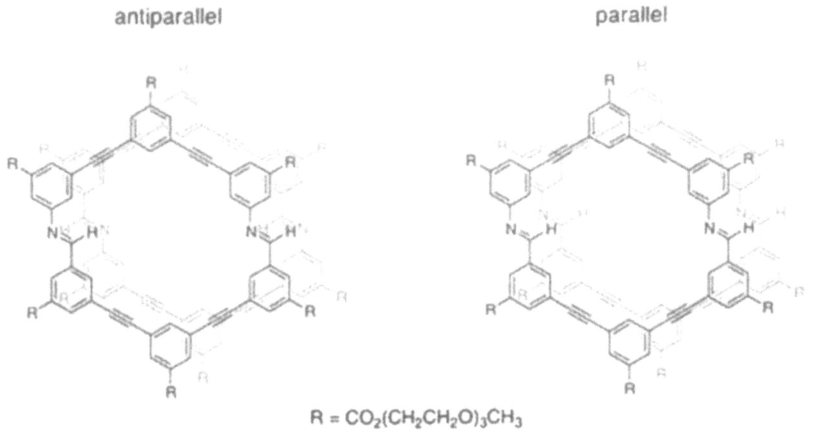

Fig. 33 Representative stacking orientation of 55 in a dimer

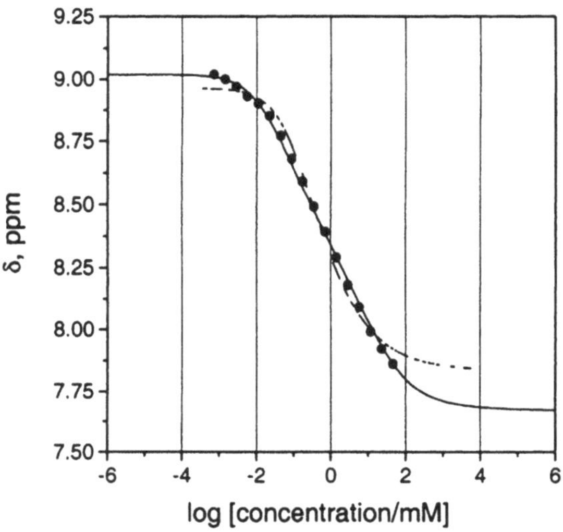

Fig. 34 Analysis of the ^{1}H NMR aggregation studies in acetonitrile using the isodesmic (- - -) and modified isodesmic model (---)

K_E of the native macrocycle, showing that the imines do not severely disrupt the phenylacetylene backbone and will allow folding to take place in a chain species.

Incorporation of an imine unit into a *m*PE began with a Monte Carlo search of dodecamer (**58e**), which indicated that the oligomer adopted a six-turn helical structure [86]. To verify these results solvent denaturation studies were performed on **58e** which showed a helix coil transition with a $\Delta G(CH_3CN)$ very similar to that of the native oligomers (3.0±0.2 for **58e** vs 3.2±0.1 for **15**). The small difference in the $\Delta G(CH_3CN)$ indicates that the imine bond has a negligible effect on the stability of the folded state of the oligomer. With this

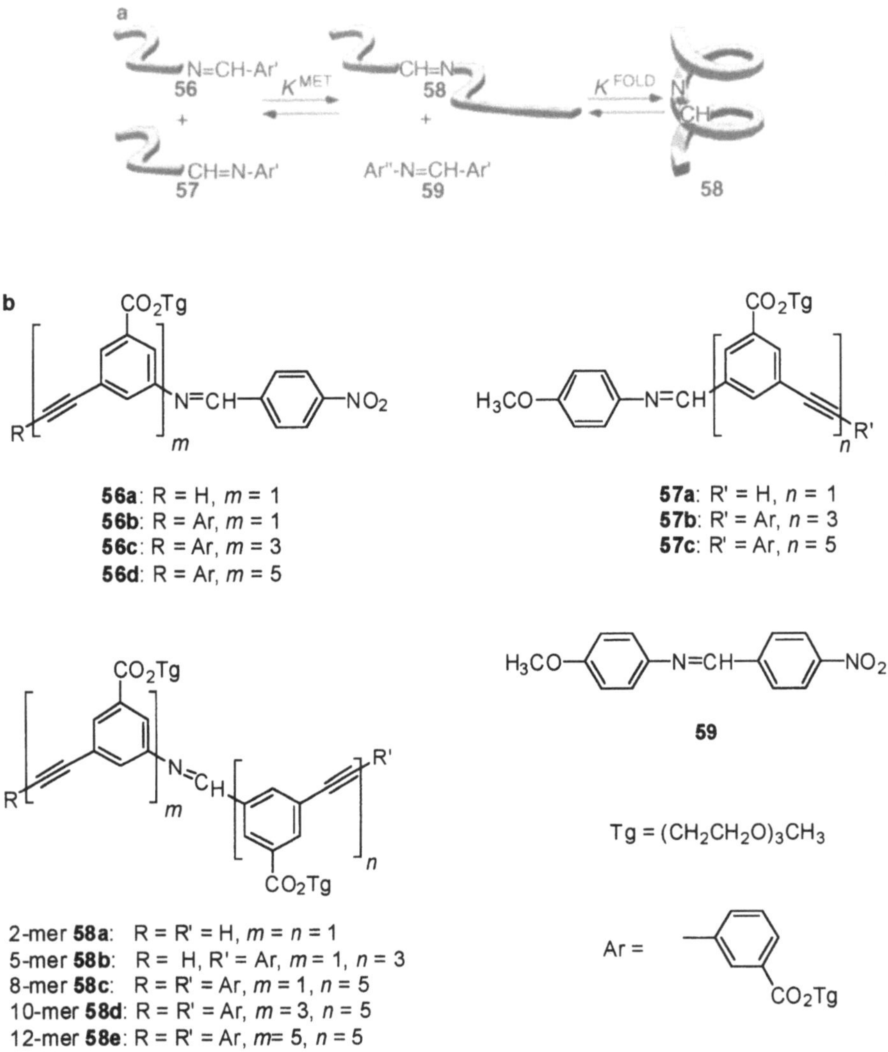

Fig. 35 The imine metathesis of *meta*-connected phenylene ethynylene oligomers. **a** Representation of the ligation reaction of oligo(*m*-phenylene ethynylene) imines. The coupled equilibria correspond to imine metathesis and helical folding. The folding equilibrium depends on the length of the ligated oligomer, the solvent, and the temperature. Equilibrium constants for the metathesis of aromatic imines are 1.0±0.2 regardless of substituents [105]. **b** Chemical structure and corresponding compound numbers of various imines used in this study. Oligomer segments of different lengths (**56a–d** and **57a–c**) were synthesized by repeated palladium/copper-catalyzed coupling of appropriate aryl halide and terminal acetylene precursors. N-terminal and C-terminal imine groups were added by condensation of the corresponding aldehydes and amines

knowledge in hand, the ability of the folding reaction to drive an imine ligation was tested [86]. In a series of experiments, imine-containing oligomer segments were allowed to equilibrate in the presence of an acid catalyst under folding conditions (Fig. 35). The equilibrium constants of these reactions showed that oligomers unable to fold (**58a, 58b**), as well as the control experiments in chloroform, showed virtually no equilibrium shifting ($K<2$) (Table 2). In larger oligomers (**58c–e**), a shift in the equilibrium was present, highlighted by a K_{eq} of 62±2 for **58e** regardless of the imine position. The shifting is attributed to the extra stabilization gained upon folding of the oligomer, which increased almost linearly from n=8–12 as additional monomer–monomer contacts are made [86]. However, once the second turn of the helix has formed, the increase in the equilibrium constant approaches an asymptotic value (Fig. 36) [87].

The demonstration of equilibrium shifting allows experiments to be performed that select for the most stable product in a dynamic situation. The first example of such a dynamic selection for the *m*PE system was the ability to preferentially form an oligomer of one length over another [86]. Under folding conditions **58e** displays a greater folding stability than **58c**, meaning that **58e** should be preferentially formed in a reaction at equilibrium. From the independently determined equilibrium constants for the formation of each product (Table 2), the ratio of [**58e**]/[**58c**] was predicted to be 5.6. To test this hypothesis, a 1:1:1 mixture of monomers **57c**, **56b**, and **56d** was combined under nonfolding and folding conditions and allowed to select between two possible products **58c** (n=8) and **58e** (n=12) [86]. The reaction was allowed to reach equilibrium and the ratio of products was determined to be 3.6 [**58e**]/[**58c**]. This ratio is slightly less than the predicted value, but much higher than the ratio observed under nonfolding conditions (1.6 [**58e**]/[**58c**]). This indicates that under equilibrium

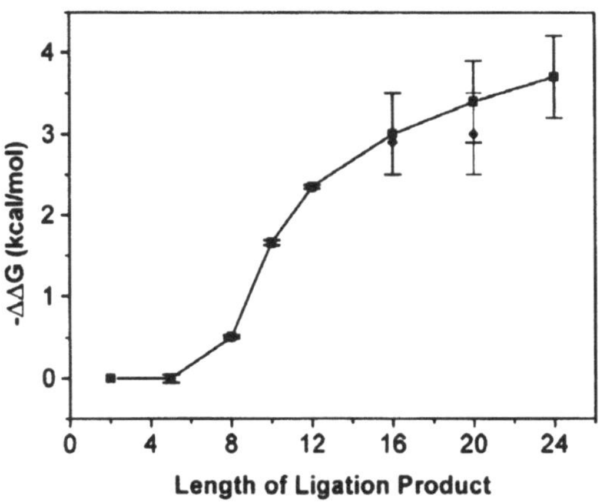

Fig. 36 Plot of the free energy change for metathesis reactions versus the chain length of the ligation product with a central imine bond (*filled square*) and with an imine bond off-center (*filled diamond*) for imine-containing *m*PE oligomers. The magnitude of the equilibrium shifting was referenced to the free energy change for dimer formation

Table 2 Equilibrium constants for various imine metathesis reactions

Entry	Reactants		Ligation product	Approx. extent of helicity[a]	K_{eq} (294 K)[b]		ΔΔG (kcal mol^{-1})[c]
					CD_3CN	$CDCl_3$	
1	**56a**	**57a**	**58a**	None	1.1	1.2±0.1	0.00
2	**56a**	**57b**	**58b**	None	1.1±0.1	1.4±0.2	0.00
3	**56b**	**57c**	**58c**	1 1/3 turns	2.6±0.1	1.6±0.1	–0.50
4	**56c**	**57c**	**58d**	1 2/3 turns	19±1	1.9±0.1	–1.66
5	**56d**	**57c**	**58e**	2 turns	62±2	1.9±0.1	–2.35

[a] Assumes six units per turn.

[b] Measured by ^{1}H NMR. Error estimates are based on the standard deviation of values measured in duplicate runs.

[c] ΔΔG is a measure of equilibrium shifting expressed as the difference in free energy for formation of entry *x* relative to dimer formation (entry 1) in CD_3CN. $\Delta\Delta G = \Delta G(\text{entry } x) - \Delta G(\text{entry } 1) = -RT \ln K_{eq} + RT \ln 1.1$.

conditions, imine metathesis of *m*PEs is able to select for structures with a more stable folded state.

The metathesis reaction of the imine bond was also able to dynamically select for oligomers that form the most stable host–guest complex [88]. It has been shown that rodlike and dumbbell-shaped guests have a higher affinity and specificity for oligomers of 20–22 repeat units [58, 61]. Imine starter sequences **60**, **61a**, and **61b**, which can potentially form 16- (**63a**), 22- (**63b**), and 28-mers (**63c**) as well as smaller molecular weight (MW) materials (**62a** and **62b**), were chosen as starter sequences for an imine ligation experiment (Fig. 37). When equilibrated in chloroform the product distribution of **63a–c** was close to 1:1:1,

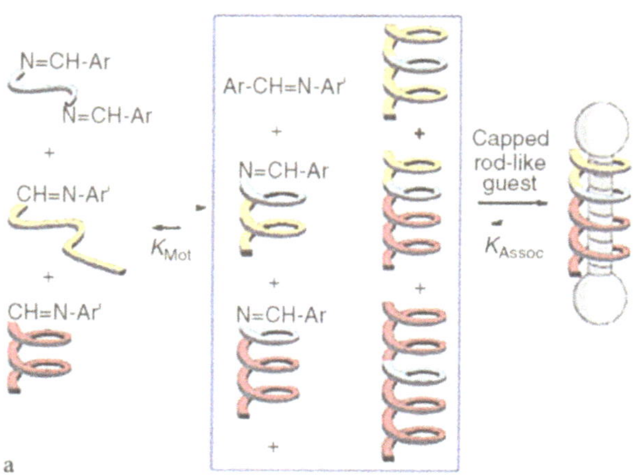

Fig. 37 **a** Schematic diagram that illustrates the equilibrium shifting driven by folding and ligand binding. The sequences are color coded. **b** Chemical structure of oligomers used in the metathesis reaction

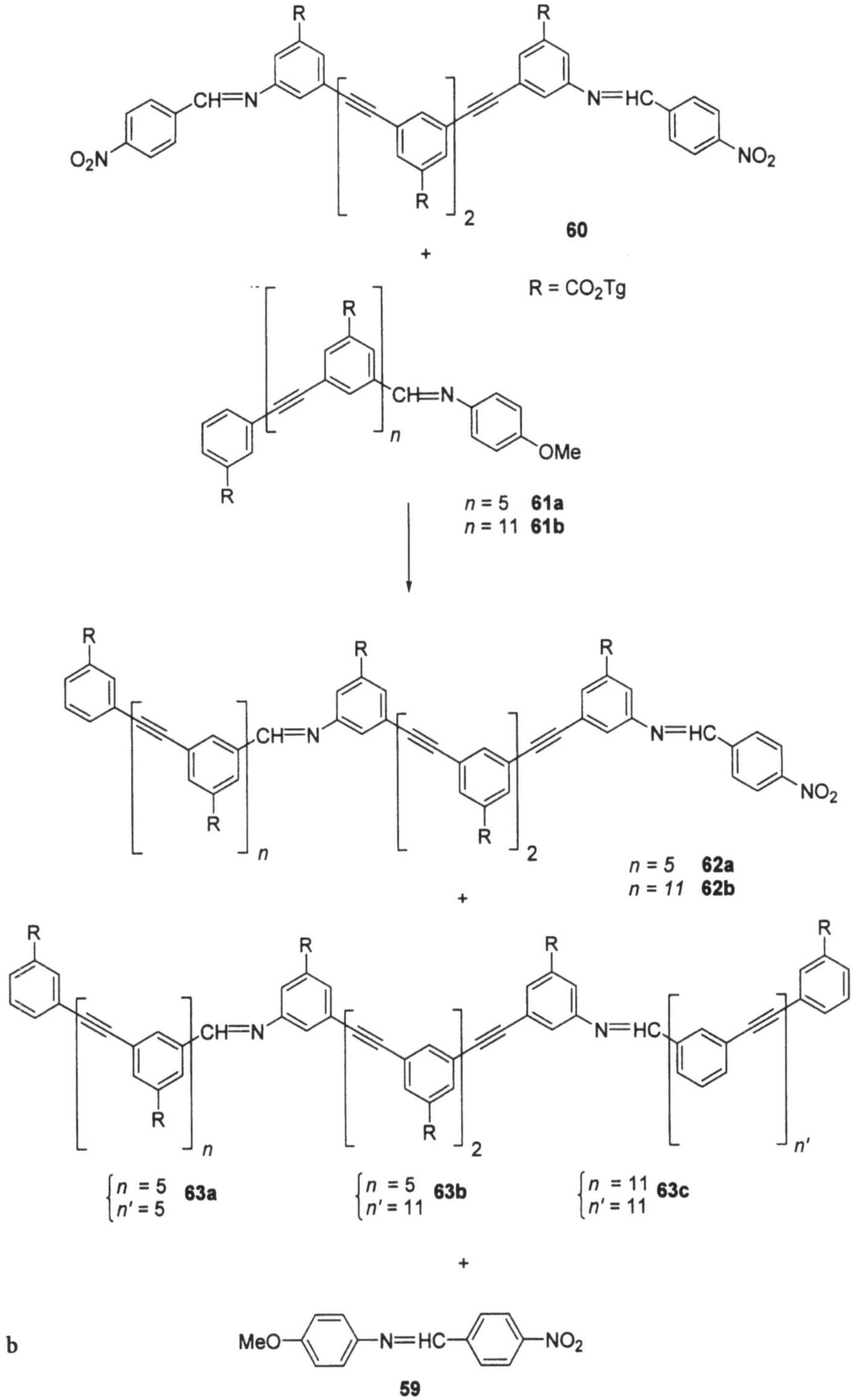

Fig. 37 (continued)

Table 3 Estimated yield of **63a–c** by HPLC

Conditions		Estimated yield (%)[a]			
Solvent	Equiv **44**[b]	**63a**	**63b**	**63c**	Total **63**
$CHCl_3$	0	6±1	8±1	5±1	19
Calcd[c]	–	8	16	8	–
CH_3CN	0	19±1	37±2	16±1	72
CH_3CN	2	10±1	66±3	9±1	85

[a] Quantitative analysis using peak areas. Error estimates are based on experimental reproducibility of duplicate runs.
[b] Equivalents to total amount of starting **60** and **61**.
[c] Assumes all metathesis equilibria have equilibrium constants of unity, as expected for $CHCl_3$ data.

with the majority of the reaction products comprised of low MW materials (Table 3). Upon changing the solvent to favor folding, the product distribution was shifted toward **63a–c**, with a statistical distribution of the three longer oligomers. The shift of the equilibrium to the higher MW species is brought on by the folding stabilization of the longer-length oligomers. Addition of the dumbbell-shaped guest, **44**, leads to the predominant formation of the complex between **63b** and **44**. Only small amounts of the other high MW species were present, reflecting the specificity of complex formation. These two experiments demonstrate the ability of imine-containing *m*PEs to be used in a dynamic combinatorial library based on folding stabilization and host–guest complexation.

2.9
Imine Metathesis Polymerization

Nucleation–elongation is a common mechanism of biopolymer formation, with helical or tubular proteins being one example [89]. This type of polymerization exhibits cooperative chain growth where a critical-sized nucleus is required before additional monomer units can add to the growing chain. Once this critical size is reached, a helical structure forms to which additional monomer units may add. Addition of monomers to this helical structure results in an increase of the stability of the growing polymer chain, driving the polymerization to higher MW (Fig. 38). As synthetic polymerizations typically join monomer units in a head-to-tail fashion without any additional stabilizing interactions, the nucleation–elongation mechanism is rarely observed [90, 91]. However, the reversible reaction of the foldable imine-containing *m*PEs makes the nucleation–polymerization mechanism possible.

The choice of starter sequences for a nucleation–elongation polymerization is very important. Starter sequences that can combine to form stable macrocycles (**64** and **65**) will result primarily in macrocycle formation (**55**), creating

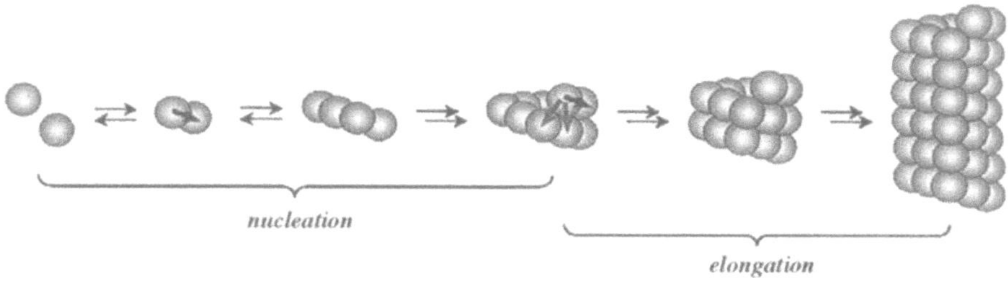

Fig. 38 General mechanism of the nucleation and elongation stages of polymerization generating a helical structure (the *arrows* represent the interactions among repeating units). Secondary interactions (*light arrows*), absent in the first turn of the helix, are the molecular origins of a less favorable nucleation event (i.e., the critical chain length) beyond which propagation becomes more favorable

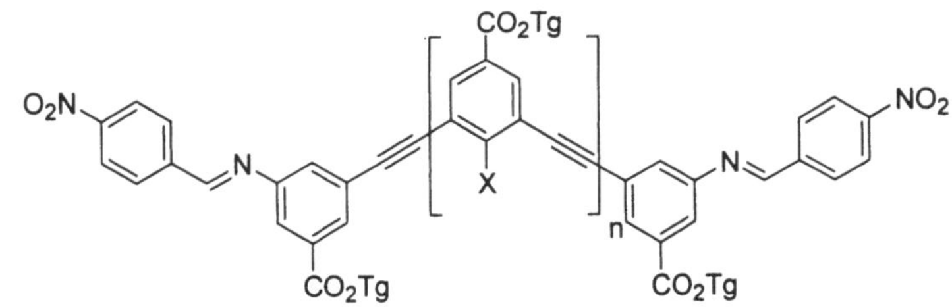

64 n = 1, X = H
69 n = 2, X = Me

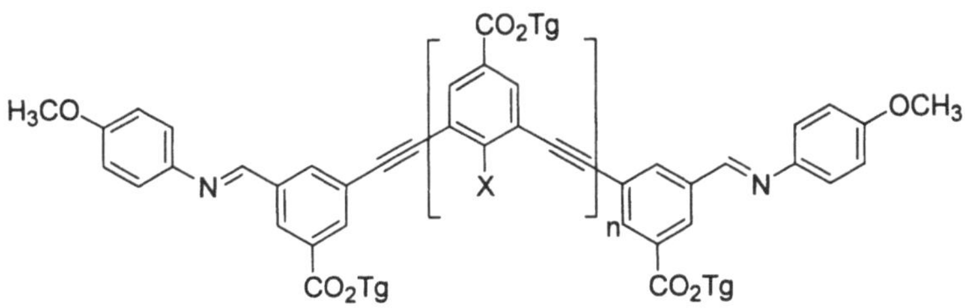

65 n = 1, X = H
66 n = 2, X = H
70 n = 2, X = Me

a thermodynamic sink that terminates the polymerization (Fig. 39) [92]. However, when starter sequences are chosen that do not favor macrocycle formation but oligomer generation, such as **60** and **66**, high molecular weight polymer may be formed. These starting materials follow the nucleation–elongation mechanism where two monomers react to form a dimer that is at or just under the nucleation size. Once the nucleation length has been obtained, the addition of more monomer units becomes energetically favorable and the polymerization proceeds.

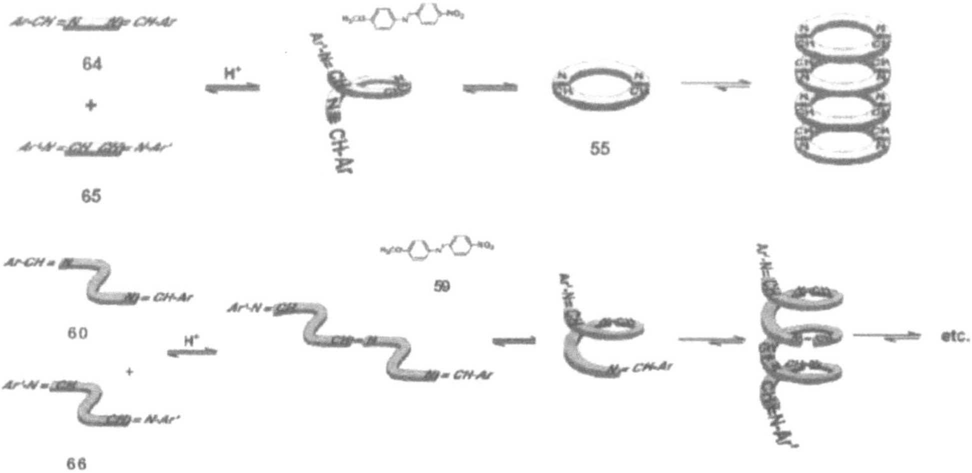

Fig. 39 Schematic illustration of the metathesis polymerization and macrocyclization

The ultimate evidence for the nucleation–elongation mechanism is the effect of imbalanced stoichiometry on the MW of the polymerization. In standard step-growth polymerizations the MW is dependent upon the ratio of the monomer concentrations where almost exact 1:1 stoichiometry is necessary to bring about high MW [90]. An irreversible imine condensation of **67** and **68** was performed under imbalanced stoichiometry ([**67**]/[**68**]=0.5) and nonfolding conditions (Fig. 40) [93]. The starting material was completely consumed and

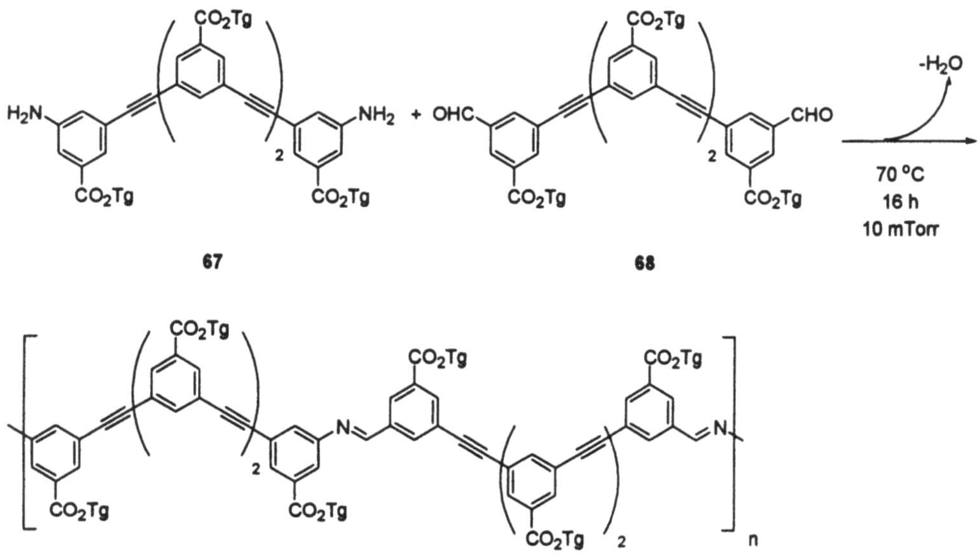

Fig. 40 Schematic representation of an imine melt polymerization

a mixture of low MW oligomeric species, mostly trimer, was formed following the Flory distribution. Using the same imbalanced monomer ratio (0.5) a metathesis polymerization was performed with **60** and **66** that gave a much higher MW. Analysis of the MW distribution showed that there was very little lower MW oligomeric material, while relatively large quantities of unreacted starting material were present (Table 4). These unusual results are readily explained by a nucleation–elongation mechanism. Upon formation of a critical-sized nucleus, the addition of subsequent monomer units to the nucleus becomes thermodynamically favored over addition to nonnucleated monomers because of the π-stacking interactions of the growing helix. The reversible nature of the imine metathesis reaction ensures monomers will then add primarily to the nucleated chains, resulting in a lack of low MW species. Upon the exhaustion of one of the starting materials the polymerization reaches an equilibrium resulting in an exclusion of the excess monomer from the polymer. To determine if this exclusion of monomer in the polymerization was general, the stoichiometry of **60** and **66** was varied and the MW and amount of remaining starting material was determined (Table 4). The results indicate that as the ratio of **60** to **66** was varied from 1/1 to 2/3 the MW of the polymerizations decreased. However, this decrease was much smaller than that predicted by the Flory distribution. Additionally, the amount of unreacted starting material present when the polymerizations had reached equilibrium was very close to the relative monomer excess, showing that the monomer had been excluded from the polymerization (Table 4).

Table 4 Polymerization conditions and molecular weight data as a function of starter sequence stoichiometry for imine- containing *m*PEs

SEC trace (Fig. 41)	[**60**][a] (mM)	[**66**][a] (mM)	Monomer molar ratio (**60**:**66**)	M_n^b (kDa)	M_w^b (kDa)	Rel. excess monomer[c]	Normalized monomer peak intensity[d]
A	5.0	5.0	1:1	31	419	0.0	0.10
B	5.0	5.5	10:11	27	283	0.10	0.16
C	5.0	6.0	5:6	21	157	0.20	0.23
D	5.0	6.3	4:5	19	94	0.25	0.28
E	5.0	6.7	3:4	17	63	0.33	0.35
F	5.0	7.5	2:3	12	34	0.50	0.48

[a] Initial monomer concentration.

[b] The molecular weight data were obtained from integrating SEC traces between 15.5 and 26.1 min (i.e., monomer peak was excluded), and calibrated on the basis of polystyrene standards (the calibrated molecular weight of starter sequences **60** and **66** is ca. 1.5 kDa; the molecular weights of polymers were underestimated).

[c] Calculated on the basis of ([**66**]–[**60**])/[**60**].

[d] Integration of the monomer peak (ca. 27 min) in the SEC traces normalized relative to the peak intensity of **66** at 5.0 mM concentration.

The MW of the imine-containing polymers, whose chain growth is described by a nucleation–elongation mechanism, is directly related to the folded-state stability of its backbone. Once a nucleus of a critical size has been formed, the addition of monomer units to the nucleus is a thermodynamically driven process. To determine if metathesis polymerizations are truly thermodynamically controlled, polymers were formed under a variety of different conditions. Metathesis polymerization of **60** and **66** under folding conditions gave high MW polymer (Fig. 41). Upon increasing the temperature of the reaction, which is known to destabilize the folded state [23], depolymerization was observed. Cooling the polymerization back to room temperature restored the original MW and demonstrated the reversible nature of the process. Dilution of the same reaction with chloroform leads to a large decrease in the MW, while diluting the reaction with acetonitrile does not effect the MW. The addition of methyl groups to the imine starter sequences (**69** and **70**) was expected to yield higher MW polymers, as the methyl group is known to stabilize the folded state. However, polymerization of **69** and **70** for 6 days gave a M_n value of 18.7 kDa, similar to that for polymerization of **60** and **66** (Fig. 42). However, when the metathesis polymerization of **69** and **70** was allowed to equilibrate for up to 19 weeks, the M_n increased with a maximum value of 40.0 kDa. The slower kinetics of polymerization for the methyl-series monomers suggests that the transition state of the imine metathesis requires a somewhat unfolded state that is not formed as readily with the methyl series in poor solvents. The ability to

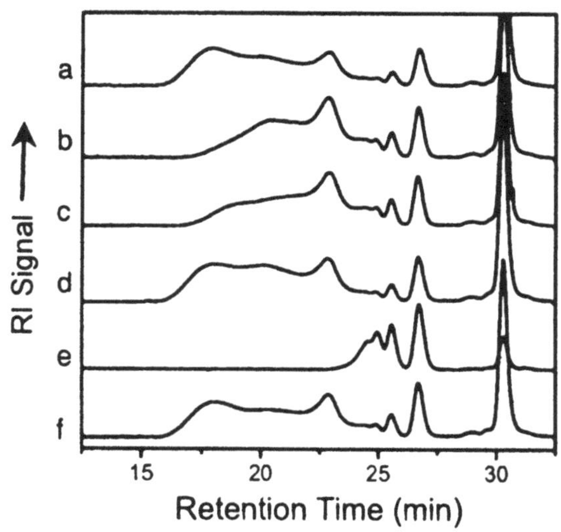

Fig. 41 Size exclusion chromatography (SEC) traces of metathesis products from **60** and **66** in CH_3CN at 5 mM rt for 6 days (a); reaction mixture from (a) was then heated at 40 °C for another 6 days (b); **60** and **66** equilibrated at 40 °C in CH_3CN at 5 mM for 6 days (c); reaction mixture from (c) was then cooled to rt and equilibrated for another 6 days (d); reaction mixture from (a) was diluted with $CHCl_3$ (e) or CH_3CN (f) to 2.5 mM and equilibrated at rt for another 6 days

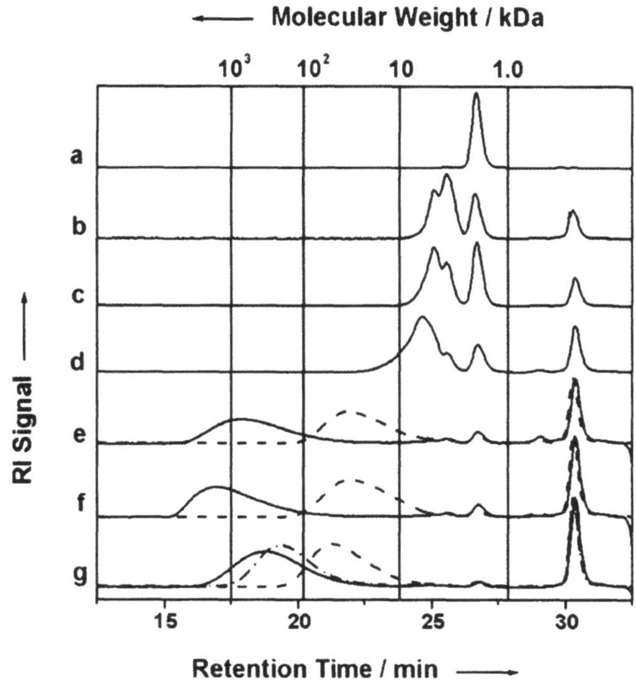

Fig. 42 SEC traces of starter sequences **69** and **70** (a) and products equilibrated at 5 mM and room temperature in chloroform (b, 6 days), tetrahydrofuran (c, 6 days), dioxane (d, 6 days), methyl acetate (e, - - - 6 days; — 19 weeks), ethyl acetate (f, - - - 6 days; — 19 weeks), and acetonitrile (g, - - - 6 days; -·- 44 days; — 25 weeks) solutions. In trace a, starter sequences **69** and **70** coelute

reversibly shift the MW of the polymerization based on the stability of the folded state shows the thermodynamic control of the polymerization.

2.10 Water-Soluble *m*-Phenylene Ethynylenes

Having the ability to study the *m*PE system in pure water would be advantageous for many reasons. Guest binding in *m*PEs has been shown to be largely based on a solvophobic driving force; however, the insolubility of oligomers in water made quantification of binding affinities at higher water concentrations impossible [57]. Also, solvophobic driving forces are at least partially responsible for the formation of the folded state of the backbone [38]. Finally, studying these oligomers in water offers the most insight into the similarities between the *m*PE backbone and biological systems. For this reason different approaches to creating water-soluble *m*PEs have been explored.

Neutral pentaethylene glycol side chains have previously shown the ability to solubilize hydrophobic cores in water [94]. A similar approach was used to bring about solubility of the *m*PE backbone in water, where hexaethylene

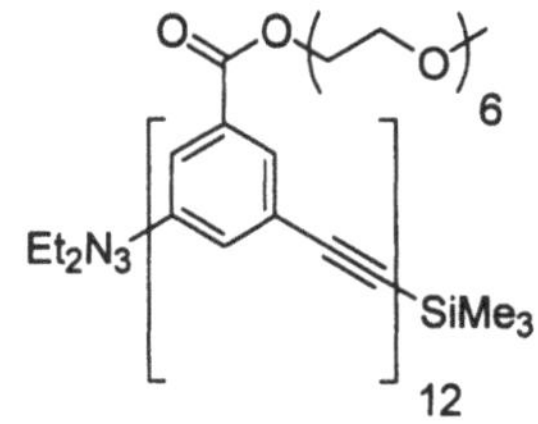

71

glycol (hexaglyme) side chains were attached through an ester linkage (**71**) [95]. It was shown that the hexaglyme side chains did not modify the folding stability, as denaturation curves were almost identical to those of the native oligomers. These studies confirmed that **71** exists in the folded state in acetonitrile and in a random coil conformation in chloroform. As the vol% water of an acetonitrile solution of **71** is increased from 0 to 90%, a blue-shifting of the maxima and a decrease in the optical density is observed in the UV spectra (Fig. 43). This hypochromic effect is believed to be caused by the tightening of the helical conformation, or by aggregation of the oligomers. In pure water, **71** exhibited a slightly larger optical density which may be a reorganization to account for the absence of acetonitrile, normally involved in solvating the oligomer backbone.

The binding of (–)-α-pinene (**39**) with **71** was studied in various compositions of water in acetonitrile. Contrary to previous results the presence of an induced CD signal was not linearly related to water concentration, as no CD

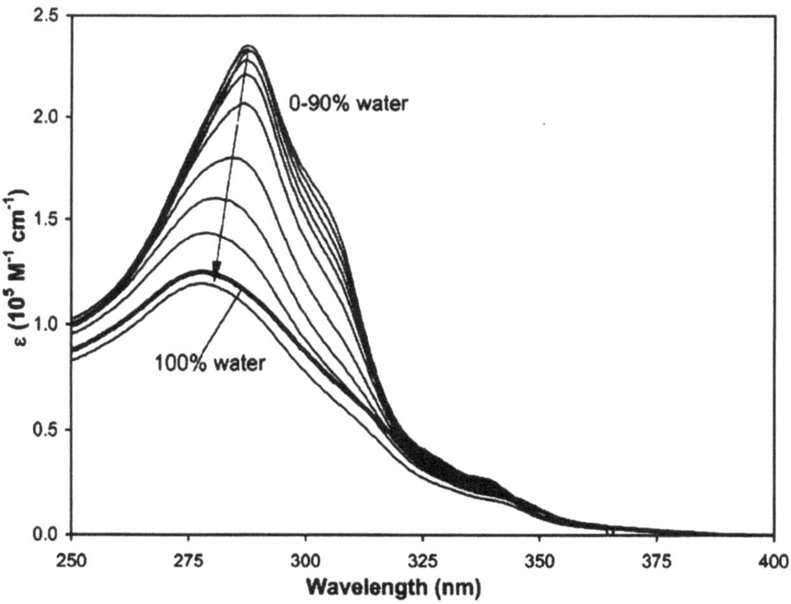

Fig. 43 UV spectra of dodecamer **71** in varying percentages of water in acetonitrile (100% water in *bold*)

signal was observed for the oligomer until the water composition was greater than 50% [95]. Monte Carlo searches of **71** showed that the longer hexaglyme side chains may be able to thread through the helical cavity (Fig. 44), such that the side chains inhibit pinene binding until sufficient driving force for binding is reached at around 50% water. Analysis of the CD binding data allowed the assignment of 1:1 stoichiometry to the complex between **71** and **39** and fitting to 1:1 binding isotherms. The association constant is plotted against solvent composition in Fig. 45, which shows that the highest binding constant occurred

Fig. 44 Minimized structure of 71 showing the ability of the terminal side chain to occupy the helical cavity

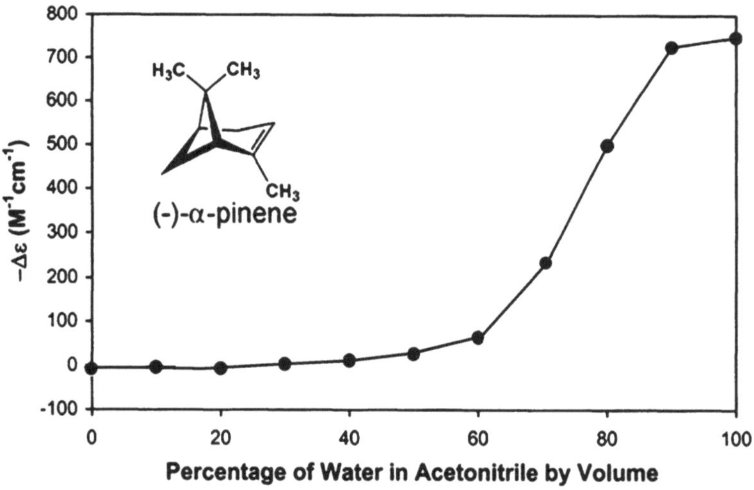

Fig. 45 Molar ellipticity at 315 nm for 4 μM dodecamer **71** and 100 equiv of (–)-α-pinene in varying compositions of water in acetonitrile

for 90% water with a value of $1.4 \pm 0.09 \times 10^6\ M^{-1}$, much higher than predicted from previous studies ($6 \times 10^4\ M^{-1}$) [57]. In 100% water the helix may be tightening in order to limit contacts between water and the solvophobic backbone, resulting in a smaller helical cavity and smaller association constant. The kinetics of the binding are definitely modified by the water composition of the solvent. In solvent compositions up to 70% water in acetonitrile the CD signal reached its maximum value in the time it took to take the measurement. However, the half-lives of the binding in 80, 90, and 100% water were on the order of seconds, minutes, and hours, respectively. This dependence of the binding kinetics on the water concentration is most likely due to the stability of the folded state in these solvent compositions.

Another water-soluble *m*PE was created by forming polymers with carboxylic acid in a basic aqueous medium (72) [96]. Treatment of the polymer with dilute sodium hydroxide deprotonates the carboxylic acid to give charged side chains that are able to "pull" the *m*PE backbone into water. Although the charged groups effectively bring about water solubility, the high charge localization leads to charge repulsion and prevents the formation of a compact structure [97, 98]. This polymer displays reversible switching from a water-soluble to hydrogel state based on the pH of the solvent.

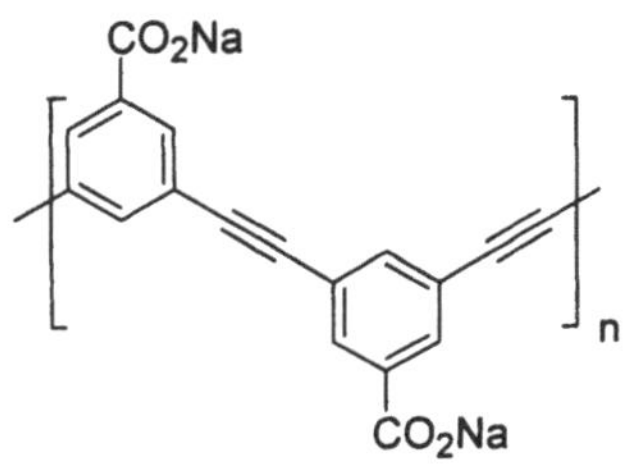

72

3 *Ortho*-Phenylene Ethynylenes

The *ortho*-phenylene ethynylene (*o*PE) backbone is a very attractive backbone for use in supramolecular chemistry. While the *m*PE helix is able to bind small molecules, the width of the structure is much larger than that typically found in peptide α-helices. By incorporating *ortho* linkages into a phenylene ethynylene backbone it may be possible to create a tighter helix, more reminiscent of the 3.1 α-helices of proteins. Synthetic methodologies for creating *o*PE backbones have been demonstrated [99, 100], leaving only the determination of the correct side chain substituents to bring about a folding reaction.

Current efforts toward *o*PE foldamers have focused on developing a computational model that will allow useful predictions of structure. This model was validated by comparing computational results to previously published experimental results [101]. For both crystal structure data and methyl- and hydrogen-series *m*PEs [25, 57], the MMFF molecular mechanics system [102, 103]

Table 5 Helix stabilization energies (kcal mol^{-1}) for substituted *o*PE hexamers calculated with MMFF

Cmpd[a]	Layer[b]	A	B	C	D	E	F	H_{ext}	H_{helix}	$-\Delta H_{helix}$
73	1	H	H	H	H	H	H	116.5	109.0	7.5
	2	H	H	H	H	H	H			
74	1	OMe	OMe	OMe	OMe	OMe	OMe	219.9	203.4	16.5[c]
	2	OMe	OMe	OMe	OMe	OMe	OMe			
75	1	OMe	CN	OMe	CN	OMe	CN	175	154	21.0
	2	CN	OMe	CN	OMe	CN	OMe			
76	1	CN	CN	CN	CN	CN	CN			8.0[d]
	2	CN	CN	CN	CN	CN	CN			
77	1	CO_2Me	CO_2Me	CO_2Me	CO_2Me	CO_2Me	CO_2Me	78.5	64.8	13.7
	2	CO_2Me	CO_2Me	CO_2Me	CO_2Me	CO_2Me	CO_2Me			

[a] Schematic of substituent placement on *o*PE backbone (to right).

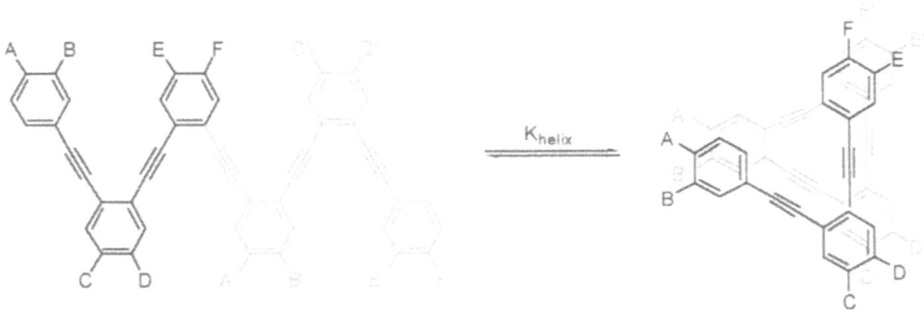

73-77

[b] Layer corresponds to the helix layer in the hexamer; substituents in column A, layer 2 are in the A′ position.
[c] The perfectly eclipsed all- *syn* helix has an H_{helix} value of 206.6 kcal mol$^{-1π}$.
[d] The helix is misshapen or poorly formed, as judged by inter-ring distances (significantly over 4 Å).

(available in Spartan [104]) gave the best agreement with experimental data. In all cases, the model was able to accurately predict values of ΔH_{helix} that agreed with the trends observed experimentally. In many cases the calculations were able to predict ΔH values that had the same magnitude change as the experimentally determined ΔG values.

Using the computational model, a series of *o*PEs with electron-rich and electron-poor substituents were evaluated as prospective synthetic targets (**73–77**) (Table 5). The computational model predicted that *o*PEs with a layer of electron-rich rings alternating with electron-poor rings (**75**) (ΔH_{helix}=–21.0 kcal mol^{-1}) would be the most stable, followed by oligomers with electron-rich rings (**74**) (ΔH_{helix}=–16.5 kcal mol^{-1}). The least stable oligomers were predicted to contain solely electron-poor rings (**76**) (ΔH_{helix}=–8.0 kcal mol^{-1}) or were unsubstituted (**73**) (ΔH_{helix}=–7.5 kcal mol^{-1}). These results are in stark contrast to classical π-stacking arguments which would predict the stability of oligomers to follow the trend alternating>electron-poor>electron-rich rings [49, 76]. To explain these unusual trends, the role of the side chain dipole orientation was examined. Calculations of intralayer dipole interaction with *syn*- and *anti*-orientated side chains (**74, 77**) indicate that the *anti* conformation is about 0.5 kcal mol^{-1} more stable than the *syn* for each pair. The inability of the rigid cyano group (**76**) to limit dipole moments by adopting alternate conformations then creates conflicting dipoles on adjacent layers, which destabilizes the helical structure. Oligomers containing ether side chains (**74**) are able to adopt a *syn* conformation, thereby limiting the dipole moment and minimizing destabilizing dipole effects, which leads to more stable helical structures than that predicted for **76**. These calculations show the importance of the side-chain dipole orientation in designing foldamers. Efforts to prove the effectiveness of this model to predict stabilities of folded structures are currently under way, and will be an exciting addition to both the fields of phenylene ethynylenes and foldamers.

4
Conclusion

Biopolymers are still the benchmark for supramolecular organization. However, through design and synthesis of foldable phenylene ethynylenes it has been possible to better understand and mimic the driving forces that biopolymers use to adopt unique, compact structures. The steps of identifying and synthesizing systems that are able to display high levels of supramolecular order have been accomplish and the basic "rules" for developing foldable systems have been determined. It is now possible to pursue more ambitious goals such as designing specific host–guest systems and performing reactions inside supramolecular reaction vessels. What will the future hold? Enzyme-like supramolecular catalysts? Highly conjugated supramolecular materials that display phenomenal mechanical properties? Using a highly modular, easily synthesized phenylene ethynylene backbone many exciting possibilities await.

Acknowledgements The authors thank the Moore group for helpful suggestions and comments. Work from the authors' laboratory was supported by the National Science Foundation (NSF CHE 0345254) and the U.S. Department of Energy, Division of Materials Sciences (DEFG02-91-ER45439).

References

1. Anfinsen CB (1973) Science 181:223
2. Hill DJ, Mio MJ, Prince RB, Hughes TS, Moore JS (2001) Chem Rev 101:3893
3. Brunsveld L, Folmer BJB, Meijer EW, Sijbesma RP (2001) Chem Rev 101:4071
4. Beer PD, Gale PA (2001) Angew Chem Int Ed 40:486
5. Nakano T, Okamoto Y (2001) Chem Rev 101:4013
6. Schmidtchen FP, Berger M (1997) Chem Rev 97:1609
7. Bunz UHF (2000) Chem Rev 100:1605
8. Giesa R (1996) JMS Rev Macromol Chem Phys 36:631
9. Martin RE, Diedrich F (1999) Angew Chem 38:1350
10. Schwab PFH, Levin MD, Michl J (1999) Chem Rev 99:1863
11. Yamamoto T (1999) Bull Chem Soc Jpn 72:621
12. Yamamoto T (2003) Synlett 4:425
13. Tour JM (1996) Chem Rev 96:537
14. Swager TM (1998) Acc Chem Res 31:201
15. Tour JM (2000) Acc Chem Res 33:791
16. Young JK, Moore JS (1995) In: Diederich F (ed) Modern acetylene chemistry. VCH, Weinheim, p 12
17. Haley MM, Pak JJ, Brand SC (1999) Top Curr Chem 201:81
18. Höger S (1999) J Polym Sci A Polym Chem 37:2685
19. Bunz UHF, Rubin Y, Tobe Y (1999) Chem Soc Rev 28:107
20. Faust R (1998) Angew Chem Int Ed 37:2825
21. Moore JS (1997) Acc Chem Res 30:402
22. Zhao D, Moore JS (2003) Chem Commun 807
23. Nelson JC, Saven JG, Moore JS, Wolynes PG (1997) Science 277:1793
24. Shetty AS, Zhang J, Moore JS (1996) J Am Chem Soc 118:1019
25. Prince RB, Saven JG, Wolynes PG, Moore JS (1999) J Am Chem Soc 121:3114
26. Hill DJ, Moore JS (2002) Proc Natl Acad Sci USA 99:5053
27. Cantor CR, Schimmel PR (1980) Biophysical chemistry. Freeman, San Francisco
28. Prince RB (2000) Phenylene ethynylene foldamers: cooperative conformational transition, twist sense bias, molecular recognition properties, and solid-state organization. PhD thesis, University of Illinois at Urbana-Champaign
29. Hanson P, Millhauser G, Formaggio F, Crisma M, Toniolo C (1996) J Am Chem Soc 118:7618
30. Jeschke G, Pannier M, Godt A, Spiess HW (2000) Chem Phys Lett 331:243
31. Mchaourab HS, Oh KJ, Fang CJ, Hubbel WL (1997) Biochemistry 36:307
32. Miick SM, Martinez GV, Fiori WR, Todd AP, Millhauser GL (1992) Nature 359
33. Matsuda K, Stone MT, Moore JS (2002) J Am Chem Soc 124:11836
34. Chan HS, Bromberg S, Dill KA (1995) Philos Trans R Soc Lond B Biol Sci 348:61
35. Chakrabartty A, Baldwin RL (1995) Adv Protein Chem 46:141
36. Saenger W (1984) Principles of nucleic acid structure. Springer, Berlin Heidelberg New York, p 141
37. Zimm BH, Bragg JK (1959) J Chem Phys 31:526

38. Prince RB, Moore JS, Brunsveld L, Meijer EW (2001) Chem Eur J 7:4150
39. Pace CN, Shriley BA, Thomson JA (1989) In: Creighton TE (ed) IRL, New York, p 311
40. Thompson PA, Eaton WA, Hofrichter J (1997) Biochemistry 36:9200
41. Elmer S, Pande VS (2001) J Phys Chem B 105:482
42. Eigen M, DeMaeyer L (1963) In: Weissberger A (ed) Techniques of organic chemistry. Interscience, New York
43. Yang WY, Prince RB, Sabelko J, Moore JS, Gruebele M (2000) J Am Chem Soc 122:3248
44. Ballew RM, Sabelko J, Reiner C, Gruebele M (1996) Rev Sci Instrum 67:3694
45. Du R, Pande VS, Grosberg AY, Tanaka T, Shakhnovich EI (1998) J Chem Phys 108:334
46. Nishio M, Hirota M, Umezawa Y (1998) The CH/π interaction: evidence, nature, and consequences. Wiley-VCH, New York
47. Lahiri S, Thompson JL, Moore JS (2000) J Am Chem Soc 122:11315
48. Burley SK, Petsko GA (1988) Adv Protein Chem 39:125
49. Hunter CA, Sanders JKM (1990) J Am Chem Soc 112:5525
50. Cary JM, Moore JS (2002) Org Lett 4:4663
51. Dill KA (1985) Biochemistry 24:1501
52. Lau KF (1989) Macromolecules 22:3986
53. Prince RB, Okada T, Moore JS (1999) Angew Chem Int Ed 38:233
54. Carlucci L, Ciani G, Prosperpio D, Sironi A (1995) J Am Chem Soc 117:4562
55. Francisco RHP, Mascarehnas YP, Lechat JR (1979) Acta Crystallogr B 35:177
56. Allenmark S (2003) Chirality 15:409
57. Prince RB, Barnes SA, Moore JS (2000) J Am Chem Soc 122:2758
58. Tanatani A, Mio MJ, Moore JS (2001) J Am Chem Soc 123:1792
59. Matsumura M, Becktel WJ, Matthews BW (1988) Nature 334:406
60. Cohen JL, Connors KA (1970) J Pharm Sci 59:1271
61. Tanatani A, Hughes TS, Moore JS (2002) Angew Chem Int Ed 41:325
62. Prince RB, Brunsveld L, Meijer EW, Moore JS (2000) Angew Chem Int Ed 39:228
63. Brunsveld L, Prince RB, Meijer EW, Moore JS (2000) Org Lett 2:1525
64. Gin MS, Yokozawa T, Prince RB, Moore JS (1999) J Am Chem Soc 121:2643
65. Gin MS, Moore JS (2000) Org Lett 2:135
66. Brunsveld L, Meijer EW, Prince RB, Moore JS (2001) J Am Chem Soc 123:7978
67. Green MM, Reidy MP, Johnson RD, Darling G, O'Leary DJ, Willson G (1989) J Am Chem Soc 111:6452
68. Kim J, McHugh SK, Swager TM (1999) Macromolecules 32:1500
69. Arias-Marin E, Arnault JC, Guillon D, Maillou T, Le Moigne J, Geffroy B, Nunzi JM (2000) Langmuir 16:4309
70. Swager TM, Gill CJ, Wrighton MS (1995) J Phys Chem 99:4886
71. McQuade DT, Kim J, Swager TM (2000) J Am Chem Soc 122:5885
72. Kim J, Swager TM (2001) Nature 411:1030
73. Arnt L, Tew GN (2003) Langmuir 19:2404
74. Arnt L, Tew GN (2002) J Am Chem Soc 124:7664
75. Venkataraman D, Lee S, Zhang J, Moore JS (1994) Nature 371
76. Hunter CA (1994) Chem Soc Rev 23:101
77. Prest P-J, Prince RB, Moore JS (1999) J Am Chem Soc 121:5933
78. Mio MJ, Prince RB, Moore JS, Kuebel C, Martin DC (2000) J Am Chem Soc 122:6134
79. Kubel C, Mio MJ, Moore JS, Martin DC (2002) J Am Chem Soc 124:8605
80. Mio MJ (2001) Hexagonally packed helical oligo(*m*-phenylene ethynylene) nanotubules via structural modification and guest templation. PhD thesis, University of Illinois at Urbana-Champaign
81. Moore JS, Zimmerman NW (2000) Org Lett 2:915

82. Trost BM (1991) Science 254:1471
83. Zhao D, Moore JS (2002) J Org Chem 67:3548
84. Martin RB (1996) Chem Rev 96:3043
85. Shetty AS, Fischer PR, Stork KF, Bohn PW, Moore JS (1996) J Am Chem Soc 118:9409
86. Oh K, Jeong K-S, Moore JS (2001) Nature 414:889
87. Oh K, Jeong K-S, Moore JS (2003) J Org Chem 68:8397
88. Nishinaga T, Tanatani A, Oh K, Moore JS (2002) J Am Chem Soc 124:5934
89. Oosawa F, Asakura S (1975) Thermodynamics of the polymerization of protein. Academic, New York
90. Flory PJ (1953) Principles of polymer chemistry. Cornell University Press, Ithaca
91. Zhao D, Moore JS (2003) Org Biomol Chem 1:3471
92. Zhao D, Moore JS (2003) Macromolecules 36:2712
93. Zhao D, Moore JS (2003) J Am Chem Soc 125:16249
94. Brunsveld L, Zhang H, Glasbeek M, Vekemans JAJM, Meijer EW (2000) J Am Chem Soc 122:6175
95. Stone MT, Moore JS (2004) Org Lett 6:469
96. Li C-J, Slaven WT IV, Chen Y-P, John VT, Rachakonda SH (1998) Chem Commun 1351
97. Walters KA, Ley KD, Schanze KS (1999) Langmuir 15:5676
98. Pinto MR, Kristal BM, Schanze KS (2003) Langmuir 19:6523
99. Orita K, Alonso E, Yaruva J, Otera J (2000) Synlett 9:1333
100. Jones TV, Blatchly RA, Tew GN (2003) Org Lett 5:3297
101. Blatchly RA, Tew GN (2003) J Org Chem 68:8780
102. Halgren TA (1996) J Comput Chem 17:490
103. Halgren TA (1999) J Comput Chem 20:730
104. Wavefunction I
105. Tóth G, Pintér I, Messmer A (1974) Tetrahedron Lett 15:735

Adv Polym Sci (2005) 177: 151–179
DOI 10.1007/b101377

Poly(arylene ethynylene)s in Chemosensing and Biosensing

Juan Zheng · Timothy M. Swager (✉)

Department of Chemistry, Massachusetts Institute of Technology, 77 Massachusetts, Avenue, Cambridge, MA, 02139 USA
tswager@mit.edu

Abstract Poly(arylene ethynylene)s (PArEs) have been used in recent years as effective transducers for a variety of sensing purposes ranging from organic molecules such as methyl viologen and TNT to biological analytes. Their superior sensitivity to minor perturbations is fundamentally governed by the energy transport properties resulting from the extended conjugation of the polymer backbone. An understanding of the underlying principles of energy transport allows the design of sensors with greater sensitivity and specificity. Pioneering work with methyl viologen as an electron-transfer quencher demonstrated that connecting receptors in series amplifies the sensing response compared to that of individual receptors. Since then, factors such as the electronic and structural nature of the polymers and their assembly architecture have proven to be important in improving sensory response. In this review, we present an overview of works to date by various groups in the field of PArE chemosensors and biosensors.

Keywords Poly(arylene ethynylene) · Chemosensor · Biosensor · Fluorescence · Energy transfer

Abbreviations and Symbols

ALF	Anthrax lethal factor
DNT	2,4-Dinitrotoluene
FRET	Fluorescence resonance energy transfer
MBL-PPV	Poly[lithium 5-methoxy-2-(4-sulfobutoxy)-1,4-phenylene vinylene]
MPS-PPV	5-Methoxy-5-propyloxy sulfonate phenylene vinylene
MV^{2+}	Methyl viologen
PArE	Poly(arylene ethynylene)
PNA	Peptide nucleic acid

PPE	Poly(phenylene ethynylene)
PPV	Poly(phenylene vinylene)
TIPS	Triisopropylsilyl
TNT	2,4,6-Trinitrotoluene

1 Introduction

An ideal sensor recognizes analytes in a sensitive, selective, and reversible manner. This recognition is in turn reported as a clear response. In recent years, conducting polymers have emerged as practical and viable transducers for translating analyte–receptor and nonspecific interactions into observable signals. Transduction schemes include electronic sensors using conductometric and potentiometric methods and optical sensors based on colorimetric and fluorescence methods [1].

Poly(arylene ethynylene)s (PArEs) are wide-gap semiconductors. They are typically insulating in their native neutral state but can be made conductive by either oxidizing or reducing the polymer's π-electron system (doping) [2, 3]. Their semiconductive nature has generated interest in developing electroluminescent polymers for device applications [4–6]. However, it is their photophysical characteristics that make PArEs good candidates for use as transducers, and they are now one of the most important classes of conducting polymers for sensing purposes. In this review, we will focus on colorimetric and fluorescence detection methods using PArE sensors. Some ingenious chemo- and biosensors that employ conducting polymers other than PArEs will also be highlighted. There are currently many research groups in the field of conjugated polymer sensors and interested readers are directed to previous reviews [1,7 and references therein].

2 PArEs for Signal Amplification

In analogy to traditional semiconductors, the extended electronic structures of PArEs can be visualized using energy bands composed of a valence band and a conduction band. Excited states consisting of electron–hole pairs, also referred to as excitons, can be generated upon photoexcitation and these travel through the polymer energy bands by Förster or Dexter mechanisms. One may consider the PArE as a molecular wire for exciton transport. Recombination of the electron–hole pairs can occur via radiative and nonradiative pathways, and the emissive properties of conjugated polymers are dominated by energy migration to and exciton recombination at the local minima of their band structures. Therefore, perturbations to the PArE will be reflected in

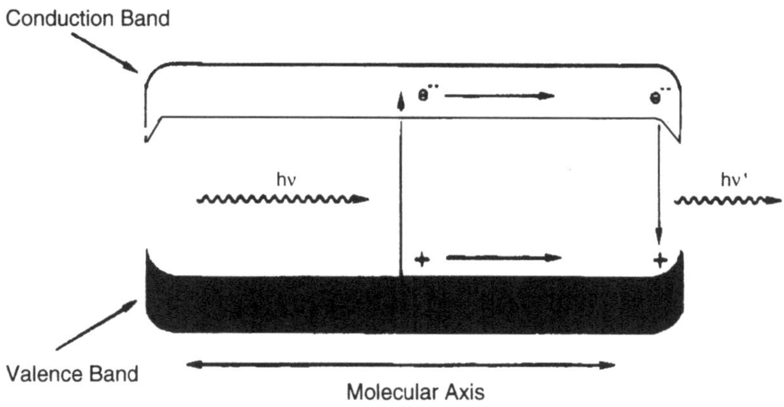

Fig. 1 Energy migration in a semiconductive molecular wire with a decrease in bandgap at the terminus. (Reprinted with permission from Ref. [8]. Copyright 1995 American Chemical Society)

its collective property. This has important implications for sensing applications.

The sensitivity of poly(phenylene ethynylene)s (PPEs) to perturbations in their band structure is illustrated by end-capping PPEs with anthracene units (**1**). The polymers act as antennae for harvesting optical energy and this is transferred to the anthracene units due to an induced localized narrowing of the bandgap (Fig. 1). Radiative recombination of the electron–hole pair results in greater than 95% of the emission occurring at the states localized at the anthracene end groups [8]. The presence of anthracene is effectively amplified.

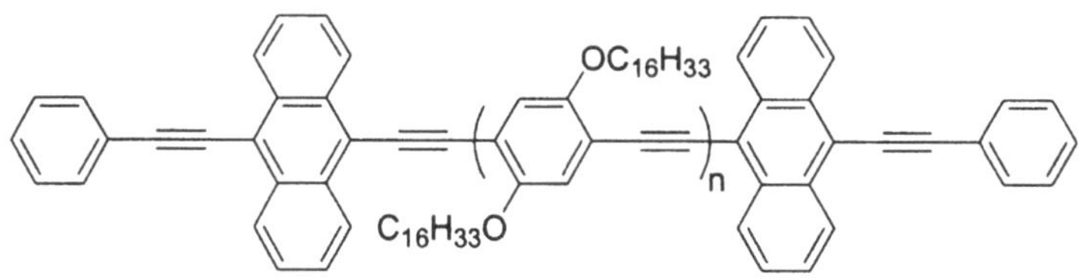

1

This amplification phenomenon can be applied to sensor design and was first demonstrated by our group in 1995 [9, 10]. In these studies, the sensing ability for a methyl viologen salt (MV^{2+} or *N,N′*-dimethyl-4,4′-bipyridinium bis(hexafluorophosphate)) is measured for a fluorescent single receptor molecule **2** and a PPE **3** where many receptors are "wired in series". In this sensing scheme, excitons were generated by photoexcitation of the polymer. When they encounter a cyclophane-bound MV^{2+}, a highly efficient electron-transfer reaction occurs to the analyte and the initially fluorescent polymer is returned to the ground state without the emission of a photon (Fig. 2). A 67-fold increase

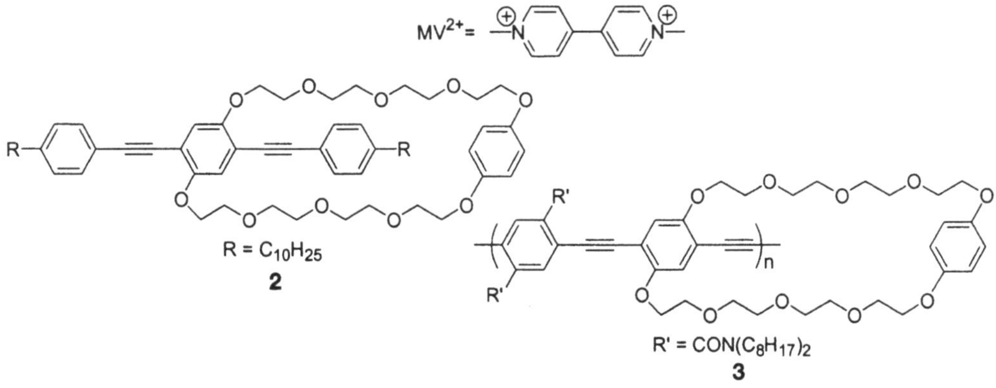

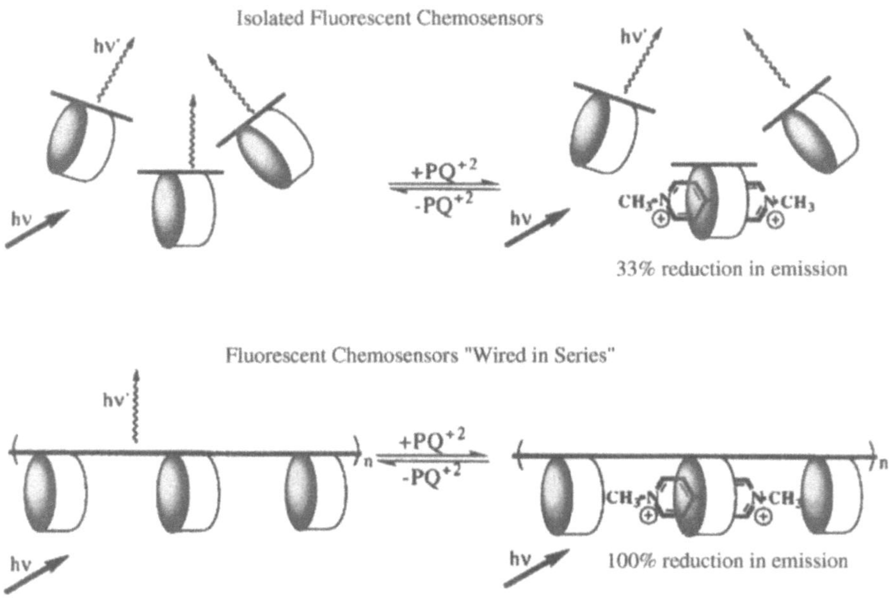

Fig. 2 Conceptual illustration of signal amplification by wiring receptors in series. (Reprinted with permission from Ref. [10]. Copyright 1995 American Chemical Society)

in quenching enhancement was obtained by comparing the Stern–Volmer quenching constants (K_{SV}) for **3** and **2**, corresponding to an average of 134 phenylene units sampled by the excitons. The quenching efficiency of MV^{2+} increased steadily with increasing polymer molecular weight, reaching a maximum at 65,000 and plateaued thereafter. The molecular weight dependence of quenching indicated that the diffusion length of the exciton is less than the length of the polymer. Polymers **4** and **5** were also synthesized and lower enhancement of quenching was observed in both cases in comparison to **3**. The *meta*-substituted comonomer in **4** resulted in a polymer with decreased delo-

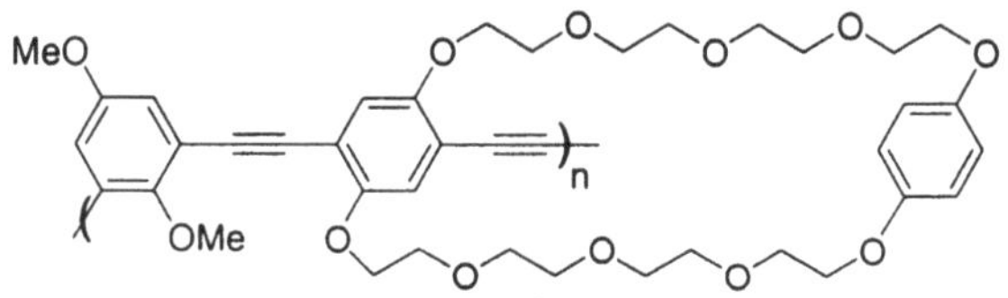

4

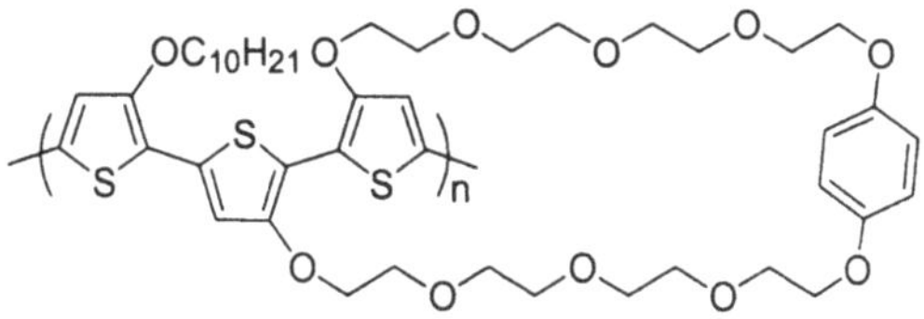

5

calization and decreased energy migration efficiency, culminating in diminished quenching. However, the greater delocalization of polythiophene **5** did not guarantee a more efficient energy migration; a decrease in its fluorescence efficiency was instead the determining factor.

The apparent binding constant K_{SV} obtained by Stern–Volmer quenching studies is the product of the number of receptors visited by the exciton and the binding constant of MV^{2+} to the cyclophane receptor. For this reason polymer **3** and its monoreceptor model **2** were designed so that the binding constant for methyl viologen to the receptor was known for both systems. This allowed the calculation of the true amplification factor of 67.

Whitten et al. have since applied the energy transport properties of conjugated polymers to amplified quenching studies [11]. In their work, the anionic conjugated polymer **6**, 5-methoxy-5-propyloxysulfonate phenylene vinylene (MPS-PPV), was quenched using the cationic methyl viologen. The Stern–Volmer quenching constant was reported to be $\sim 10^7\ M^{-1}$ and more than a million-fold amplification relative to the neutral small molecule model **7** was suggested. However, these numbers are misleading, firstly because the K_{SV} is a combination of the amplification factor and the binding constant of MV^{2+} with the anionic PPV, and secondly because the small molecule model is not a reasonable analog of the receptor–analyte recognition and so an accurate binding constant cannot be obtained. Since the MV^{2+} is reported to quench at a level of 95% when the viologen concentration is 1/100 that of the repeat unit, the amplification is likely overestimated by as much as 10^4–10^5.

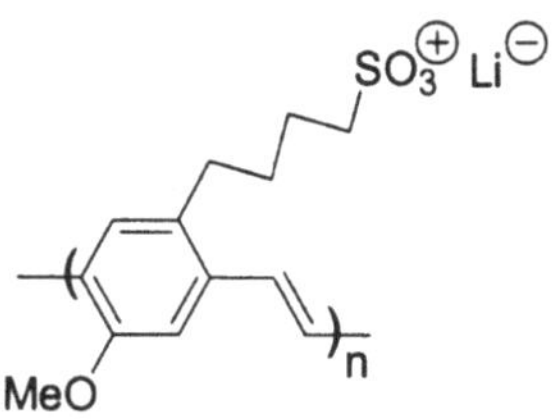

6

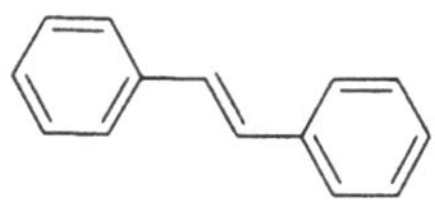

7

Anionic PPEs with higher quantum yields than MPS-PPV that are also quenched by MV^{2+} have been synthesized by Schanze et al. In methanol, sulfonated PPE **8** and its small molecule analog **9** expressed K_{SV} values of 1.4×10^7 M^{-1} and 2.2×10^4 M^{-1}, respectively. In water, the K_{SV} values were 2.7×10^7 M^{-1} and 7.0×10^3 M^{-1} for the two species [12]. The polymer was aggregated in water but not in methanol and greater amplification was seen in the former solvent. Since aggregation of the polymer is induced by the addition of only a few quenchers per chain, this may contribute to the increased response due to self-quenching. Phosphonated PPEs such as **10** have also been prepared by the same group [13]. In this case, the amplification was ~100-fold compared to a small molecule model **11** with the MV^{2+} quencher. Interestingly, similar quenching amplification factors were observed with cationic dyes such as ethydium bromide and rhodamine 6G, which shunt the polymer emission via a singlet–singlet Förster energy-transfer mechanism. As MV^{2+} precludes a Förster mechanism for quenching the polymer, such long-range transfer from a polymer to a dye acceptor may not be necessary.

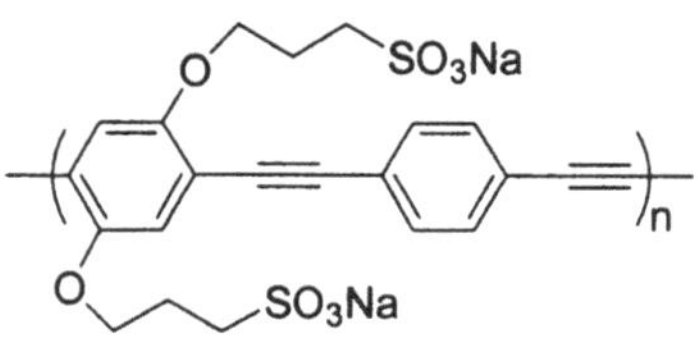

8

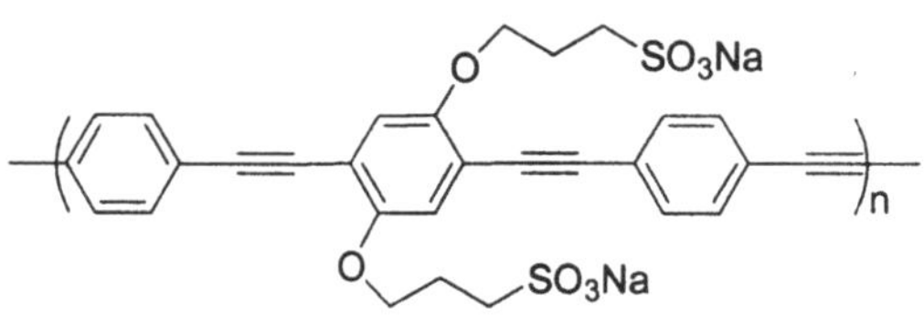

9

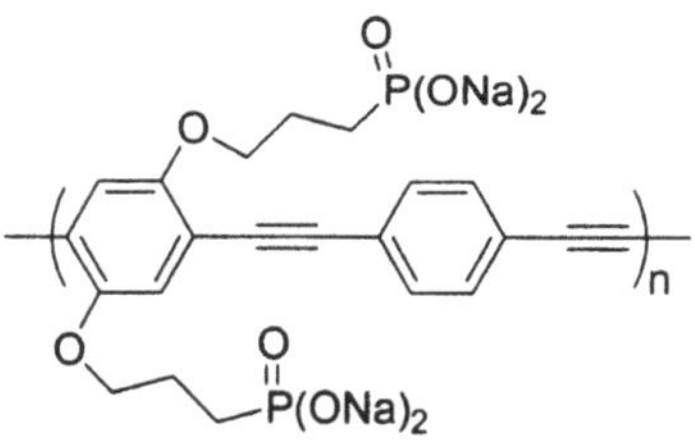

10

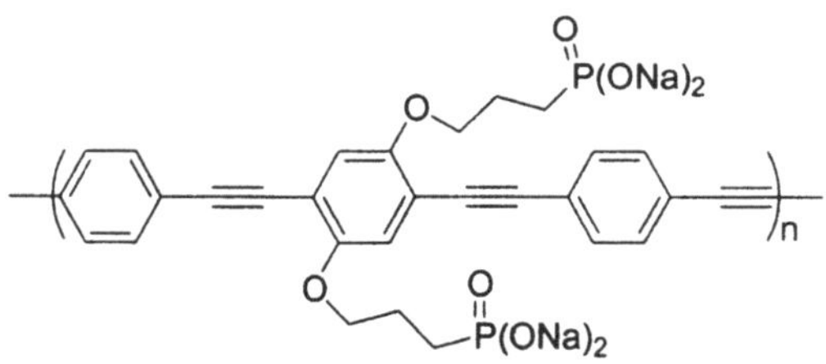

11

The 67-fold amplification obtained for polymer **3** is restricted by an inherent limitation of the "wired in series" design. As the exciton travels in a one-dimensional random walk process down the polymer chain, it has equal opportunity to visit a preceding or an ensuing receptor. This represents 134^2 random stepwise movements for 134 phenylene ethynylene units, and so much of the receptor sampling by the exciton is redundant. Increasing the efficiency of receptor sampling requires maximization of the number of different receptors that an exciton can visit throughout its lifetime. To achieve this end we extended the polymer sensor into two dimensions by use of a thin film and thereby increased the sensitivity.

PPEs often π-stack and form excimers in the solid state [14–17]. To circumvent this problem we designed PPE films that incorporate rigid three-dimensional pentiptycene scaffolds in the polymer backbone [18, 19]. These polymers form porous films and discriminately bind to various analytes of suitable size and electronic properties. Strongly electron-deficient analytes such as 2,4,6-trinitrotoluene (TNT) and 2,4-dinitrotoluene (DNT) cause fluorescence quenching by an electron-transfer mechanism. Films of polymer **12** were quenched by 50% within 30 s of exposure to TNT and by 75% within 60 s, despite the low equilibrium vapor pressure of 7 ppb for the analyte (Fig. 3 and Fig. 4).

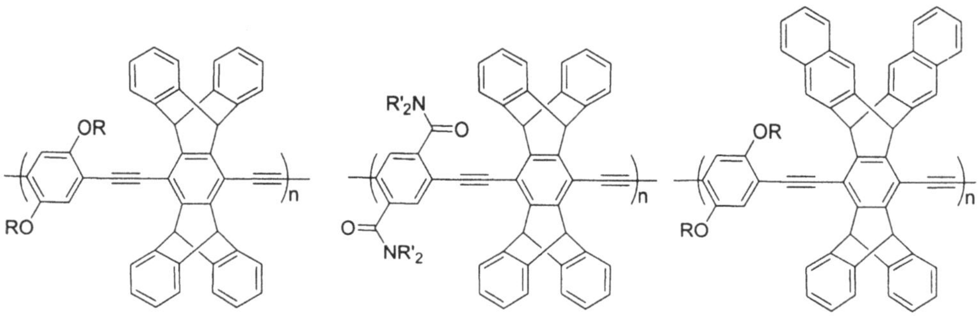

12 **13** **14**

$R = C_{14}H_{29}$
$R' = C_8H_{17}$

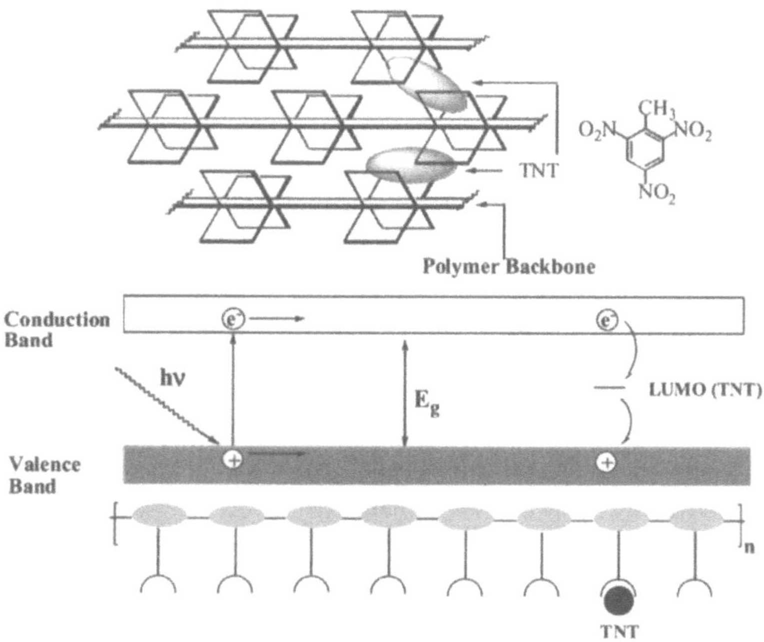

Fig. 3 *Top*: Schematic representation of porous polymer films allowing for analyte docking. *Bottom*: Band diagram depicting quenching resulting from electron transfer from PPE to TNT.

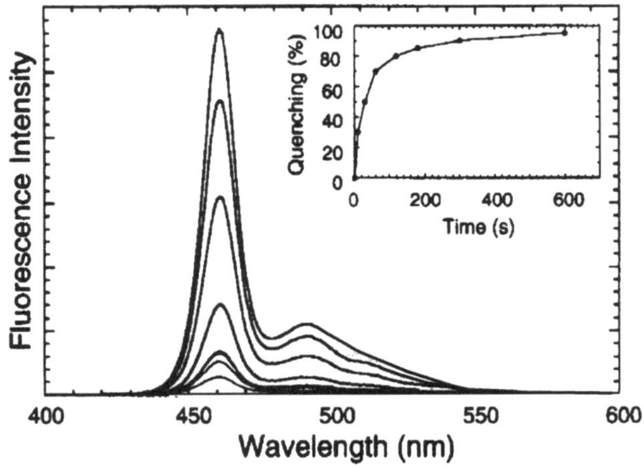

Fig. 4 Time-dependent fluorescence intensity of **12** upon exposure to TNT vapor at 0, 10, 30, 60, 120, 180, 300, and 600 s (*top to bottom*), and fluorescence quenching (%) as a function of time (*inset*). (Reprinted with permission from Ref. [18]. Copyright 1998 American Chemical Society)

Films of varying thicknesses were investigated. In thick films of **12** at 200 Å, analytes such as duroquinone that had fewer electrostatic interactions with the polymers diffused to greater depths and so better cumulative quenching was observed. In the case of electron-deficient nitro compounds, the stronger electrostatic interactions limited their diffusion in thicker films, resulting in a smaller quenching response. However, in thin films of 25 Å, nitroaromatic compounds displayed superior quenching effects to those of quinones due to stronger film interactions, despite their lower vapor pressures and thermodynamically less favorable electron-transfer reactions.

The polymer composition was varied in order to probe the effects of polymer electronic and steric properties on analyte binding (structures **13** and **14**). As expected, the electron-deficient polymer **13** was less sensitive to oxidative quenchers. Steric factors were also important, as larger molecules had slower rates of diffusion through film cavities that were obstructed by the nonplanar structure and the double alkyl chains of the amide groups in **13**. Interpolymer interactions between the naphthalenoid pentiptycenes and adjacent polymer backbones in **14** also resulted in smaller cavity sizes and slower analyte diffusion. A good balance of electrostatic interactions and film porosity is therefore crucial to the design of a sensitive optical sensor for TNT. Detectors based on this technology are currently being manufactured by Nomadics Inc. and have been shown to be effective in detecting landmines in the field [20].

As greater sensitivity was realized in 2D compared to 1D, to further maximize the sensitivity of the polymer we investigated energy migration in three dimensions. The Langmuir–Blodgett technique was used to construct layers of aligned PPE in order to facilitate dipolar Förster-type processes for efficient intermolecular energy transfer from the PPE to surface acridine orange ac-

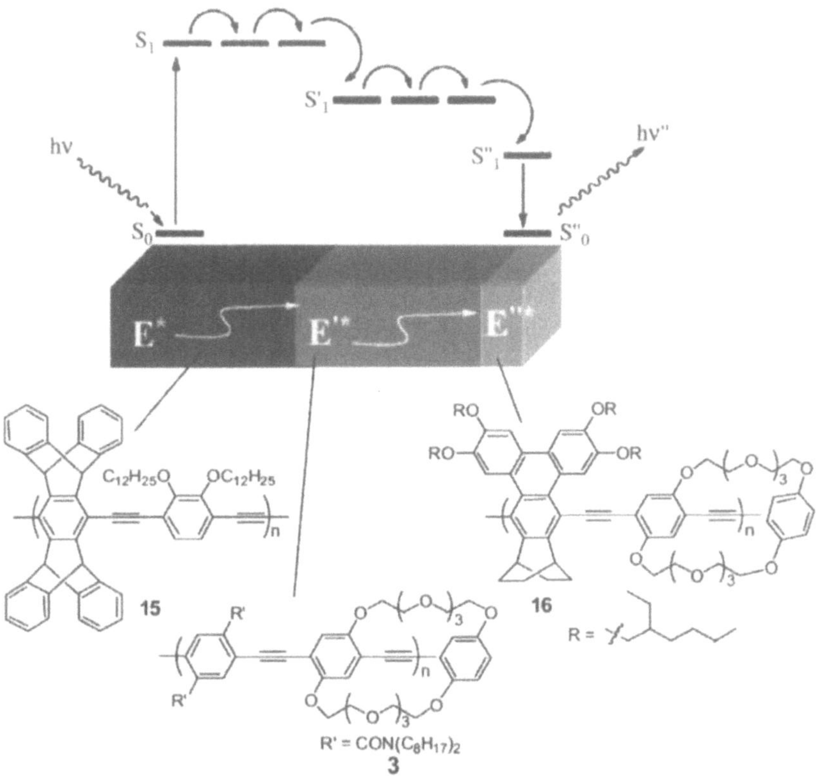

Fig. 5 Energy transfer from polymers **15** to **3** to **16**. The films have decreasing bandgaps moving from the *bottom* to the *top*. Polymer **15** (abs./em. max. 390/424 nm) overlaps with **3** (abs./em. max. 430/465 nm) overlaps with **16** (abs./em. max. 495/514 nm). (Reprinted with permission from Ref. [22]. Copyright 2001 American Chemical Society)

ceptors [21]. Increasing the number of polymer layers steadily enhanced the acridine orange emission with the energy transfer peaking at 16 layers. This may at first glance seem counterintuitive since the relative amount of acridine orange acceptors to polymer donor decreases as the polymer thickness increases. However, in analogy to the amplification observed for the two-dimensional process discussed earlier, the energy trapping efficiency is maximized in 3D as the exciton does not retrace its steps. Enhanced acridine orange emission is therefore observed with thicker films. These insights have been applied to directed energy transfer with PPEs by layering polymer films with decreasing bandgaps on top of one another (Fig. 5) [22]. Energy is preferentially transferred to the surface of the thin film, where the bandgap is the smallest. This ability to control the exciton pathway has important implications for the design of chemosensors.

Polymers with extended lifetimes can have excitons with higher diffusion lengths and hence a greater probability of encountering a receptor-bound analyte. Polymers **17** and **18** were synthesized to this effect. Incorporation of triphenylenes into **17** enhanced the lifetimes of the polymer by about 30% compared to those of the PPE analog. Dexter energy transfer was determined to be the dominant intramolecular energy transport process in these poly(triphenylene ethynylene) materials [23]. PArE **18** and related polymers based on dibenzo[*g,p*]chrysene also display long excited lifetimes in the range of 1.4–2.6 ns. These polymers are promising materials for sensing applications.

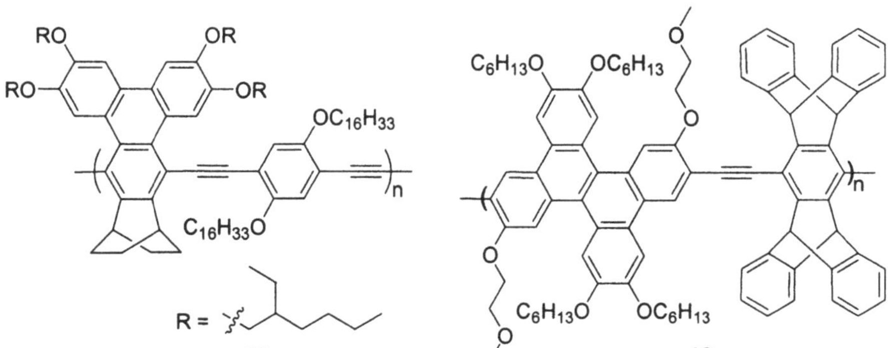

17 **18**

3 PArEs with Specific Receptors

The excellent amplifying abilities of PArEs are advantageous to sensor design. However, adequate and specific receptors are also of crucial importance. Conjugated polymers with custom receptors, for use in the detection of heavy-metal contaminants, were presented by Wasielewski et al. in 1997. In this study, the polymer incorporated 2,2′-bipyridyl ligands in the backbone, which could coordinate to a wide variety of metal ions. Metal sensing is realized by inducing changes in the degree of conjugation in poly(phenylene vinylene)s, thereby influencing their bandgap and affecting their absorption and fluorescence spectra [24].

In its *transoid* form, a 20° dihedral angle exists between the two pyridine rings in 2, 2′-bipyridine and as a result, polymers **19** and **20** are not fully conjugated. When chelated to a metal ion, the coordination between the metal ion

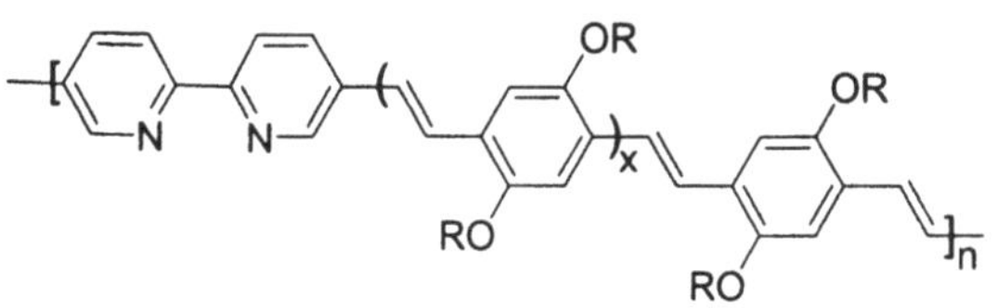

19: x = 0
20: x = 2
R = $C_{10}H_{21}$

and the bipyridine would forcefully planarize the initially twisted conformation, thereby increasing the conjugation. Metals such as Zn^{2+}, Cd^{2+}, Hg^{2+}, Ag^{+}, Al^{3+}, and lanthanide ions caused a significant redshift in the polymers' absorption and fluorescence with minimal fluorescence quenching. This redshift can be explained by a conjugation enhancement due to metal binding and to a lesser extent, due to the electron density variations on the polymer backbone by coordinating to electron-deficient metal ions. Other metal ions such as Pb^{2+}, Fe^{2+}, Fe^{3+}, Cu^{+}, Sb^{3+}, and certain lanthanide ions caused a blueshift in the fluorescence spectra with significant quenching. This was explained by a monodentate binding of metal ion to bipyridyl ligands, which induced a more twisted backbone and hence a decrease in the conjugation of the polymer. Cu^{2+}, Ni^{2+}, Co^{2+}, Mn^{2+}, Sn^{2+}, and Pd^{2+}, on the other hand, quenched the fluorescence quantitatively due to energy- or electron-transfer reactions between the phenylene vinylene segments and the metal complexes. Absorption and fluorescence spectral profiles were dependent on the metal complex and the ions may be removed by treatment with competing ligands such as ammonium and cyanide ions.

In accordance with Wasielewski's report, Huang et al. synthesized three polymers with different linkers between the chromophores (**21–23**), including a PArE derivative, to study the effect of polymer backbone extension and rigidity on the sensitivity and selectivity of bipyridyl-based metal sensors [25]. The polymers were very responsive to transition metal ions (typically in the micromolar range, 10^{-6} M) and insensitive to alkali and alkaline earth metals with the exception of Mg^{2+}. Decreasing sensitivity was observed as the polymer was varied from **21** to **23**. For transition metal ions in the 2nd and 3rd groups, the least amount of analyte was required to quench **21**, while **22** and **23** required progressively larger amounts of analyte to effect quenching.

21
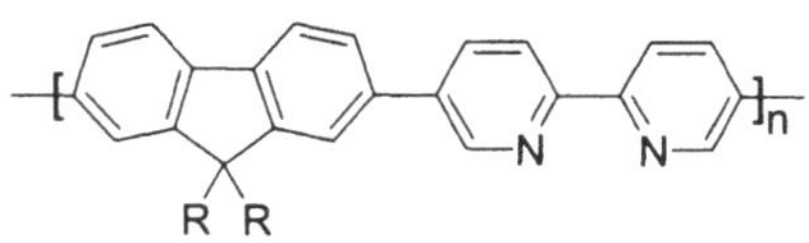

22
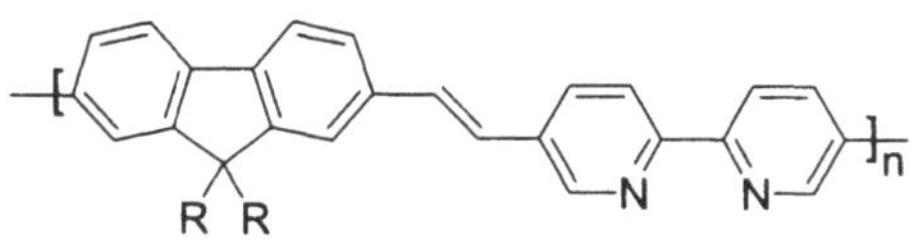

23
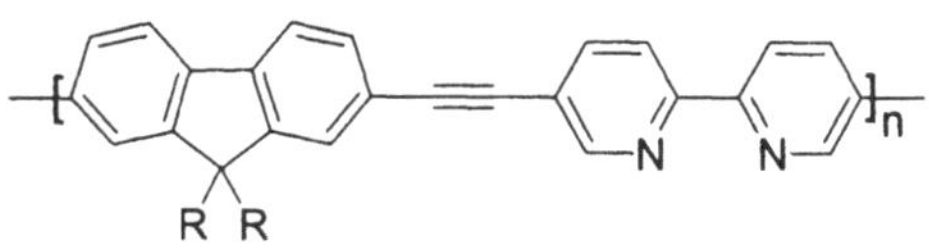

PArEs with pendant bipyridyl or terpyridyl ligands have also been reported [26]. At 3.08×10^{-6} M polymer concentration, a Ni^{2+} concentration of 1.0×10^{-6} M quenched **24** and **25** to 10.9 and 38.2% of their respective initial fluorescence

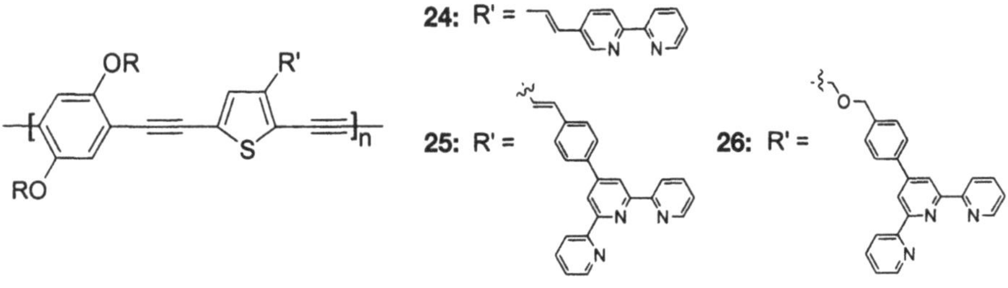

emissions. Polymers with ligands linked via a vinylene linkage were therefore more sensitive than those where the ligands were connected via an ether linkage, highlighting the importance of electronic communication for increased sensitivity. Selectivity toward different transition metals was mediated by the chelating ability of the terpyridine ligand. PArE **24** showed response to Cr^{6+}, Cd^{2+}, Ni^{2+}, and Mn^{2+}, while Na^{+}, Ca^{2+}, Pb^{2+}, and Hg^{2+} did not elicit any sensory response (Fig. 6). Consistent with the tridentate nature of terpyridine, polymers substituted with this ligand were more sensitive than the bidentate bipyridine-substituted PArE **26**.

The same group has synthesized well-defined porphyrin-containing PArEs **27** and **28** [27]. The central metal cation could complex to various solvent ligands. Depending on the electron-donating ability of the solvent ligand, complexation of solvent to metal ion changes the electron density on the porphyrin

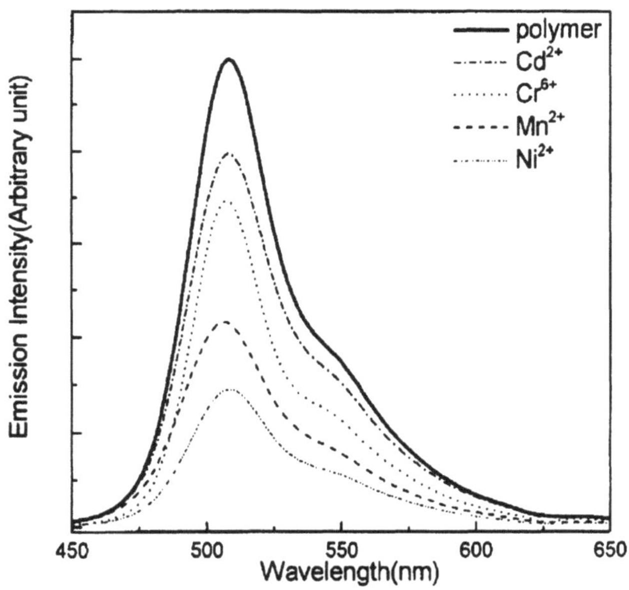

Fig. 6 Emission quenching of polymer **25** by different transition metal ions. The polymer concentrations are held fixed at 3.08×10^{-6} M corresponding to receptor unit. Transition metal ions are 1.54×10^{-6} M. (Reprinted with permission from Ref. [26]. Copyright 2002 American Chemical Society)

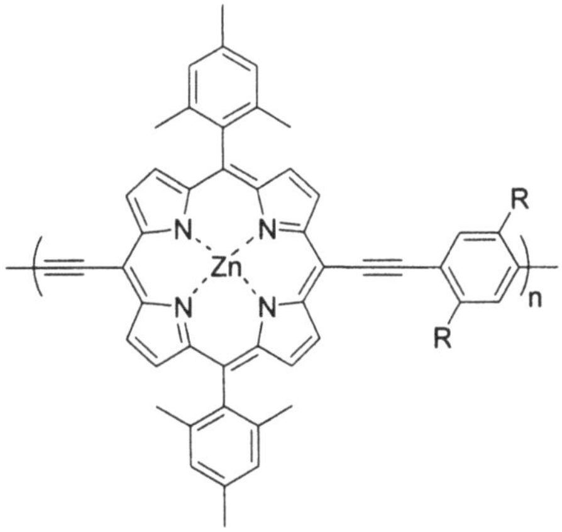

27: R = $O_{15}H_{31}$

28: R = $CON(C_8H_{17})_2$

units. For basic solvents, an increase in electron density leads to a redshift in the absorption and emission bands of the polymers. For instance, on going from chloroform to *n*-butylamine, the emission shifted from 704 to 741 nm. The coordinating ability of the central metal cation and the tunability of the PArE polymer could potentially be used for optical sensors.

Other polymers with coordinating abilities such as quinoline-containing PArEs **29** and **30** have also been synthesized by Bunz et al. [28]. In this case, the polymers were sensitive to metal ions such as Pd^{2+}, La^{3+}, and Ag^{+}.

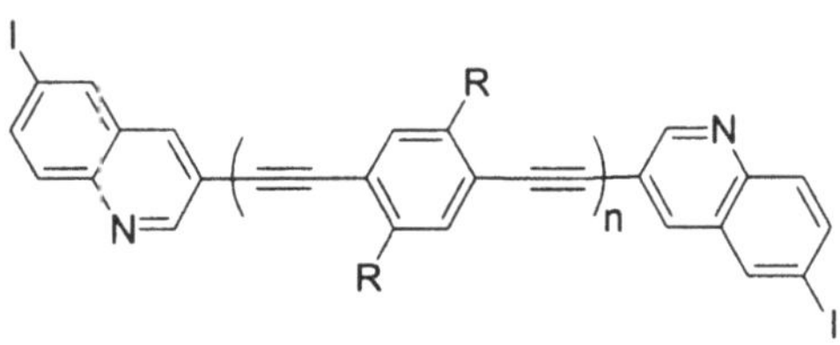

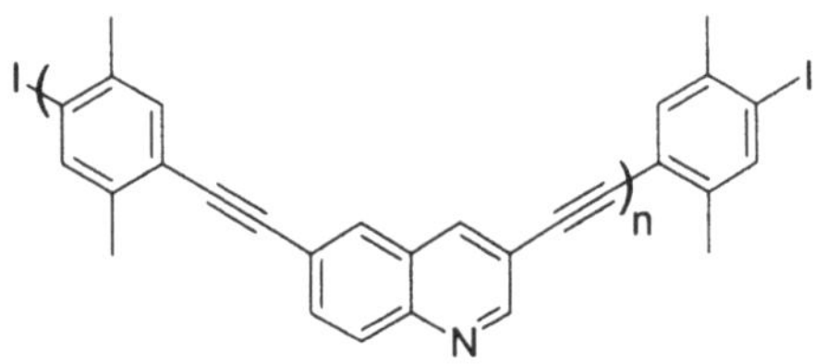

29 R = 2-ethylhexyl **30**

While aggregation is usually an undesired characteristic for PArEs, we have seized the accompanying spectral changes of this phenomenon for sensing purposes. This was demonstrated by the detection of potassium ions with use of a 15-crown-5-substituted PPE [29]. K^{+} induces a 2:1 complex with 15-crown-5 while Li^{+} and Na^{+} form 1:1 complexes with the same crown ether. As a consequence, aggregation between polymer chains occurs only with K^{+} and this was observed at a polymer-to-ion ratio of 0.5:1 (Fig. 7). This was manifested by a diminished emission and appearance of a bathochromic band in the absorption spectrum upon addition of K^{+} to a solution of **31**, consistent with spectra obtained by compressing the polymer film at the air–water interface (Fig. 8). No change in the absorbance and emission spectra was observed even at 1,500-fold excess of Na^{+} and Li^{+}. The comonomer's steric and electronic properties were important in facilitating π-stacking interactions between poly-

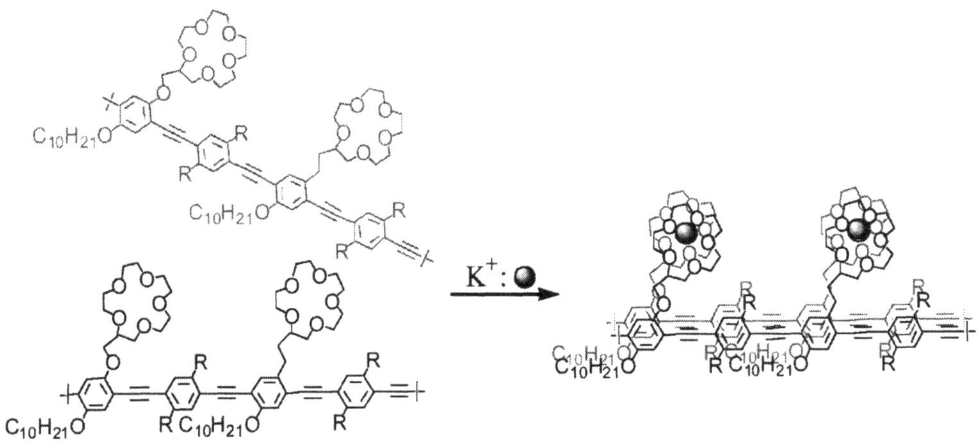

Fig. 7 Schematic representation of K^+ ion-induced aggregation. (Reprinted with permission from Ref. [29]. Copyright 2000 Wiley-VCH Verlag GmbH)

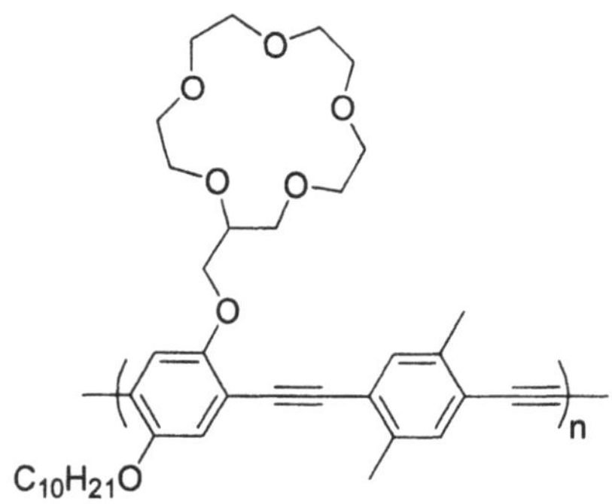

31

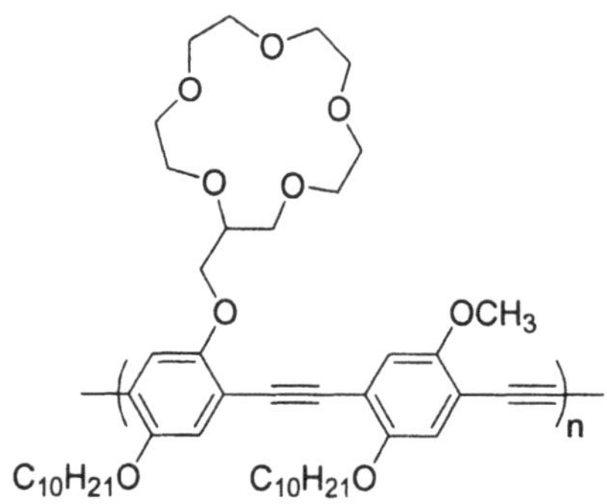

32

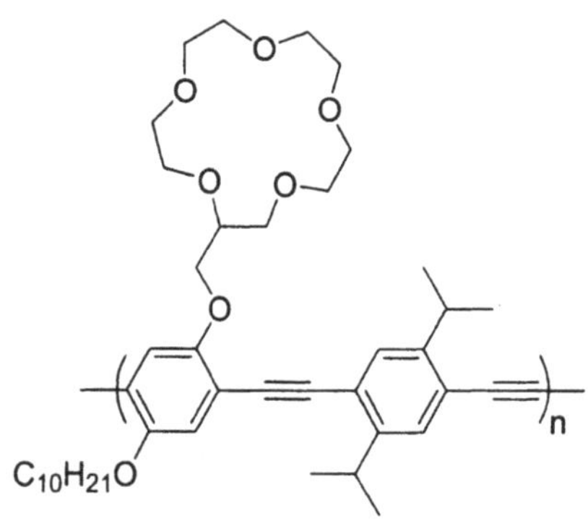

33

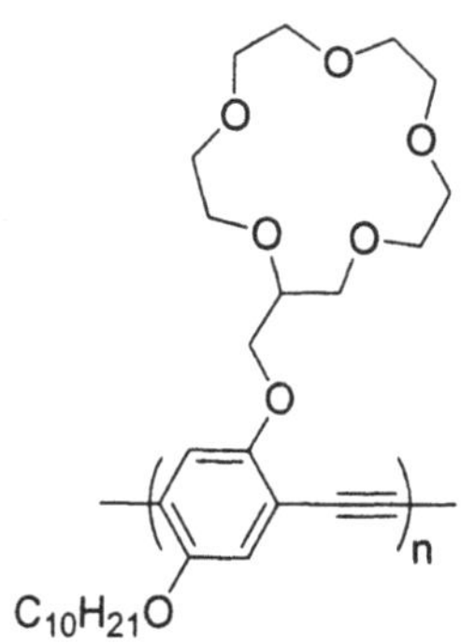

34

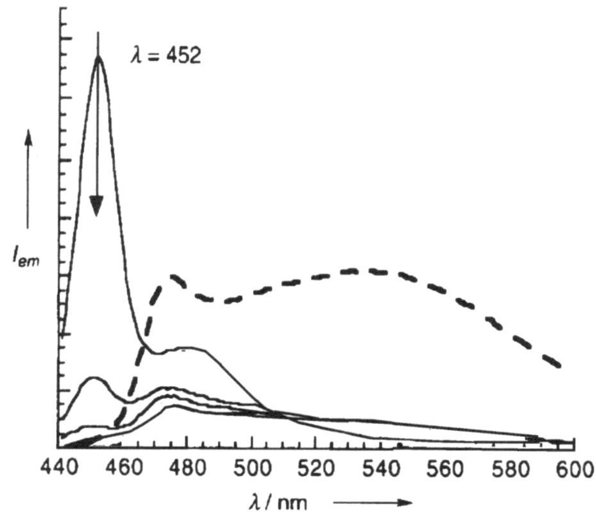

Fig. 8 Fluorescence spectra of polymer **29** with various mole ratios of K^+ ions. The *arrow* indicates the changes that result from increasing the concentration ratio of K^+ to 15-crown-5: 15-crown-5:K^+=1:0, 1:0.5, 1:1, 1:2.5, 1:5. *Dashed line* is data from a monolayer Langmuir–Blodgett film. (Reprinted with permission from Ref. [29]. Copyright 2000 Wiley-VCH Verlag GmbH)

mer chains. For instance, PPE **32** showed diminished aggregation efficiency due to competition from possible Lariat-type complexation between the crown ether and the adjacent methoxy oxygen. In **33**, the isopropyl side chains on the comonomer prevented π-stacking due to increased steric bulk. PPE **34** with crown ether moieties on every repeat unit could form intrachain complexes with K^+, so ion-induced interchain aggregation was absent.

In addition to metal ions, detection of halides such as fluoride is important, as the latter is often present in nerve gases and nuclear weapons manufacture. Our group utilized the unique reactivities of F^- with Si to specifically sense for fluoride ions [30]. PArE **35** was constructed with a masked precursor to a flu-

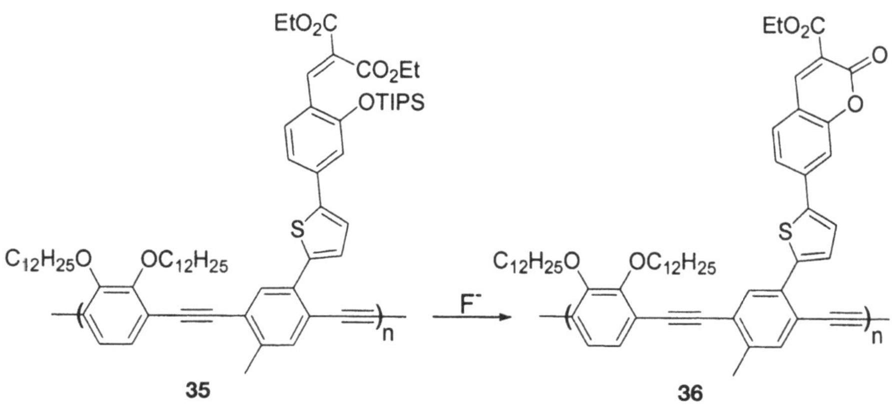

35 **36**

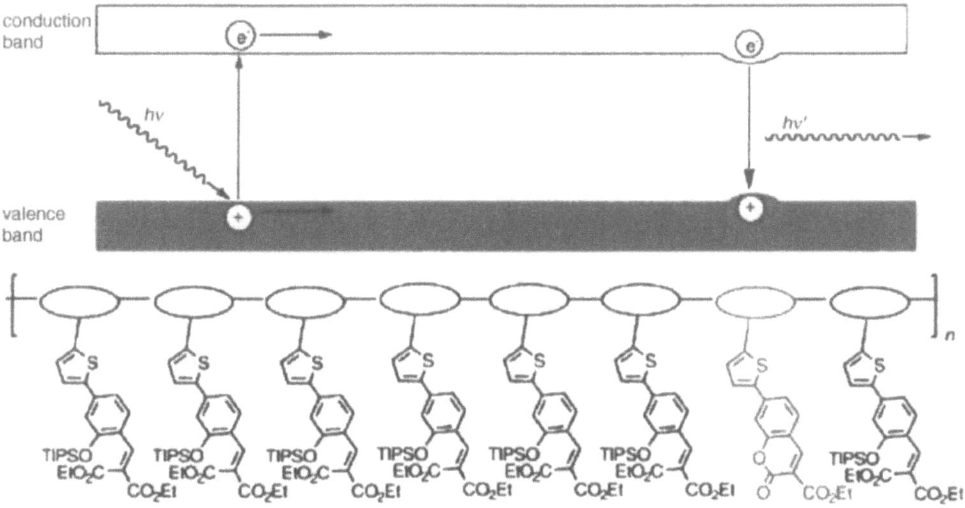

Fig. 9 Schematic band diagram depicting the mechanism by which a semiconductive polymer can produce an enhancement in PArE **34**. (Reprinted with permission from Ref. [30]. Copyright 2003 Wiley-VCH Verlag GmbH)

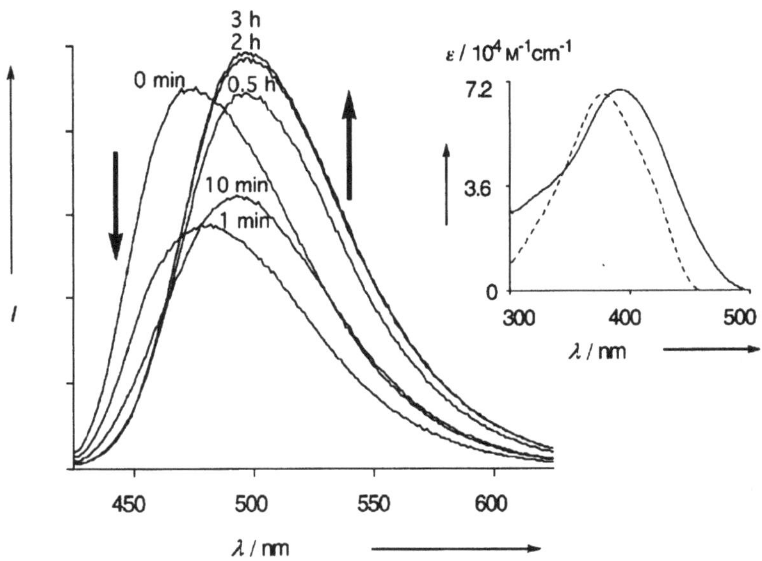

Fig. 10 Emission changes of polymer **33** upon addition of tetrabutylammonium fluoride (TBAF). *Inset* are the respective absorption spectra of the initial solution (0 min, ---) and after exposure to TBAF (—). 1.5×10^{-6} M in CH_2Cl_2 (repeating group) and TBAF, 1.6×10^{-7} M. (Reprinted with permission from Ref. [30]. Copyright 2003 Wiley-VCH Verlag GmbH)

orescent coumarin dye that is conjugated to the PArE. Upon exposure to fluoride ions, reaction with TIPS occurs and the coumarin dye forms to generate **36**. The electronic character of the dye is strongly coupled to the band structure of the PArE and locally perturbs the bandgap (Fig. 9). As a result, energy transfer occurs via the Dexter mechanism to the dye (Fig. 10). Under identical experimental conditions, the polymer proved to be 100-fold more sensitive to F^- compared to a simple masked coumarin, once again demonstrating amplification.

4 PArEs as Biosensors

The amplification ability of conjugated polymers can be used in biosensors and this capability becomes especially relevant considering the often minute quantities of biological analytes. Biological interactions such as those between proteins and their ligands, DNA strands, carbohydrates, and cells could all be potentially amplified using conjugated polymers. To be compatible with biological systems the polymer should be hydrophilic, and this is usually accomplished by installing ionic groups onto the polymer backbone. The transducer should also be highly selective, as it should only amplify the desired signal while minimizing the response due to nonspecific interactions.

As a proof of principle, the conjugated polymer MPS-PPV **6** with pendant sulfonate groups was used for sensing the interaction between biotin and avidin [11]. The strong association (K_a~10^{15} M) between the small molecule biotin and the tetrameric proteins avidin or streptavidin lends itself to a well-suited model system for biological recognition. As methyl viologen (MV^{2+}) is a good electron transfer quenching agent for conjugated polymers, a biotin-conjugated viologen was synthesized and added to a solution of MPS-PPV in water, resulting in a quenched polymer. Upon addition of avidin, the comparatively bulkier protein binds to biotin–viologen and removes it from the polymer, effecting an "unquench" (Fig. 11). Amounts of avidin on the order of 10^{-8} M can be detected. Limitations of this bioassay lie in the requirements for

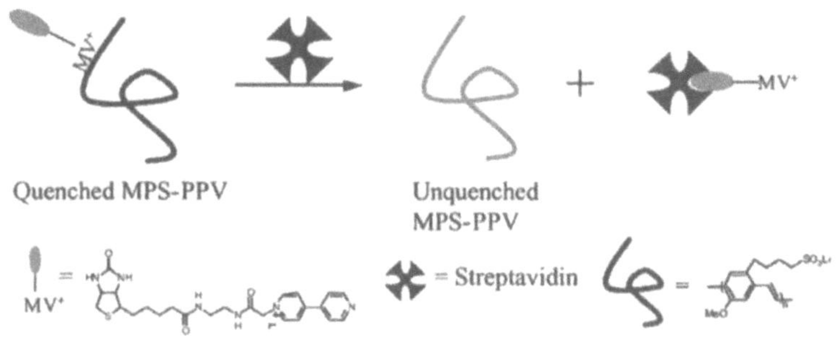

Fig. 11 Removal of a biotin-labeled quencher by avidin results in recovery of the MPS-PPV fluorescence.

the analyte. A protein that is too small may be unable to fully remove and segregate the quencher from the polymer. Changes in the charge of the quencher and in the protein may also affect the sensitivity of the ionic polymer.

Fluorescence recovery-type bioassays have also been designed by Heeger et al., utilizing a charge-neutral complex formed by the anionic sulfonated PPV **37**, MBL-PPV, and a cationic polyelectrolyte, poly(*N*,*N*-dimethylammonio-ethylene iodide) [31]. Both positively and negatively charged quenchers affect the photoluminescence of the polymer complex in phosphate-buffered saline (PBS), although at a decreased sensitivity corresponding to about 2 orders of magnitude compared to only the anionic PPV in water. The decreased sensitivity was attributed to a combination of ionic screening in the buffer solution and to the absence of static quenching for the neutral polymer complex. As a demonstration of the possible biosensing applications, a negatively charged dinitrophenol (DNP) derivative quenches the complex and this is reversed upon addition of anti-DNP IgG antibody.

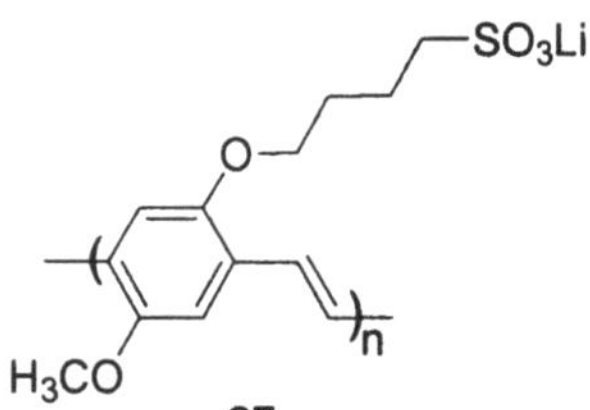

37

Electron transfer quenching analogous to that between MV^{2+} and PArEs is also possible with proteins. Fluorescence quenching of **37** was carried out with cytochrome c, a cationic protein, at neutral pH (pI=9.6 at room temperature) that can undergo rapid electron transfer [32]. An apparent K_{SV} for quenching of 3.2×10^8 M^{-1} at pH 7.4 was obtained and compared to the previously reported amplification factor of 67 [10]. However, the K_{SV} is a composite of the amplification factor and the binding constant of the protein to the polymer and so a fair comparison cannot be made. As can be expected in electrostatic complexations, the K_{SV} value was sensitive to the pH of the solution and dropped by up to 6 orders of magnitude when the pH was increased to >10 (where the protein was slightly anionic). Control experiments were conducted with myoglobin and lysozyme. No electron transfer could occur in the case of myoglobin; however, at a pH where it has the same surface charge as cytochrome c (at pH 7.4), a K_{SV} on the order of 10^6 M^{-1} could still be obtained. At pH 7.4, quenching of the polymer was also observed upon addition of the cationic lysozyme. There exists certainly some dependence of quenching efficiency on the electron transfer ability of the protein analyte. However, in this case the charge of the protein also plays a significant role in causing nonspecific quenching of the polymer, possibly by inducing aggregation and subsequent self-quenching.

Sulfonated **8** and biotinylated polymer **38**-coated streptavidin-derivatized polystyrene microspheres were useful as a platform for the detection of DNA

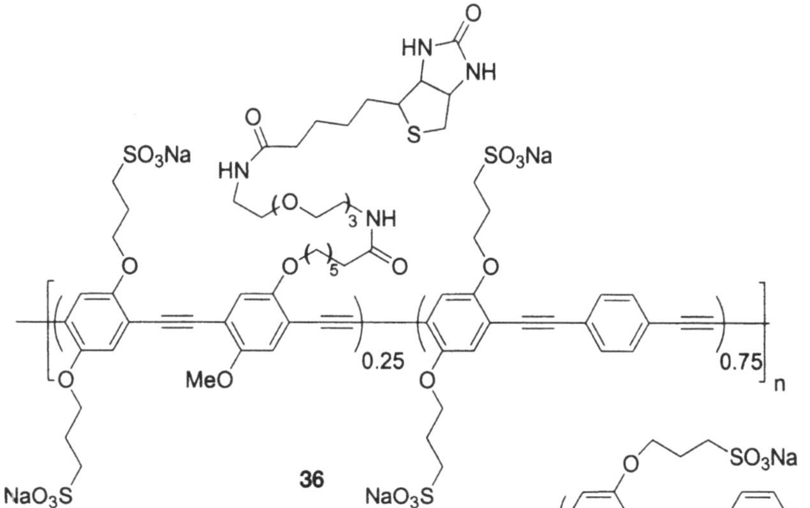

36

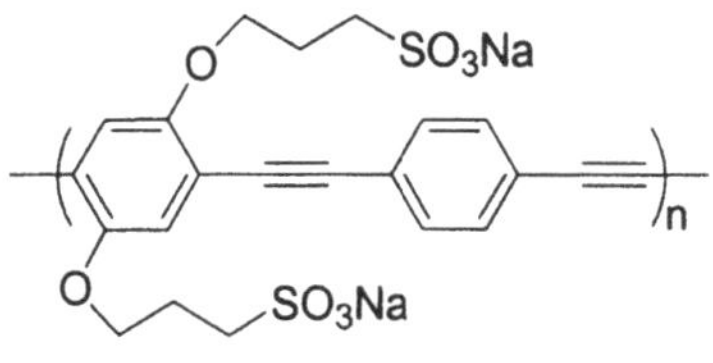

8

hybridization in competition assays [33]. A 20-base biotinylated capture DNA strand for anthrax lethal factor (ALF) was first bound to the microsphere. To this was added the quencher (QSY-7)-labeled complementary target strand. Quenching of the PPEs occurred upon hybridization of the target strand to the capture strand as the energy-accepting quencher was now brought into close proximity to the polymers (Fig. 12a). A series of competition assays were carried out. In the first case, the microspheres were subjected to a mixture of quencher-labeled target DNA and a nonlabeled ALF target strand. The QSY-7 DNA was kept constant while the concentration of the target strand was varied; it was expected that the quenching due to QSY-7 would be attenuated with increasing amounts of ALF target strand, as the latter would compete with the labeled DNA strand for the capture strand. However, little attenuation was observed even at higher concentrations of the target strand. The authors attribute this to a hydrophobic interaction between the quencher molecules with the polymers on the microsphere that enhanced preferential association of the

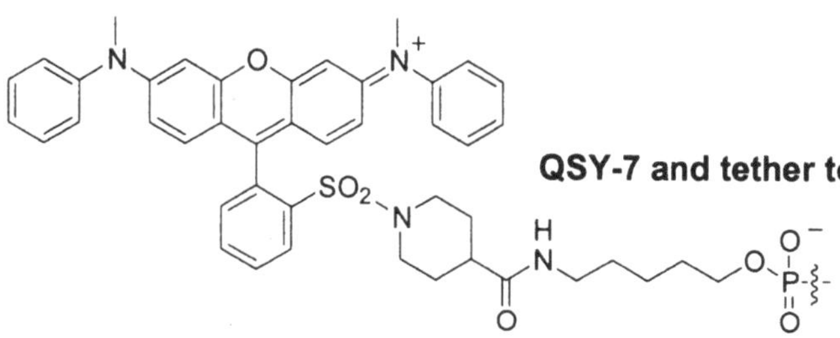

QSY-7 and tether to DNA

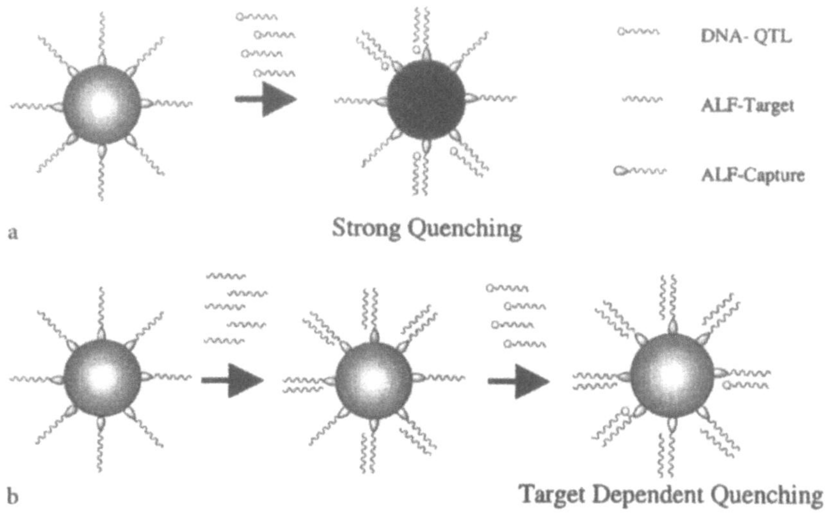

Fig. 12 **a** Quenching assay where QSY-7-labeled DNA strand (DNA-QTL) is presented to microsphere and microsphere-bound biotinylated ALF capture strand. **b** Competition assay for detection of target DNA. Variable amounts of ALF target are added to microsphere and bound ALF capture strand. This is then subjected to a fixed amount of QSY-7-labeled DNA. The amount of quenching is dependent on the amount of ALF target already bound on the microsphere. (Reprinted with permission from Ref. [33]. Copyright 2002 American Chemical Society)

QSY-7 conjugate. To circumvent this, variations on the competition assay were carried out. By a two-step sequential incubation of microsphere with varying amounts of ALF target strand followed by the QSY-7-labeled target DNA, attenuation of the level of quenching by the latter is observed (Fig. 12b). Alternatively the biotinylated capture strand, QSY-7 DNA, and variable amounts of ALF target could be incubated together. This mixture could then be added to the microspheres and similar attenuation of quenching was observed. By plotting the quenching ratio vs. the molar concentration of the ALF target, quenching attenuation curves were obtained. In a 200-μl assay containing 3.7×10^7 microspheres, 3 pmol of ALF capture strand, and 10 pmol of QSY-7-labeled complementary DNA, an attenuation of fluorescence can be observed at 0.5 pmol of ALF target. Thus, fluorescence recovery can be correlated to the amount of analyte oligonucleotide.

To elaborate upon this scheme, the authors replaced the DNA-based capture strand with one that is based on peptide nucleic acid (PNA) in order to distinguish single nucleotide mismatches [34]. Since PNA does not have a phosphodiester backbone, a PNA–DNA duplex is more thermodynamically stable than DNA–DNA or RNA–RNA helices due to a lack of electrostatic repulsion between the two chains. PNAs have also been shown to be more selective. However, PNAs are more difficult to synthesize and PNA–DNA interactions are not as well understood as DNA–DNA binding. In this study, quaternary ammo-

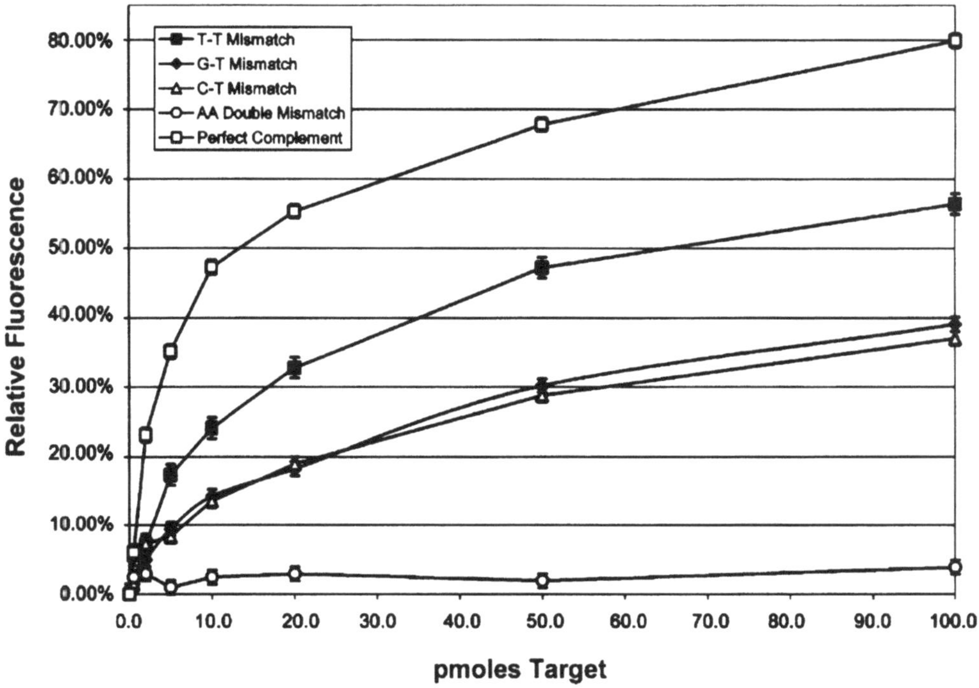

Fig. 13 Mismatch analysis at 40 °C using a microsphere sensor loaded with a PNA-based capture strand. T-T, G-T, C-T are single mismatch sequences. (Reprinted with permission from Ref. [34]. Copyright 2003 American Chemical Society)

nium-functionalized polystyrene beads were coated with a precomplexed solution of polymer **38** and neutravidin. By a similar competition assay to one previously described (Fig. 12b) using target sequences (mismatched or perfect complement) in variable amounts, a plot of the fluorescence ratio (before and after exposure to QSY-7-labeled target DNA) vs. the amount of ALF–DNA target sequence revealed the quenching attenuation as increasing amounts of the ALF–DNA competed for the hybridization sites. To improve the selectivity of the single mismatch in the 20-mer sequence, the assays were conducted at an elevated temperature of 40 °C, and mismatched sequences were expected to yield lower rates of quenching attenuation compared to that of the perfect complement. For a suspension containing 1×10^7 microspheres, 2 pmol bound PNA capture sequence, and 10 pmol QSY-7-labeled DNA target sequence, a 15% difference in quenching attenuation could be observed between the perfect complement and the mismatched sequences for 0.5 pmol of the various target ALF–DNA strands (Fig. 13). When the identical assay was performed at 25 °C with a DNA-based capture strand, poor resolution of the mismatched oligonucleotides was observed.

As a simple and practical alternative to coated microspheres, submicron particles of pendant amine-functionalized PPE **39** could be fabricated by phase

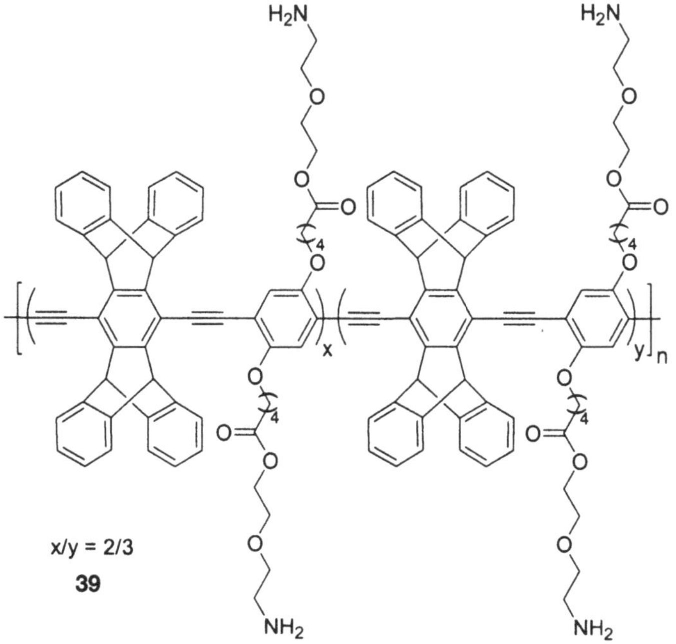

x/y = 2/3

39

inversion and used for detecting Cy-5-labeled oligonucleotides [35]. While Cy-5 is well known as a fluorophore, the authors of this report use the dye as a quencher. Cy-5 concentrations in the range 10^{-10} to 10^{-7} M produced measurable quenching. A large K_{SV} of 8.8×10^{7} M^{-1} was obtained. The fluorescence quenching was 2 orders of magnitude more sensitive than direct excitation of the Cy-5 dye, and the platform could potentially be applied to DNA sensing arrays.

Carbohydrate detection is important for applications such as glucose monitors; these are arguably one of the most successful and relevant biosensors. An interesting fluorescence recovery-type saccharide sensor based on the reactivity of carbohydrates with boronic acids was reported in 2002 [36]. Specifically, modification of the cationic viologen-linked boronic acid derivative **40** to a zwitterionic species **41** upon covalent and reversible reaction of boronic acid with monosaccharides (Scheme 1) can cause the dissociation of the ion-pair in-

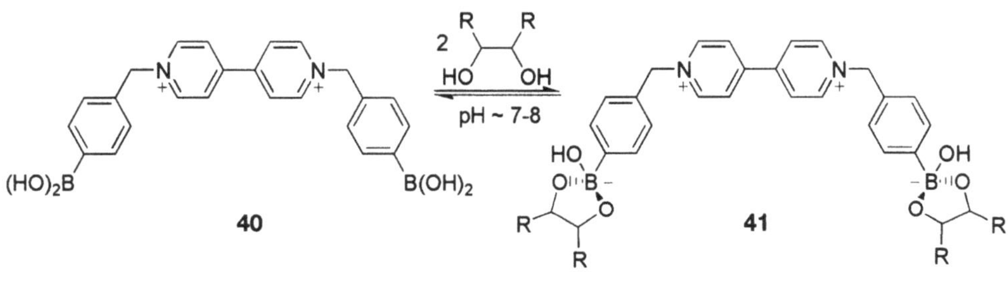

Scheme 1 Interaction between 4,4′-*N,N*′-bis(benzyl-4-boronic acid)-bipyridinium dibromide with a carbohydrate at near-neutral pH

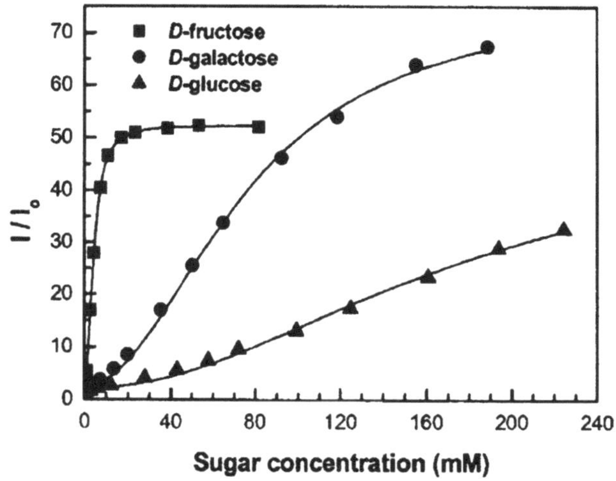

Fig. 14 Titration curves against sugar for PPE **8** (2.5×10^{-6} M)/**38** (8×10^{-7} M), measured in PBS (6 mM) of pH 7.4. (Reprinted with permission from Ref. [36]. Copyright 2002 American Chemical Society)

teraction between the viologen quencher and polymer **8**. This in turn leads to fluorescence recovery from the fluorescent polymer. In the case of galactose, up to 70-fold increase in fluorescence intensity was observed (Fig. 14).

PArE-induced aggregation of biological agents could potentially be used as a method of detection, considering the prevalence of multivalent interactions in biology. This was demonstrated by Bunz et al., who used a biotin-functionalized PPE and streptavidin-functionalized microspheres as a primitive model system for the recognition that occurs in cells and bacteria [37]. Upon incubation of the microspheres with PPE **42**, the solution fluorescence showed aggregation and the disappearance of a blue shoulder compared to the uncomplexed biotinylated polymer. Agglomeration of the microspheres was observed by scanning electron microscopy and fluorescence microscopy.

While quenching experiments are useful and provide a sensitive response to analytes, a turn-on sensor offers advantages such as improved sensitivity and selectivity to the species to be detected, as well as diminished response to non-

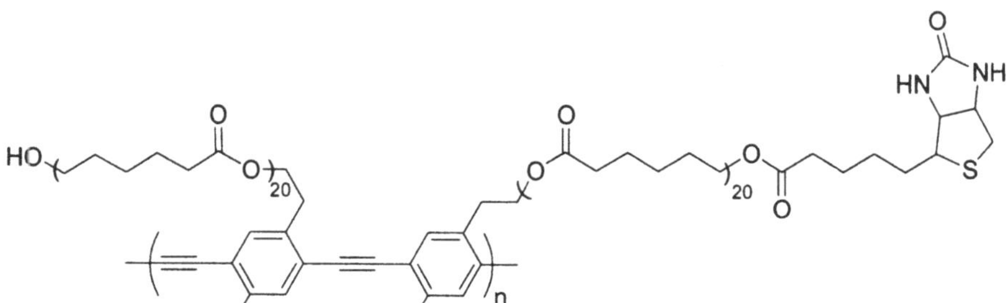

42

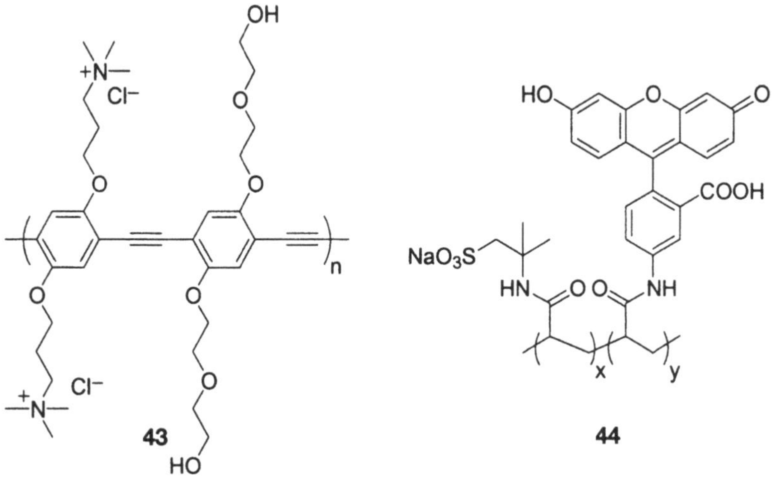

specific interactions. An obvious method is to use fluorescence resonance energy transfer (FRET) to transduce the recognition event. For this purpose our group demonstrated the efficient energy transfer from a PPE to a fluorescent pH-sensitive dye [38]. Films of cationic PPE **43** and the anionic fluoresceinamine-appended polyacrylate **44** were coated onto a glass substrate using layer-by-layer deposition. The absorption cross section, energy migration efficiency, and emission efficiency of the pendant fluoresceinamine dye could change as

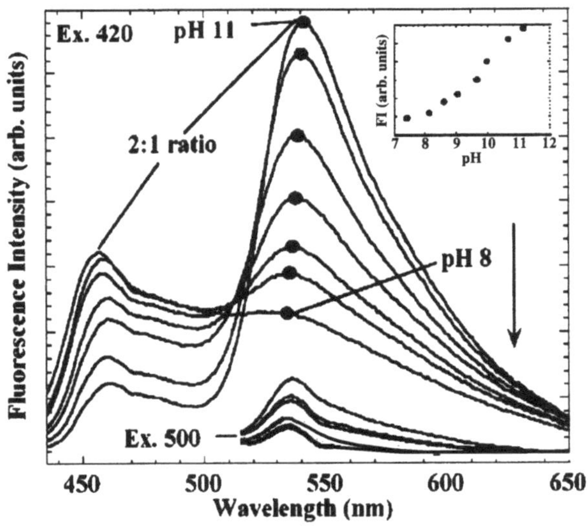

Fig. 15 A film composed of **42** sandwiched between two layers of **41**. The fluorescence spectra spanning 435 to 650 nm and the spectra beginning at 515 nm were excited at 420 and 500 nm, respectively. *Inset*: The emission maximum of the fluoresceinamine band after excitation at 420 nm plotted against pH. (Reprinted with permission from Ref. [38]. Copyright 2000 American Chemical Society)

a function of pH. At high pH, the dye is highly absorptive and fluorescent, acting as a shunt and withdrawing energy from the light-harvesting conjugated polymer. At low pH, the absorbance decreases and there is no fluorescence. Excitation of the PPE at 420 nm resulted in a tenfold increase in emission of the dye relative to its emission obtained by direct excitation. At pH 11, ~90% of the PPE's emission is transferred to the dye (Fig. 15). The architecture of the film is important; when **44** was sandwiched between two layers of **43**, more energy was transferred to the dye acceptor, as there was more energy-harvesting PPE available.

Based on these observations for FRET between PPE and fluorescent dyes, our group reported a model biosensor using biotinylated PPE **45** and dye-labeled streptavidin [39]. Upon binding of rhodamine B or Texas Red-X-labeled streptavidin to **45** in Tris buffer at pH 7.4, energy transfer to the dyes was observed. Surprisingly, Texas Red-X–streptavidin exhibited greater emission intensity, even though it has less spectral overlap (compared to rhodamine B–streptavidin) with the polymeric donor under identical experimental conditions (Fig. 16). This finding did not obey the requirements for Förster-type energy transfer, where diminished spectral overlap between the polymer donor's emission and the dye acceptor's absorption would result in decreased energy transfer. As Texas Red-X possesses a more planar structure and greater hydrophobic character compared to rhodamine B, once brought into close proximity to

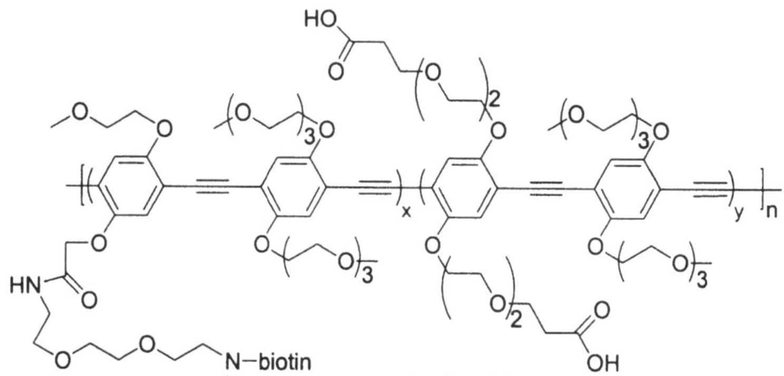

45 x/y = 1/4

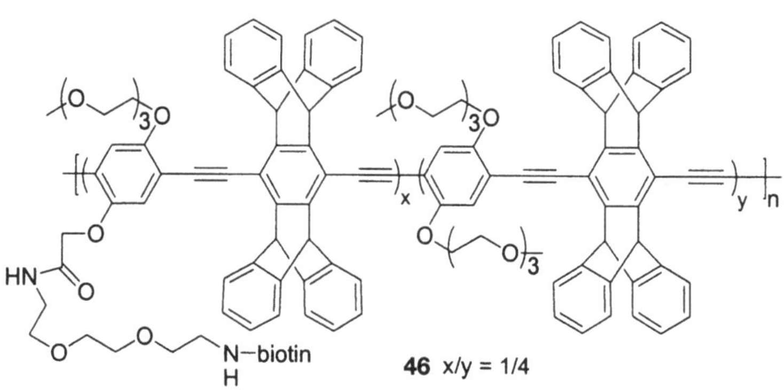

46 x/y = 1/4

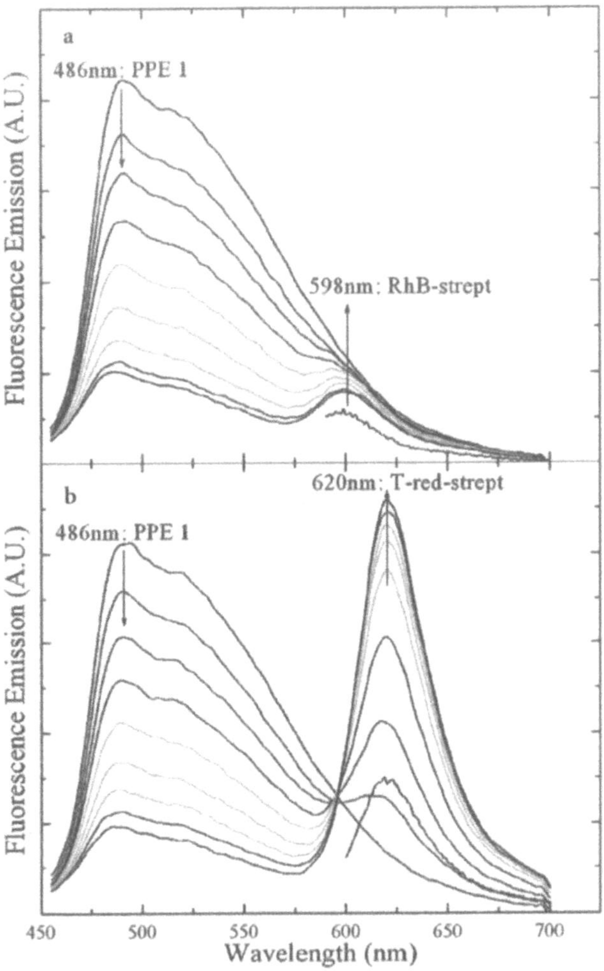

Fig. 16 Addition of 0.017-nmol aliquots of **a** rhodamine B-labeled streptavidin and **b** Texas Red-X-labeled streptavidin to 1.51 nmol of **43**. Energy transfer observed in both cases with amplified emission of the dyes to the light-harvesting conjugated polymers. Direct excitation of the dyes at 575 and 585 nm correspond to 0.100 nmol of streptavidin.

the polymer by the streptavidin–biotin recognition it may be able to interact more intimately with the polymer's phenylene ethynylene backbone and undergo greater orbital mixing. We proposed that a greater Dexter energy-transfer mechanism may be operative in this model system.

The interactions of the dye acceptors to polymers could also be witnessed in the case of solid films. In this case, the sterically more restrictive cavities of the polymeric film **46** allowed better orbital interaction with the smaller and more flexible rhodamine B dye, and accordingly higher energy transfer with rhodamine B-labeled streptavidin was observed compared to Texas Red-

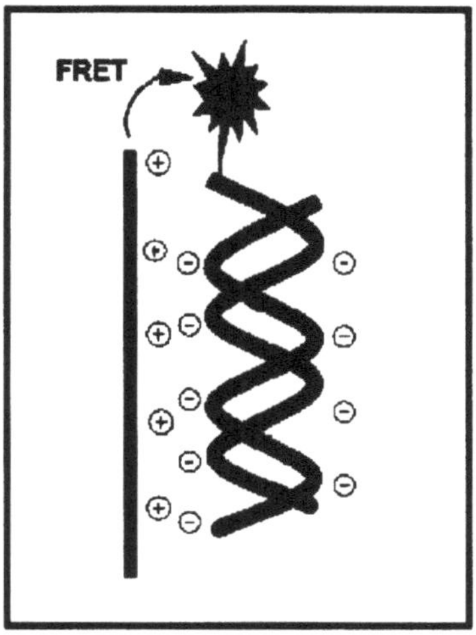

Fig. 17 Schematic representation of a DNA–DNA duplex bound electrostatically to polymer **45**. (Reprinted with permission from Ref. [44]. Copyright 2004 American Chemical Society)

X–streptavidin. The intricate interplay between the steric and electronic properties of the acceptor and the polymeric donor may have important impact for the design of future biosensors.

Some non-PArE based colorimetric and fluorometric sensors also deserve mention. Bazan et al. have used energy-transfer platforms in the design of sensors for detecting negatively charged PNA–DNA [40, 41], DNA–DNA [42–45] (Fig. 17), and RNA–peptide [46] duplexes. Typically these assays use cationic polymers such as **47**, a nonlabeled single-stranded capture DNA/RNA and a fluorophore-labeled complementary DNA/PNA/peptide. Upon formation of the recognition duplex, the fluorophore is brought into close proximity with the polymer by electrostatic interactions between the anionic duplex and cationic polymer; subsequent emission from the dye occurs due to energy transfer. A three-tiered energy-transfer assay has also been constructed [44] where energy transfer occurs from the conjugated polymer to a fluorescein-labeled DNA, which in turn transfers energy to an intercalated ethydium bromide. This could potentially improve selectivity and optical resolution of the biosensor.

Sensitive, cationic, water-soluble polythiophenes have been utilized for biosensing. Leclerc et al. have created affinitychromic sensors with such polymers for transducing a variety of recognition events such as those between

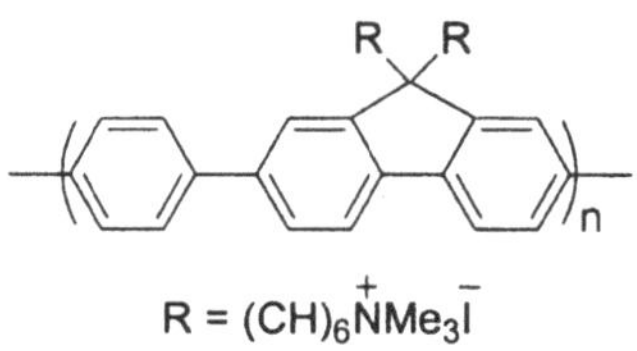

47

biotinylated polythiophene–avidin, DNA–DNA, and aptamer–protein. Electrostatic and conformational differences that occur upon recognition disrupt the planarization and aggregation of the polymer backbone, leading to visible absorption and fluorescence changes. Remarkably, analyte detection levels down to 10^{-21} mol could in some cases be achieved [50]. Interested readers are referred to relevant publications [46–50].

Biosensors based on poly(diacetylene)s have also been investigated by other groups [51–54]. These have proved successful in detecting biologically relevant agents such as the influenza virus and cholera toxin.

5 Summary

Poly(arylene ethynylene)s have proven to be sensitive transducers capable of amplifying binding events ranging from detection of TNT and ions to detecting single mismatches in DNA. The ability to manipulate the properties of PArEs by changing their reactivities, assembly architecture, and emissive properties creates wide-ranging possibilities for practical real-life sensing purposes.

References

1. McQuade DT, Pullen AE, Swager TM (2000) Chem Rev 100:2537
2. Ofer D, Swager TM, Wrighton MS (1995) Chem Mater 7:418
3. Zotti G, Schiavon G, Zecchin S, Berlin A (1998) Synth Met 97:245
4. Greiner A (1998) Polym Adv Technol 9:371
5. Pschirer NG, Miteva T, Evans U, Roberts RS, Marshall AR, Neher D, Myrick ML, Bunz UHF (2001) Chem Mater 13:2691
6. Kocher C, Montali A, Smith P, Weder C (2001) Adv Funct Mater 11:31
7. Bunz UHF (2000) Chem Rev 100:1605
8. Swager TM, Gil CJ, Wrighton MS (1995) J Phys Chem 99:4886
9. Zhou Q, Swager TM (1995) J Am Chem Soc 117:7017
10. Zhou Q, Swager TM (1995) J Am Chem Soc 117:12593
11. Chen L, McBranch DW, Wang H-L, Helgeson R, Wudl F, Whitten DG (1999) Proc Natl Acad Sci U S A 96:12287
12. Tan C, Pinto MR, Schanze KS (2002) Chem Commun 446
13. Pinto MR, Kristal BM, Schanze KS (2003) Langmuir 19:6523
14. Jenekhe SA, Osaheni JA (1994) Science 265:765
15. Osaheni JA, Jenekhe SA (1995) J Am Chem Soc 117:7389
16. Kim J, Swager TM (2001) Nature 411:1030
17. (a) Levitus M, Schmieder K, Ricks H, Shimizu KD, Bunz UHF, Garcia-Garibay MA (2001) J Am Chem Soc 123:4259; (b) Levitus M, Schmieder K, Ricks H, Shimizu KD, Bunz UHF, Garcia-Garibay MA (2002) J Am Chem Soc 123:8181
18. Yang J-S, Swager TM (1998) J Am Chem Soc 120:5321
19. Yang J-S, Swager TM (1998) J Am Chem Soc 120:11864
20. Cumming JC, Aker C, Fisher M, Fox M, la Grone MJ, Reust D, Rockley MG, Swager TM, Towers E, Williams V (2001) IEEE Trans Geosci Remote Sens 39:1119

21. Levitsky IA, Kim J, Swager TM (1999) J Am Chem Soc 121:1466
22. Kim J, McQuade DT, Rose R, Zhu Z, Swager TM (2001) J Am Chem Soc 123:11488
23. Rose A, Lugmair CG, Swager TM (2001) J Am Chem Soc 123:11298
24. Wang B, Wasielewski MR (1997) J Am Chem Soc 119: 12
25. Liu B, Yu W-L, Pei J, Liu S-Y, Lai Y-H, Huang W (2001) Macromolecules 34:7932
26. Zhang Y, Murphy CB, Jones WE Jr (2002) Macromolecules 35:630
27. Jiang B, Yang S-W, Barbini DC, Jones WE Jr (1998) Chem Commun 213
28. Bangcuyo CG, Ranpey-Vaughn ME, Quan LT, Angel SM, Smith MD, Bunz UHF (2002) Macromolecules 35:1563
29. Kim J, McQuade DT, McHugh SK, Swager TM (2000) Angew Chem Int Ed 39:3868
30. Kim T-H, Swager TM (2003) Angew Chem Int Ed 42:4803
31. Wang D, Gong X, Heeger PS, Rininsland F, Bazan GC, Heeger AJ (2002) Proc Natl Acad Sci U S A 99:49
32. Fan C, Plaxco KW, Heeger AJ (2002) J Am Chem Soc 124:5642
33. Kushon SA, Ley KD, Bradford K, Jones RM, McBranch D, Whitten D (2002) Langmuir 18:7245
34. Kushon SA, Bradford K, Marin V, Suhrada C, Armitage BA, McBranch D, Whitten D (2003) Langmuir 19:6456
35. Moon JH, Deans R, Krueger E, Hancock LF (2003) Chem Commun 104
36. DiCesare N, Pinto MR, Schanze KS, Lakowicz JR (2002) Langmuir 18:7785
37. Wilson JN, Wang Y, Lavigne JJ, Bunz UHF (2003) Chem Commun 1626
38. McQuade DT, Hegedus AH, Swager TM (2000) J Am Chem Soc 122:12389
39. Zheng J, Swager TM (2004) Chem Commun 2798
40. Gaylord BS, Heeger AJ, Bazan GC (2002) Proc Natl Acad Sci U S A 99:10954
41. Liu B, Bazan GC (2004) J Am Chem Soc 126:1942
42. Gaylord BS, Heeger AJ, Bazan GC (2003) J Am Chem Soc 125:896
43. Liu B, Gaylord BS, Wang S, Bazan GC (2003) J Am Chem Soc 125:6705
44. Wang S, Gaylord BS, Bazan GC (2004) J Am Chem Soc 126:5446
45. Wang S, Bazan GC (2003) Adv Mater 15:1425
46. Leclerc M, Ho H-A (2004) Synlett 2:380
47. Leclerc M (1999) Adv Mater 11:1491
48. Bernier S, Garreau S, Béra-Abérem M, Gravel C, Leclerc M (2002) J Am Chem Soc 124:12463
49. Ho H-A, Leclerc M (2004) J Am Chem Soc 126:1384
50. Doré K, Dubus S, Ho H-A, Lévesque I, Brunette M, Corbeil G, Boissinot M, Boivin G, Bergeron MG, Boudreau D, Leclerc M (2004) J Am Chem Soc 126:4240
51. Charych DH, Nagy JO, Spevak W, Bednarski MD (1993) Science 261:585
52. Reichert A, Nagy JO, Spevak W, Charych D (1995) J Am Chem Soc 117:829
53. Pan JJ, Charych D (1997) Langmuir 13:1365
54. Song J, Cheng Q, Zhu S, Stevens RC (2002) Biomed Microdevices 4:213

Adv Polym Sci (2005) 177: 181–208
DOI 10.1007/b101378

PAEs with Heteroaromatic Rings

Takakazu Yamamoto (✉) · Isao Yamaguchi · Takuma Yasuda

Chemical Resources Laboratory, Tokyo Institute of Technology, 4259 Nagatsuta, Midori-ku, Yokohama 226-8503, Japan
tyamamot@res.titech.ac.jp

Abstract This article reviews the synthesis and chemical properties of poly(aryleneethynylene)s, PAEs, consisting of sulfur-containing heteroaromatic rings (e.g., thiophene), nitrogen-containing heteroaromatic rings (e.g., pyridine), and silicon-containing heteroaromatic rings (e.g., silole). The polymers are usually prepared via Pd-catalyzed C–C coupling between diethynyl compounds and dihalo organic compounds. However, synthesis using other monomers such as distannylacetylenic compounds is also possible. This article describes the chemical and optical properties of PAEs, which are affected by the basic electronic properties of the heteroaromatic ring and the molecular structure of the PAE.

Keywords Poly(aryleneethynylene) · Heteroaromatic ring · Palladium-catalyzed coupling · Optical properties · Chemical reactivity

1 Introduction

The following Pd-catalyzed organometallic C–C coupling reaction [1–3] is an important coupling reaction in organic synthesis [4–7].

$$RX + R'C{\equiv}CH + R''_3N \xrightarrow{\text{Pd-Cu}} RC{\equiv}CR' + [R''_3HN]X \qquad (1)$$

The coupling reaction is considered to involve the following elementary reactions.

a) Oxidative addition:

$$\mathrm{RX + \underset{\substack{\text{zerovalent}\\ \text{Pd complex}}}{Pd(0)Lm} \longrightarrow Pd(R)(X)Lm} \tag{2}$$

b) Formation of copper acetylide:

$$\mathrm{R'C{\equiv}CH + CuX + R''_3N \longrightarrow R'C{\equiv}C\text{-}Cu + [R''_3HN]X} \tag{3}$$

c) Intermetal transfer of alkynyl group:

$$\mathrm{Pd(R)(X)Lm + R'C{\equiv}C\text{-}Cu \longrightarrow Pd(R)(C{\equiv}CR')Lm + CuX} \tag{4}$$

d) Reductive elimination (C–C coupling on Pd):

$$\mathrm{Pd(R)(C{\equiv}CR')Lm \longrightarrow RC{\equiv}CR' + Pd(0)Lm} \tag{5}$$

The total of the reactions 2–5 results in reaction 1. The oxidative addition and formation of copper acetylide are well known. The intermetal transfer of the alkynyl group from copper acetylide (e.g., from Cu(I) to Pd(II) [8]) has been revealed by organometallic chemistry [8–11].

Reductive elimination on transition metal complexes seems to be enhanced by coordination of electron-withdrawing π-acids such as cyanobenzene and cyanoethylene. For example, the reductive elimination reaction of NiR_2(bpy) (R=alkyl or aryl group; bpy =2,2′-bipyridyl) is enhanced by electron-withdrawing olefinic and aromatic compounds [12–16] (Scheme 1).

The Ni–R group is considered to be polarized as $Ni^{2\delta+}$–$R^{\delta-}$, whereas the reductive elimination produces electrically neutral R–R. Consequently the reductive elimination is accompanied by partial migration of electrons from R to Ni, which is enhanced by coordination of the π-acid to Ni. The coordination of the electron-withdrawing olefin to Ni accelerates the reductive elimination by a factor of 10^{10} or larger [16]. This situation is similar to the enhancement effect of electron-withdrawing substituent on acid dissociation [17]:

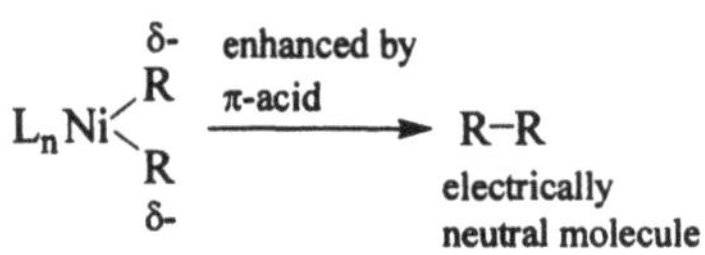

(6)

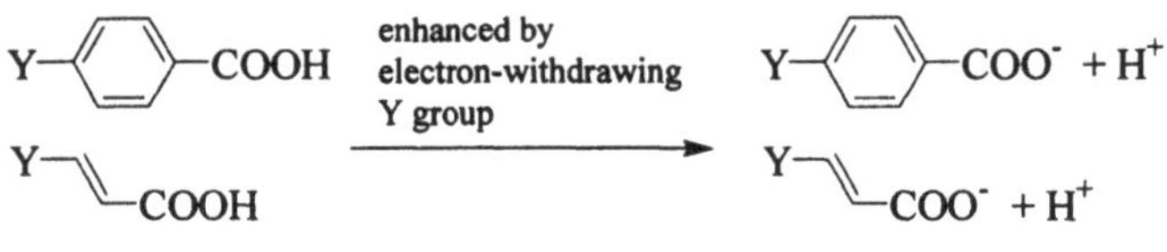

(7)

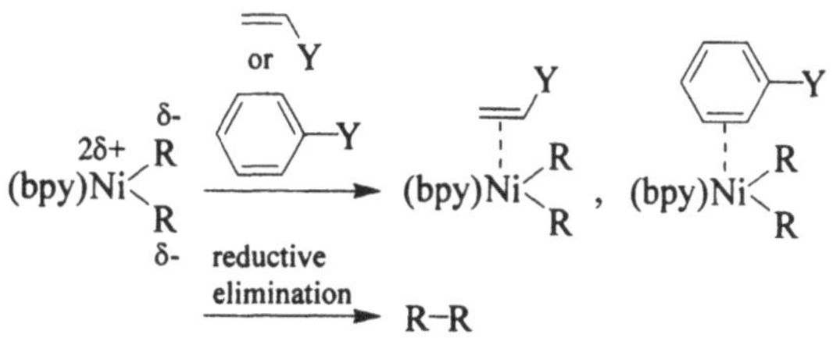

Scheme 1 Enhancement of reductive elimination reaction of $NiR_2(bpy)$ by coordination of π-acid

In the case of the acid dissociation, however, it is considered to involve migration of the –COO–H bonding electron to the –COO unit to generate $-COO^-$ and H^+. The following Ni-catalyzed synthetic C–C coupling reaction between RMgX and R′X [18–21] is strongly influenced by the kind of R′X, whereas there is no restriction about the kind of RMgX. The reaction proceeds well with olefinic R′X and aromatic R′X, whereas the reaction cannot be carried out well with aliphatic R′X. On the other hand, there is no restriction about RMgX as described above.

$$RMgX + R'X \xrightarrow{NiLm} R\text{–}R' \tag{8}$$

The C–C coupling reaction between RMgX and R′X is considered to proceed though an Ni(R)(R′)Lm intermediate, and acceleration of the reductive elimination of R–R′ by coordination with olefinic or aromatic R′X to Ni(R)(R′)Lm is necessitated for a smooth catalytic reaction [15, 16]. On these bases Ni-promoted dehalogenative polycondensation of dihalo organic compounds is suited to the preparation of π-conjugated aromatic and olefinic polymers.

The controlling factor of the reductive elimination on Pd(R)(C≡CR′)Lm (Eq. 5) may be different from that observed with the nickel complex. However, participation of a similar activation process by coordination of electron-withdrawing RX and R′C≡CH is conceivable. The Pd(R)(C≡CR′)Lm-type complex can be isolated, and it has been shown that isolated Pd(R)(C≡CR′)Lm undergoes the reductive elimination exhibited in Eq. 5 [8]. The reductive elimination seems to be enhanced by addition of CuI. CuI may interact with the Pd complex, and an acceleration effect of Lewis acids on the reductive elimination reaction of $NiR_2(bpy)$ has been shown [22]. The X-ray crystallographic structure of an isolated Pd(R)(C≡CR′)Lm ($R=C_6H_4Me$-*p*; $R'=C_6H_5$) has been determined [8].

The Pd-catalyzed C–C coupling (Sonogashira coupling) was applied to polymer synthesis about 20 years ago [23–25], and has especially been developed for the synthesis of π-conjugated poly(aryleneethynylene)s (PAEs) (for reviews, see refs. [16, 26–33]). Recently other synthetic routes for PAEs were also developed, e.g., the alkyne metathesis method [28, 34] and the coupling reaction of $\equiv C\text{–}MR_3$ with R′X (M=Si [35, 36] or Sn [37, 38]). In this review, we are concerned with the synthesis and chemical properties of PAEs with heteroaromatic rings.

2 PAEs with Sulfur-Containing Heteroaromatic Rings

2.1 Synthesis

Table 1 summarizes PAEs with sulfur-containing heteroaromatic rings. Some PAEs with thiophene rings are included in Table 2 and Table 3.

When the PAE has long side chains such as hexyl and hexyloxy groups, the polymer is soluble in organic solvents, and data from NMR and elemental analysis indicate that PAEs prepared by the Sonogashira coupling usually have halogenated groups at the polymer ends [41, 58, 59]. As shown in Table 1, various palladium complexes with various phosphine ligands have been used. The dppf ligand used in Nos. 16–18 has attracted recent attention, because it often gives effective transition metal catalysts [60, 61]. As the copper cocatalyst, CuI is most widely used; however, other group 11 metal compounds such as copper(II) acetate (No. 40 in Table 1) and Ag_2O (No. 12) can also be used. When a monomer of type X–Ar–C≡CH is used as indicated in Nos. 21, 23, and 29 (as well as No. 3 in Table 2), the polymerization gives a regioregular head-to-tail-type PAE.

2.2 Chemical Properties

The –C≡C– bond undergoes an addition reaction with HBr [54] and a chlorofluorination reaction [41].

$$\text{-C}\equiv\text{C- + HBr} \longrightarrow \text{-CH=CBr-} \qquad (9)$$

$$\text{-C}\equiv\text{C-} \xrightarrow{\text{NCS, [PyH}^+\text{][(HF)}_X\text{F}^-\text{]}} \text{-CCl}\equiv\text{C F-} \qquad (10)$$

NCS = *N*-Chlorosuccinimide, Py = pyridine

The C–Br group formed can be converted into an ester group via organometallic reactions [54]. For one of the PAE described later (No. 16 in Table 2), *trans*-hydrogenation of the –C≡C– bond has also been reported [41].

$$\text{-C}\equiv\text{C-} \xrightarrow{\text{SMEAH or DIBAL}} \text{-CH=CH-} \qquad (11)$$

SMEAH = $Na[CH_3OCH_2CH_2O)_2AlH_2]$
DIBAL = $Al(i\text{-}C_4H_9)_2H$

The –C≡C– bond did not show good reactivity toward halogenation with Br_2 and I_2, presumably due to the sterically condensed structure around the –C≡C– bond.

Table 1 PAEs with sulfur-containing heteroaromatic rings (C_nH_{2n+1} is n-alkyl group)

no.	monomer		conditions		molecular weight[a]	absorption / nm		emission / nm		ref.
	diyne	dihalide	catalyst	solvent		solution	film	solution	film	
1		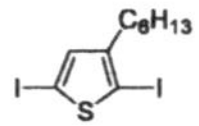	$Pd(PPh_3)_4$ CuI	Et_3N	$M_w = 96000$[b]	403	not measured	455, 483	not measured	39-41
2		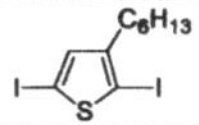	$Pd(PPh_3)_4$ CuI	toluene Et_3N	$M_n = 32000$ $M_w = 51000$	418	not measured	459, (485)	not measured	39-41
3	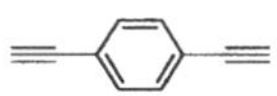	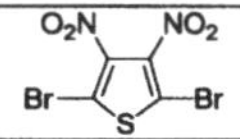	$Pd(PPh_3)_4$ CuI	Et_3N	not measured	not measured	not measured	not measured	not measured	42
4	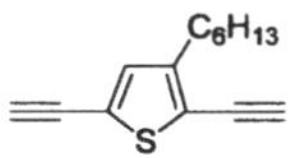	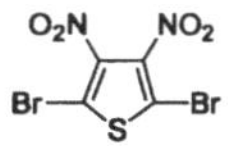	$Pd(PPh_3)_4$ CuI	Et_3N	not measured	not measured	not measured	not measured	not measured	42
5	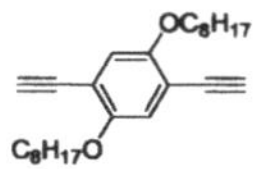	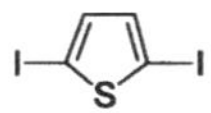	$PdCl_2(PPh_3)_2$ CuI	toluene Et_3N	$M_w = 43900$ PDI = 2.12	444	not measured	482, 514	not measured	43
6	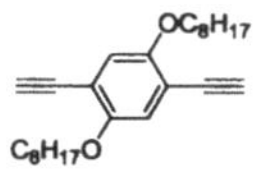	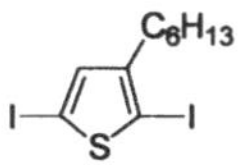	$PdCl_2(PPh_3)_2$ CuI	toluene Et_3N	$M_w = 61600$ PDI = 2.15	460	not measured	489, 520	not measured	43
7	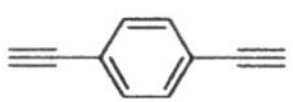	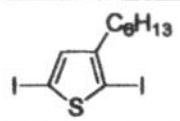	$PdCl_2(PPh_3)_2$ CuI	toluene Et_3N	$M_w = 78500$ PDI = 2.27	414	not measured	451, 478	not measured	43
8	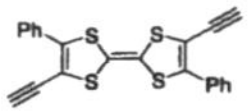	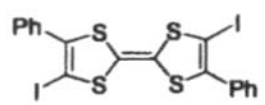	$Pd(PPh_3)_4$ CuI	THF Et_3N	$M_n = 2320$ $M_w = 4700$	502	not measured	not measured	not measured	44
9	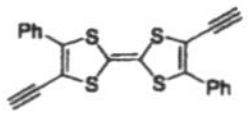	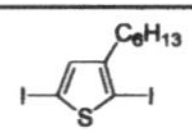	$Pd(PPh_3)_4$ CuI	THF Et_3N	$M_w = 610000$[b]	not measured	not measured	not measured	not measured	44

[a] Determined by GPC; [b] Determined by light scattering method; [c] dppf=1,1′-Diphenylphosphinoferrocene.

Table 1 (continued)

no.	monomer		conditions		molecular weight[a]	absorption / nm		emission / nm		ref.
	diyne	dihalide	catalyst	solvent		solution	film	solution	film	
10	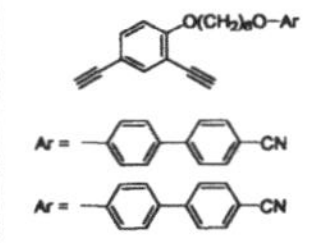	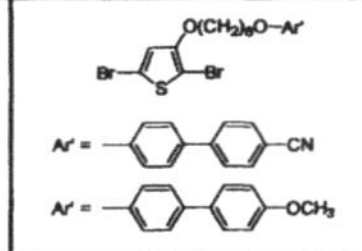	$Pd(PPh_3)_4$ CuI	toluene Et_3N	$M_n = 15000$ $M_n = 14500$	not measured	not measured	not measured	not measured	45
11	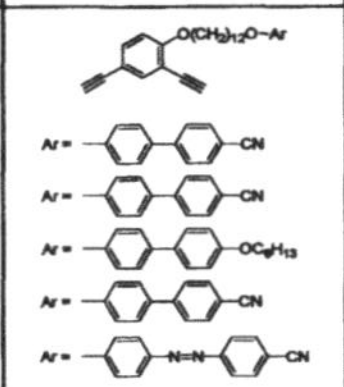	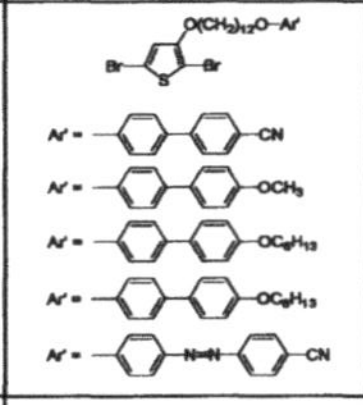	$Pd(PPh_3)_4$ CuI	toluene Et_3N	$M_n = 10000$ $M_n = 10800$ $M_n = 26800$ $M_n = 18500$ $M_n = 6100$	not measured	not measured	not measured	not measured	45
12	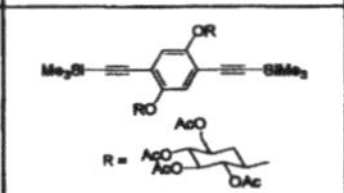	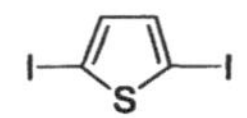	$Pd(PPh_3)_4$ Ag_2O	THF	$M_n = 2600$ $M_w = 16000$	427	not measured	472	not measured	46
13	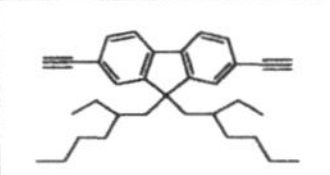	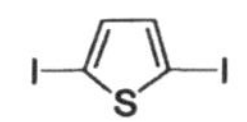	$Pd(PPh_3)_4$ CuI	toluene Et_3N	$M_w = 19700$ PDI = 2.26	312	311	447	519	47
14	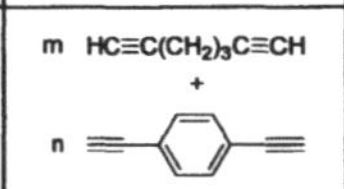	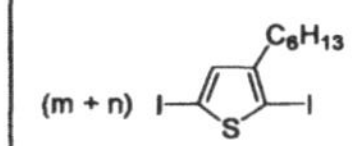	$Pd(PPh_3)_4$ CuI	Et_3N	m, M_n 0.5, 12000 0.25, 11000	383, 400 (383), 402	not measured	453, (484) 458, (486)	not measured	48
15	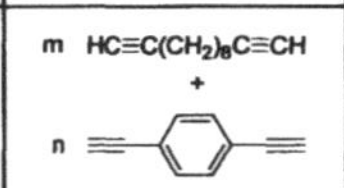	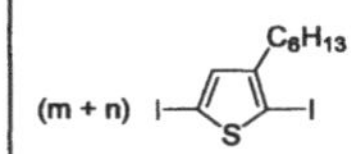	$Pd(PPh_3)_4$ CuI	Et_3N	m, M_n 0.5, 12000 0.25, 11000	382, 402 (382), 403	not measured	not measured	not measured	48

Table 1 (continued)

no.	monomer		conditions		molecular weight[a]	absorption / nm		emission / nm		ref.
	diyne	dihalide	catalyst	solvent		solution	film	solution	film	
16	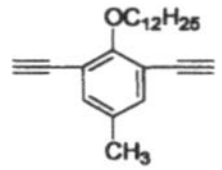	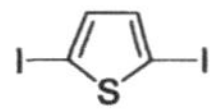	$PdCl_2(dppf)^c \cdot CH_2Cl_2$ CuI	$^iPr_2NH$	$M_n = 3560$ PDI = 2.35	365	not measured	410	not measured	49
17	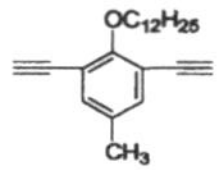	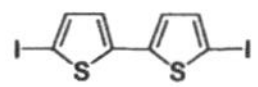	$PdCl_2(dppf)^c \cdot CH_2Cl_2$ CuI	$^iPr_2NH$	$M_n = 8360$ PDI = 3.29	388	not measured	451	not measured	49
18	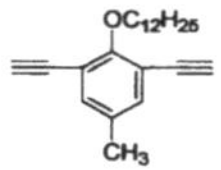	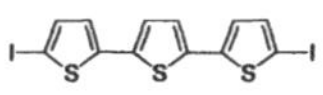	$PdCl_2(dppf)^c \cdot CH_2Cl_2$ CuI	$^iPr_2NH$	$M_n = 4460$ PDI = 5.95	407	not measured	485	not measured	49
19	$Bu_3Sn-C \equiv C-H$	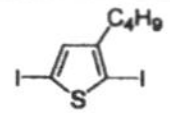	1) $Pd(PPh_3)_4$ 2) LDA	dioxane	$M_n = 5420$ $M_w = 15200$	414	405	550	630	37,38
20	$Bu_3Sn-C \equiv C-H$	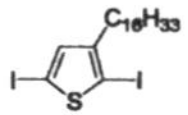	1) $Pd(PPh_3)_4$ 2) LDA	dioxane	$M_n = 10200$ $M_w = 24500$	413	413	538	595	37,38
21	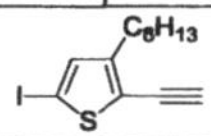		$PdCl_2(PPh_3)_2$ CuI	toluene Et_3N	$M_w = 20100$ PDI = 1.9	440	486	506, 536	575	50
22	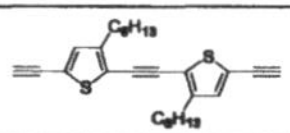	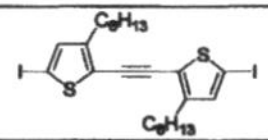	$PdCl_2(PPh_3)_2$ CuI	toluene Et_3N	$M_w = 116400$ PDI = 7.8	436	488	505, 535	575	50
23	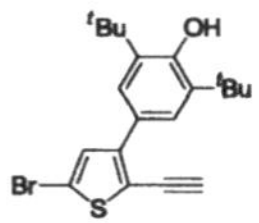		$Pd(PPh_3)_4$ CuI	Et_3N	$M_n = 7700$	460	not measured	535	not measured	51

Table 1 (continued)

no.	monomer		conditions		molecular weight[a]	absorption / nm		emission / nm		ref.
	diyne	dihalide	catalyst	solvent		solution	film	solution	film	
24	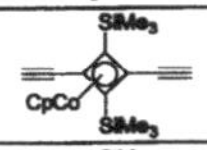	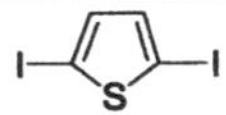	$PdCl_2(PPh_3)_2$ CuI	piperidine	$M_n = 8100$ $M_w = 12000$	not measured	not measured	not measured	not measured	52
25	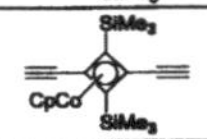	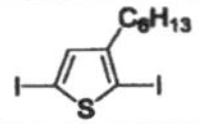	$PdCl_2(PPh_3)_2$ CuI	piperidine	$M_n = 14000$ $M_w = 30000$	not measured	not measured	not measured	not measured	52
26	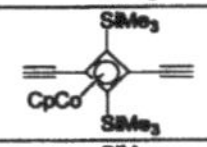	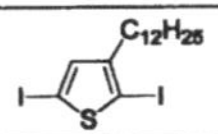	$PdCl_2(PPh_3)_2$ CuI	piperidine	$M_n = 5180$ $M_w = 11350$	not measured	not measured	not measured	not measured	52
27	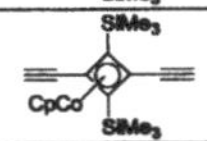	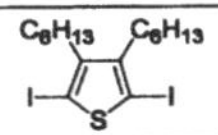	$PdCl_2(PPh_3)_2$ CuI	piperidine	$M_n = 9600$ $M_w = 29000$	not measured	not measured	not measured	not measured	52
28	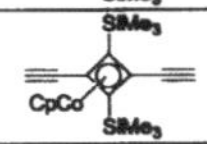	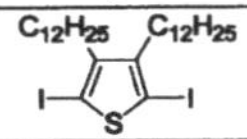	$PdCl_2(PPh_3)_2$ CuI	piperidine	$M_n = 13400$ $M_w = 35300$	not measured	not measured	not measured	not measured	52
29	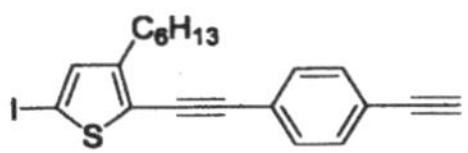		$PdCl_2(PPh_3)_2$ CuI	Et_3N	$M_w = 37900$ PDI = 1.43	414	408	453, 479	525	53
30	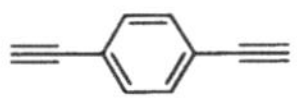	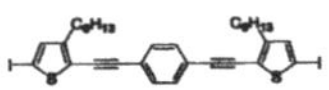	$PdCl_2(PPh_3)_2$ CuI	Et_3N	$M_w = 47200$ PDI = 1.98	415	431	453, 478	534	53
31	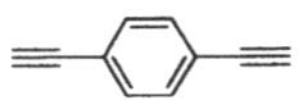	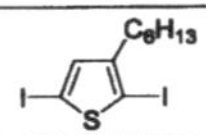	$PdCl_2(PPh_3)_2$ CuI	Et_3N	$M_w = 31100$ PDI = 2.10	414	421	451, 478	530	53
32	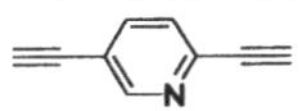	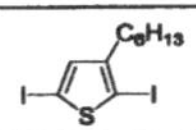	$Pd(PPh_3)_4$ CuI	Et_3N	$M_w = 480000$[b]	426	432	465, 495	592	40,54
33	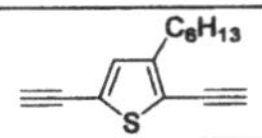	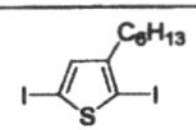	$Pd(PPh_3)_4$ CuI	Et_3N	$M_w = 190000$[b]	438	446	510, 538	529, 558	54
34	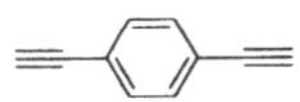	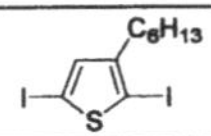	$Pd(PPh_3)_4$ CuI	Et_3N	$M_w = 96000$[b]	403	410	455, 485	582	40,54

Table 1 (continued)

no.	monomer		conditions		molecular weight[a]	absorption / nm		emission / nm		ref.
	diyne	dihalide	catalyst	solvent		solution	film	solution	film	
35		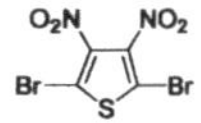	$Pd(PPh_3)_4$ CuI	Et_3N	$M_w = 690000$[b]	360	350	430	not measured	54
36	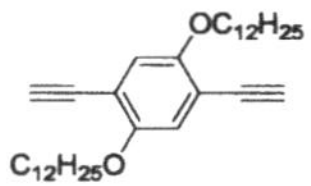	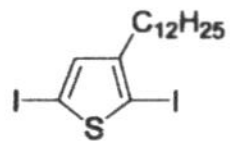	$Pd(PPh_3)_4$ CuI	THF $^iPr_2NH$	$M_n = 320000$ PDI = 2.8	448	not measured	496	not measured	55
37	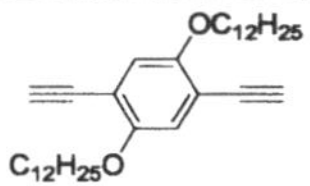	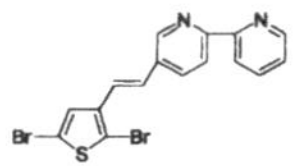	$Pd(PPh_3)_4$ CuI	THF $^iPr_2NH$	$M_n = 18000$ PDI = 2.5	344, 462	not measured	508	not measured	55
38	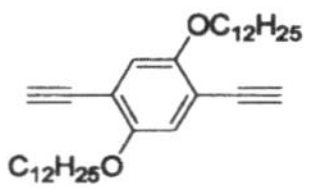	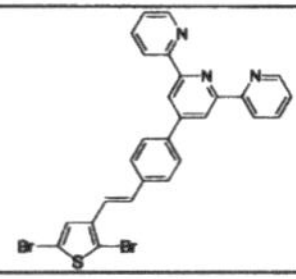	$Pd(PPh_3)_4$ CuI	THF $^iPr_2NH$	$M_n = 170000$ PDI = 1.4	338, 454	not measured	508	not measured	55
39	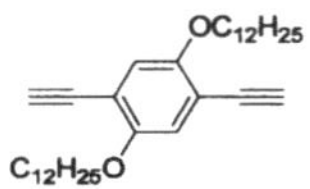	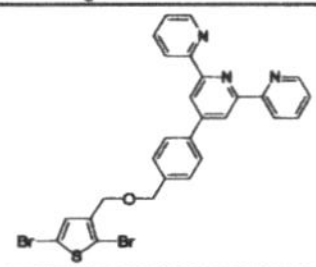	$Pd(PPh_3)_4$ CuI	THF $^iPr_2NH$	$M_n = 190000$ PDI = 1.5	276, 444	not measured	488	not measured	55
40	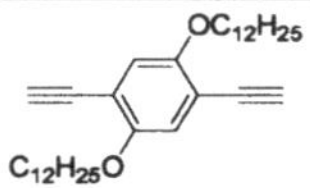	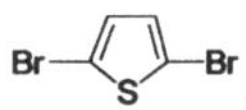	$PdCl_2$ PPh_3 $Cu(OAc)_2$	THF Et_3N	not measured	433	450	not measured	not measured	56
41	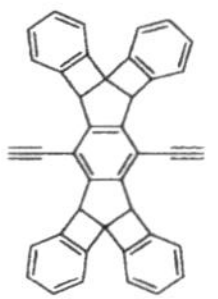	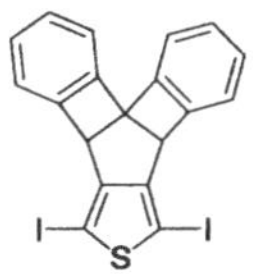	$Pd(PPh_3)_4$ CuI	toluene $^iPr_2NH$	$M_n = 22000$ $M_w = 37000$	405	420	458	460	57

The –C≡C– group is a typical electron-accepting group in organic chemistry [17]. Consequently, PAEs are active in electrochemical reduction at about –2 V vs Ag^+/Ag [40, 54].

$$\text{PAE} + xe^- + xC^+ \longrightarrow \text{n-doped PAE} \qquad (12)$$

C^+ = cation

The electronic properties of π-conjugated polymers reflect well the basic electron-withdrawing or electron-donating properties of the components of the π-conjugated polymer [62]. In view of the electrochemical reduction potential, the thiophene unit and tetrathiafulvalene unit (Nos. 8 and 9 in Table 1) have a similar electronic effect in PAEs. It is reported that poly(arylenevinylene)s are also susceptible to electrochemical reduction [63, 64]. Due to the electron-accepting properties, PAEs are usually inert in electrochemical and chemical (e.g., by I_2 [54]) oxidation.

Pang, Li, and Barton indicated, based on viscosity data, that PAEs had a stiff structure in solution [43]. A PAE shows a Mark–Houwink constant as high as α=1.92 [43], revealing that it has a very stiff structure, similar to rodlike poly(pyridine-2,5-diyl) [65, 66]. Liquid crystalline behavior of PAEs with a mesogen has been reported [45]. The chemical properties of PAEs have also been investigated with oligomeric model compounds [67–76].

2.3 Optical Properties

As shown in Table 1, π-conjugated PAEs usually show a UV–vis absorption peak in the range of 400–460 nm in solution. In the case of π-conjugated poly(arylene)s, the UV–vis absorption peak is strongly influenced by the presence or absence of *o*-H or *o*-substituent repulsion. Poly(*p*-phenylene) receives the *o*-hydrogen repulsion to form a twisted main chain and gives a peak at about 380 nm, whereas the *o*-H repulsion is weak in poly(thiophene-2,5-diyl). Molecular modeling indicates that poly(thiophene-2,5-diyl) can form a coplanar structure, and this polymer gives the UV–vis peak at a longer wavelength (about 450 nm) [16]. Poly(pyrazine-2,5-diyl), which receives no *o*-H repulsion, shows a UV–vis peak at 439 nm [77].

In the case of PAEs, the *o*-H and *o*-substituent repulsion does not seem to be important due to the presence of the spacing –C≡C– group. When a 3,4-dinitrothiophene unit is incorporated, PAEs may receive the *o*-substituent repulsion, and the polymer exhibits the UV–vis peak at a shorter wavelength (e.g., No. 35). Non-π-conjugated PAEs, such as Nos. 16 and 17 in Table 1, also exhibit the UV–vis peak at a shorter wavelength. PAEs are usually photoluminescent, and show an emission peak agreeing with the onset position of the UV–vis peak in solution, similar to the cases of aromatic compounds and aromatic polymers. Pang, Li, and Barton reported high quantum yields of 37–48% for photoluminescence of a PAE consisting of *p*-phenylene and thiophene-2,5-diyl units [43]. Weder reported that photoenergy captured by photosensitizers was transferred to PAEs, and PAEs emitted light polarized along the direction of the PAE main chain [78]. In the solid state, PAEs show excimer-like emission [54], which is shifted to a longer wavelength, due to the presence of a strong intermolecular interaction in the solid.

When a unit with a shorter π-conjugation length and a unit with a longer π-conjugation length are connected through a non-π-conjugated unit (e.g., Nos. 14 and 15), the photoenergy captured by the unit with the shorter π-conjugation length is transferred to the unit with the longer π-conjugation length and a smaller π–π* transition energy [48] (Scheme 2). Similar photoenergy transfers in energy gradient systems [79, 80] and a light-harvesting model system [81] have been reported. Swager pointed out the applicability of the photoluminescent property of PAEs for sensing nitro compounds such as TNT [82].

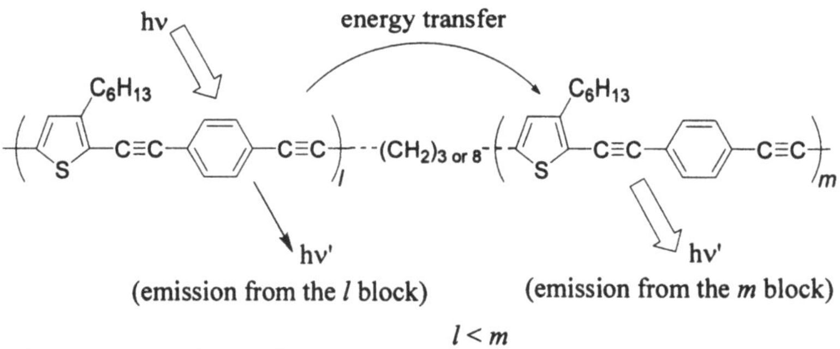

Scheme 2 Energy transfer in a PAE

Due to the presence of a highly expanded π-conjugation system, π-conjugated PAEs show a large optical third-order nonlinear susceptibility, $\chi^{(3)}$, of $4\text{–}5\times10^{-11}$ esu as measured by third-order harmonic generation (THG) [54]. The rapid switching and effective manipulation of light using π-conjugated polymers will be the key to constructing future optical communication systems and devices such as optical computers. π-Conjugated polymers can undergo refractive index changes at a timescale of about 10^{-12} s under irradiation with laser pulses. The effective manipulation of communicating light (or larger changes in the refractive index with weaker pumping light) requires polymers with large $\chi^{(3)}$ values, which is related to a degree of change in the refractive index per power of pumping (or manipulating) light [83, 84]. The first waveguide based on π-conjugated polymer was constructed with a PAE, and the waveguide worked as expected [85]. Electroluminescent devices using π-conjugated polymers are the subject of recent interest, and the devices using PAEs have been shown in papers [47, 86, 87] and Bunz's review [27].

3
PAEs with Nitrogen-Containing Heteroaromatic Rings

Table 2 shows PAEs with nitrogen-containing heteroaromatic rings. Le Moigne's paper (Nos. 25 and 26 in Table 2) mainly reports PAEs with nitrogen-containing heteroaromatic rings; however, it also reports a PAE with a thiophene ring [100]. This paper describes the morphology of the PAEs. In organic chemistry, the imine –CH=N– group is known as an electron-withdrawing group. Consequently, the heteroaromatic ring containing the imine group is an electron-accepting ring, and π-conjugated homopolymer constituents of the ring undergo chemical and electrochemical reduction or n-type doping. The number of imine groups affects the ease of the n-doping (or the reduction potential E_{red}) [33, 62] (Scheme 3). In

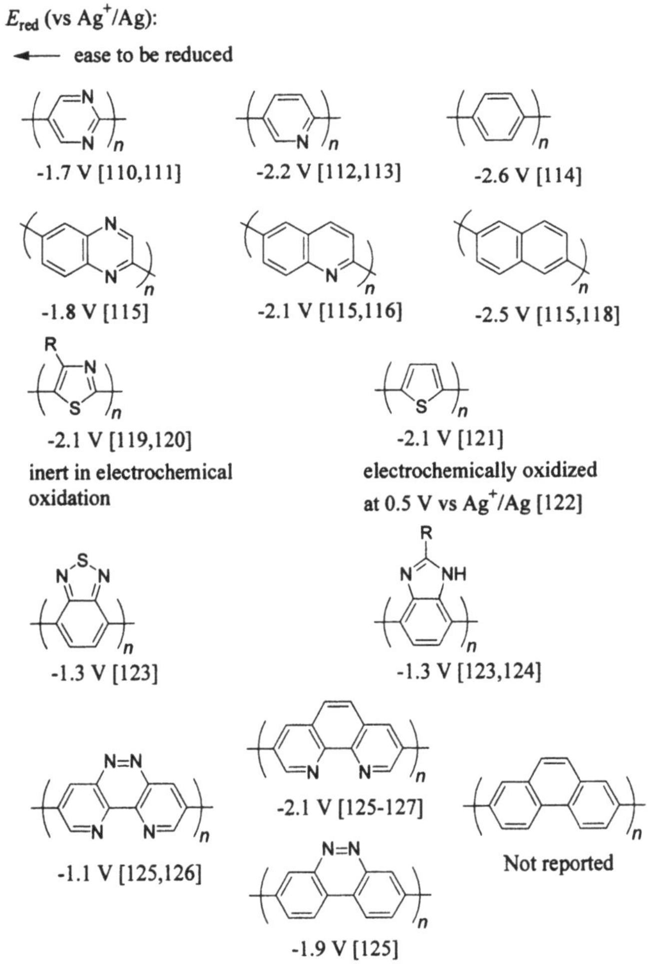

Scheme 3 Electrochemical reduction (n-doping) of π-conjugated heteroaromatic polymers. Effect of the imine group

Table 2 PAEs with nitrogen-containing heteroaromatic rings

no.	monomer		conditions		molecular weight[a]	absorption / nm		emission / nm		ref.
	diyne	dihalide	catalyst	solvent		solution	film	solution	film	
1	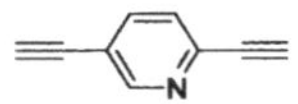	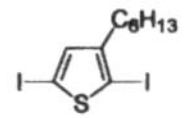	$Pd(PPh_3)_4$ CuI	toluene Et_3N	$M_n = 32000$ $M_w = 51000$	418	not measured	459, (485)	not measured	39-41, 54
2	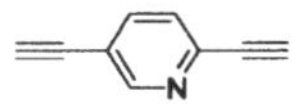	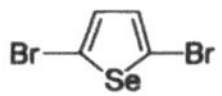	$Pd(PPh_3)_4$ CuI	toluene Et_3N	insoluble	not measured	not measured	not measured	not measured	39-41, 54
3	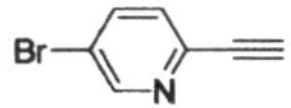		$Pd(PPh_3)_4$ CuI	toluene Et_3N	$M_w = 210000$[b]	not measured	not measured	not measured	not measured	39-41, 54
4	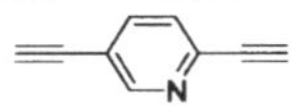	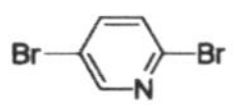	$Pd(PPh_3)_4$ CuI	toluene Et_3N	not measured	340	not measured	430	not measured	39-41, 54
5	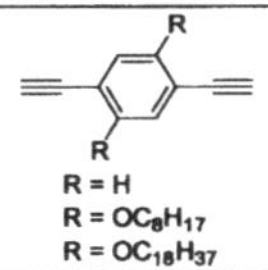	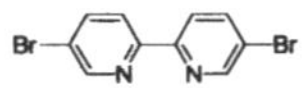	$Pd(PPh_3)_4$ CuI	toluene Et_3N	M_n, PDI 1250[d] - 6900, 1.95 8540, 2.70	357 413 420	not measured	388 456 454	not measured	88,106
6	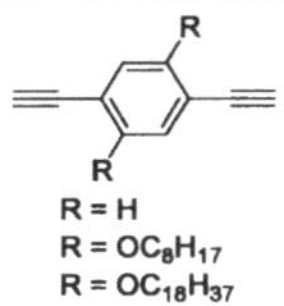	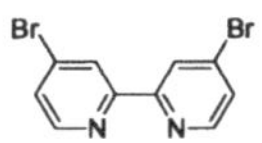	$Pd(PPh_3)_4$ CuI	toluene Et_3N	M_n, PDI 1360[d] - 12300, 2.57 29800, 2.80	329 387 386	not measured	389 446 445	not measured	106
7	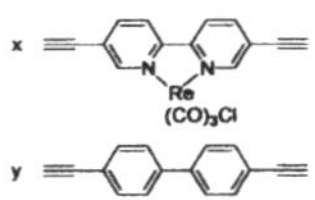	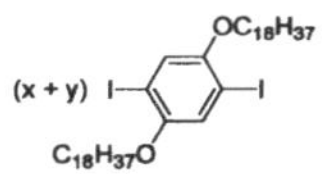	$Pd(PPh_3)_4$ CuI	THFe $^iPr_2NH$	x, M_n, PDI 0, 13500, 2.7 0.1, 8800, 2.3 0.25, 7900, 2.1 0.5, 7800, 2.0	400 400, 465 400, 469 388, 469	not measured	435 435 435 432	not measured	107, 108

[a] Determined by GPC; [b] Determined by light scattering method; [c] DMF-soluble part; [d] Determined by elemental analysis.

Table 2 (continued)

no.	monomer		conditions		molecular weight[a]	absorption / nm		emission / nm		ref.
	diyne	dihalide	catalyst	solvent		solution	film	solution	film	
8	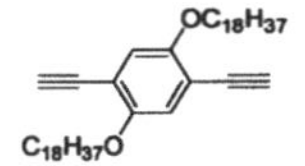	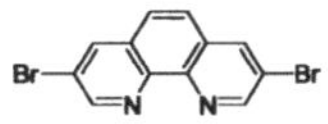	$Pd(PPh_3)_4$ CuI	toluene $^iPr_2NH$	$M_n = 8700$	413	not measured	452	not measured	90
9	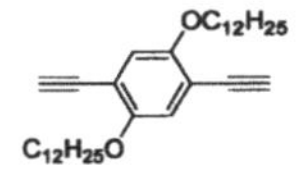	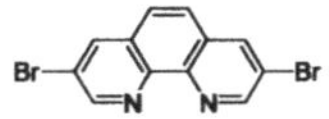	$Pd(PPh_3)_4$ CuI	toluene $^iPr_2NH$	$M_n = 6900$	414	not measured	460	not measured	90
10	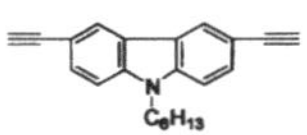	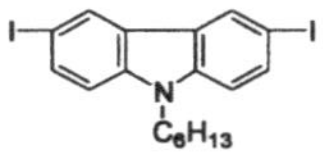	$PdCl_2$ PPh_3 $Cu(OAc)_2$	$^iPr_2NH$	$M_w = 9700$	not measured	not measured	not measured	not measured	89
11	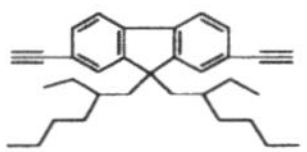	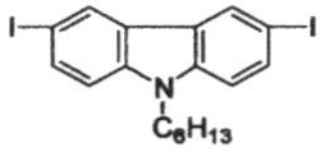	$Pd(PPh_3)_4$ CuI	toluene Et_3N	$M_w = 22500$ PDI = 2.04	297	300	470	525	47
12	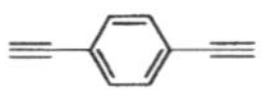	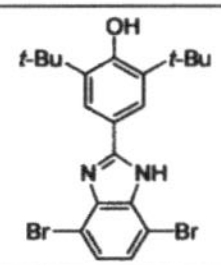	$Pd(PPh_3)_4$ CuI	DMF Et_3N	$M_n = 15100$[c]	363	not measured	463	not measured	91
13	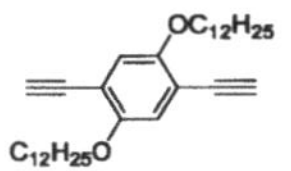	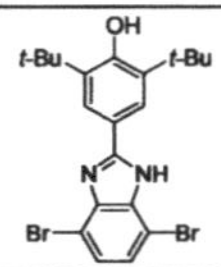	$Pd(PPh_3)_4$ CuI	DMF Et_3N	$M_n = 3900$[c]	410	not measured	511	not measured	91
14	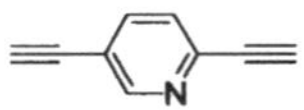	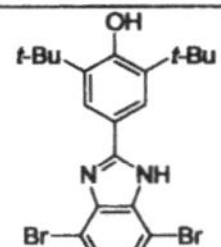	$Pd(PPh_3)_4$ CuI	DMF Et_3N	$M_n = 5100$[c]	397	not measured	478	not measured	91

Table 2 (continued)

no.	monomer		conditions		molecular weight[a]	absorption / nm		emission / nm		ref.
	diyne	dihalide	catalyst	solvent		solution	film	solution	film	
15	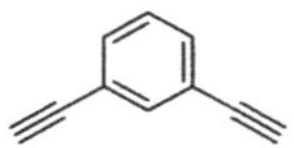	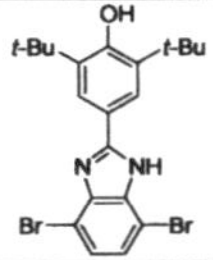	$Pd(PPh_3)_4$ CuI	DMF Et_3N	M_n = 33000	353	362	463	not measured	91
16	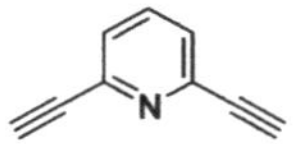	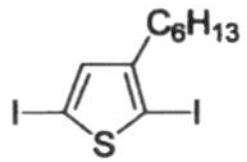	$Pd(PPh_3)_4$ CuI	toluene Et_3N	M_n = 5700	not measured	not measured	not measured	not measured	41
17	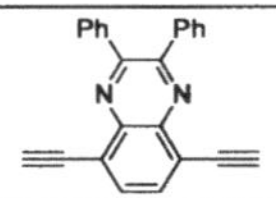	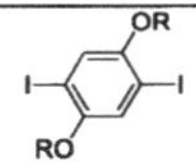 R = 2-ethylhexyl	$PdCl_2(PPh_3)_2$ CuI	piperidine	DP = 110	462	534	538	not measured	92
18	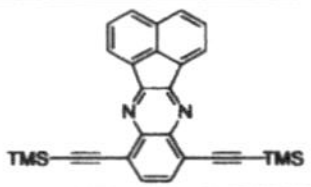	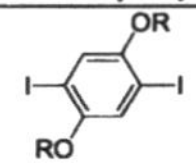 R = 2-ethylhexyl	$PdCl_2(PPh_3)_2$ CuI	piperidine THF/KOH in EtOH	DP = 36	474	not measured	540	not measured	92
19	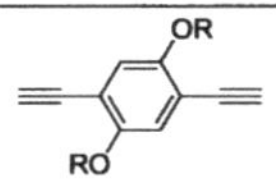 R = C_6H_{13} R = C_9H_{19} R = $C_{12}H_{25}$ R = $C_{18}H_{37}$	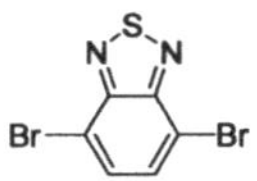	$Pd(PPh_3)_4$ CuI	toluene Et_3N	M_n, PDI 12600, 3.4 30000, 5.3 18400, 2.7 13200, 2.9	512 514 513 512	564 564 561 560	564 565 565 565	673 673 673 673	93-95
20	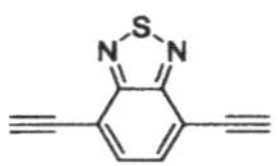	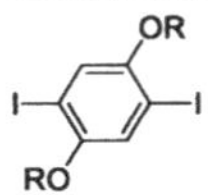 R = 2-ethylhexyl	$PdCl_2(PPh_3)_2$ CuI	piperidine	DP = 18 PDI = 2.78	498	508, 553	562	not observed	96,97
21	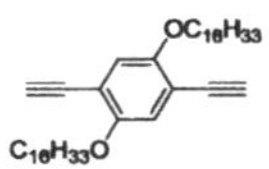	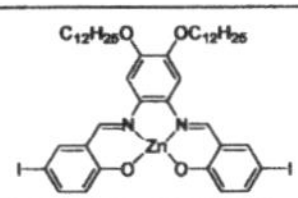	$Pd(PPh_3)_4$ CuI	THF $^iPr_2NH$	M_n = 12300 M_w = 37000	407, 475	not measured	546	not measured	98

Table 2 (continued)

no.	monomer		conditions		molecular weight[a]	absorption / nm		emission / nm		ref.
	diyne	dihalide	catalyst	solvent		solution	film	solution	film	
22	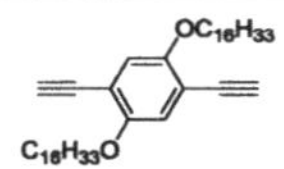	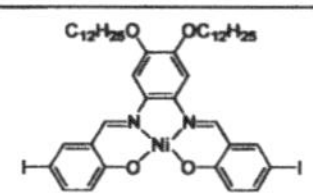	$Pd(PPh_3)_4$ CuI	THF $^iPr_2NH$	$M_n = 8100$ $M_w = 17000$	407, 509	not measured	not observed	not measured	98
23	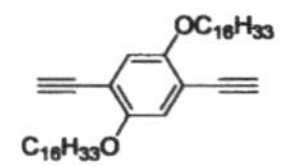	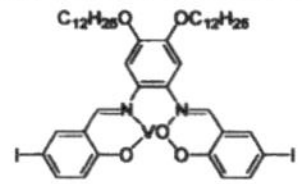	$Pd(PPh_3)_4$ CuI	THF $^iPr_2NH$	$M_n = 21500$ $M_w = 84000$	396, 481	not measured	not observed	not measured	98
24	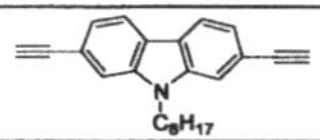	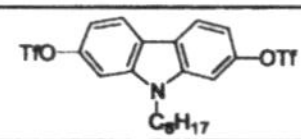	$Pd(PPh_3)_4$ CuI	toluene Et_3N	$M_n = 2700$ $M_w = 5400$	380	not measured	not measured	not measured	99
25	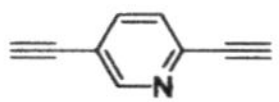	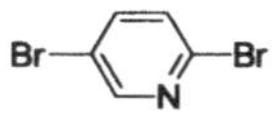	$PdCl_2$ PPh_3 CuI	Et_3N	$M_w = 3760$ PDI = 1.21	327	362	not measured	not observed	100
26	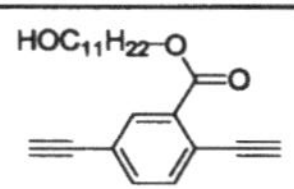	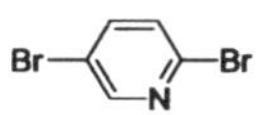	$PdCl_2$ PPh_3 CuI	Et_3N	$M_w = 14500$ PDI = 2.96	371	390	not measured	502	100
27	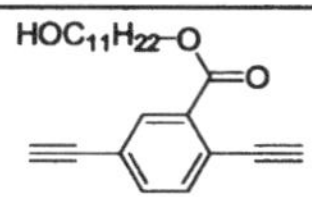	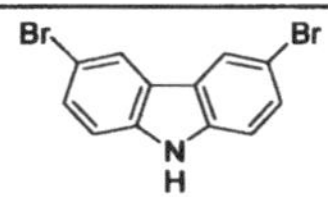	$PdCl_2$ PPh_3 CuI	Et_3N	$M_w = 4570$ PDI = 1.42	378	404	not measured	563	100
28	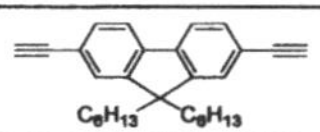	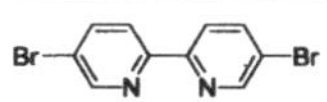	$Pd(PPh_3)_4$ CuI	toluene $^iPr_2NH$	$M_w = 113500$ PDI = 1.92	409, (392)	425, (395)	424, 444	477, 503, (439)	101
29	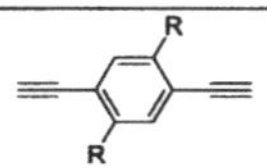 R = 2-ethylhexyl	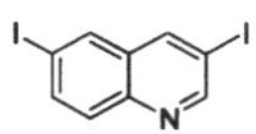	$PdCl_2(PPh_3)_2$ CuI	piperidine	DP = 37 PDI = 3.3	380	not measured	399, 417	not measured	102

Table 2 (continued)

no.	monomer		conditions		molecular weight[a]	absorption / nm		emission / nm		ref.
	diyne	dihalide	catalyst	solvent		solution	film	solution	film	
30	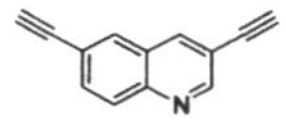	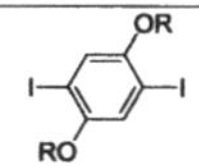 R = 2-ethylhexyl	$PdCl_2(PPh_3)_2$ CuI	piperidine	DP = 95 PDI = 3.1	401	not measured	437	not measured	102
31	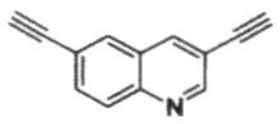	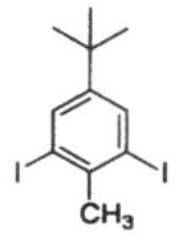	$PdCl_2(PPh_3)_2$ CuI	piperidine	DP = 11 PDI = 1.8	310, 368	not measured	381	not measured	102
32	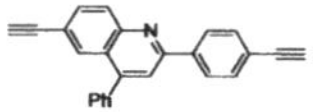	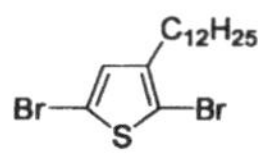	$PdCl_2(PPh_3)_2$ CuI	THF $^iPr_2NH$	M_n = 5200 M_w = 9700	430	449	508	605	103
33	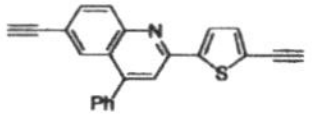	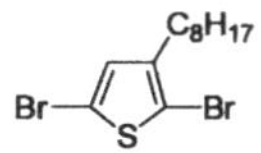	$PdCl_2(PPh_3)_2$ CuI	THF $^iPr_2NH$	M_n = 7500 M_w = 14600	412	424	470	560	103
34	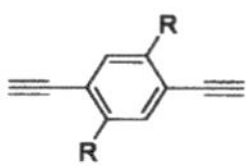 R = H R = OC_6H_{25} R = $OC_{12}H_{25}$ R = $OC_{18}H_{37}$	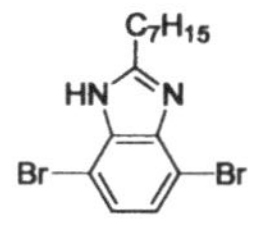	$Pd(PPh_3)_4$ CuI	dioxane Et_3N	not measured	362 396 402 367	- 401 439, (463) 376, (410)	not measured	not measured	105
35	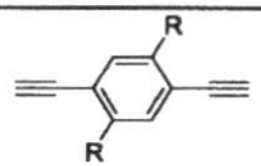 R = H R = OC_6H_{25} R = $OC_{12}H_{25}$ R = $OC_{18}H_{37}$ R = $C_{12}H_{25}$	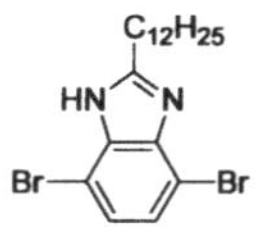	$Pd(PPh_3)_4$ CuI	dioxane Et_3N	not measured	364 409 401 381 390	- 423, (460) 438, (465) 385 403, (462)	not measured	not measured	105

Table 2 (continued)

no.	monomer		conditions		molecular weight[a]	absorption / nm		emission / nm		ref.
	diyne	dihalide	catalyst	solvent		solution	film	solution	film	
36	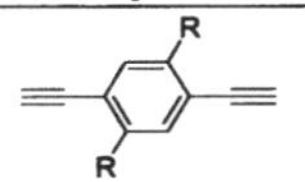 R = H R = OC_6H_{25} R = $OC_{12}H_{25}$ R = $OC_{18}H_{37}$	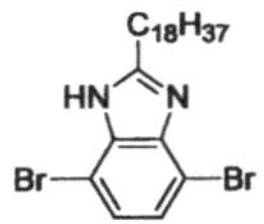	$Pd(PPh_3)_4$ CuI	dioxane Et_3N	not measured	362 422 421 418	- - - 419, (463)	not measured	not measured	105
37	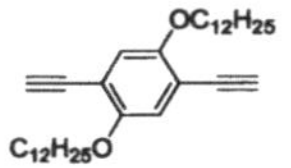	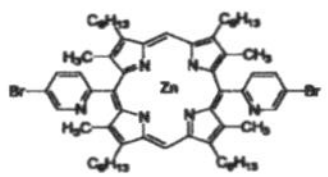	$Pd(PPh_3)_4$ CuI	toluene Et_3N	not measured	B:415,427 Q:544,602	not measured	591 643	not measured	104
38	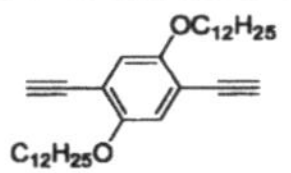	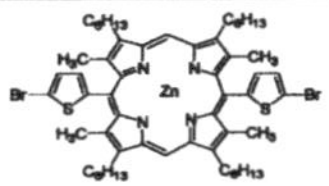	$Pd(PPh_3)_4$ CuI	toluene Et_3N	$M_n = 12200$ $M_w = 21500$	B:417,431 Q:547,585	B:418,433 Q:551,590	591 646	not measured	104
39	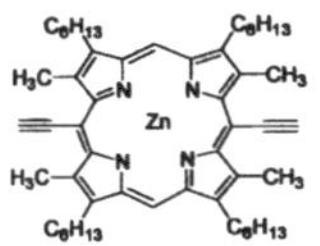	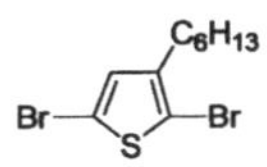	$Pd(PPh_3)_4$ CuI	toluene Et_3N	$M_n = 4800$ $M_w = 5800$	B:454,490 Q:697	not measured	724	not measured	104
40	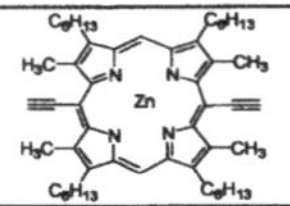	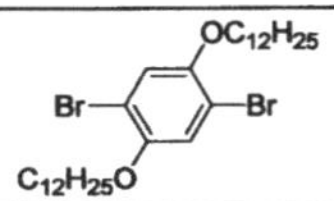	$Pd(PPh_3)_4$ CuI	toluene Et_3N	$M_n = 4600$ $M_w = 6300$	B:448,490 Q:645	B:434,500 Q:670	695	not measured	104
41	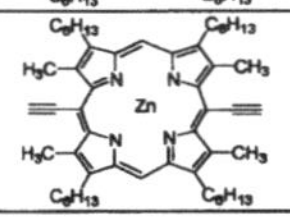	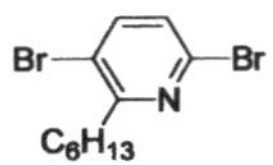	$Pd(PPh_3)_4$ CuI	toluene Et_3N	$M_n = 6500$ $M_w = 9300$	B:440 Q:645	B:440 Q:655	693	not measured	104
42		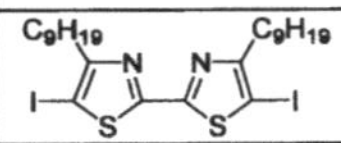	$Pd(dba)_2$ PPh_3	toluene	$M_n = 2100$	475	503, (553)	532	600	109

contrast to the imine-containing rings, the carbazole unit such as those shown in Nos. 10, 11, and 24 in Table 2 are considered to be an electron-donating unit.

As depicted in Scheme 3, the 2,1,3-benzothiadiazole is a typical electron-accepting unit, whereas *p*-dialkoxy benzene rings such as those shown in Nos. 8, 19, and 20 are considered to have electron-excessive properties due to the electron-donating OR group. On these bases the PAEs shown in Nos. 19 and 20 have a strong tendency to stack in the solid state as well as in colloidal solutions [93–97]; the cast film of the polymer shines like a film of gold (Scheme 4). Analogous stacking of a PAE with electron-donating *p*-dialkoxybenzene and electron-accepting benzoquinone and dicyanobenzoquinone units was also reported [133].

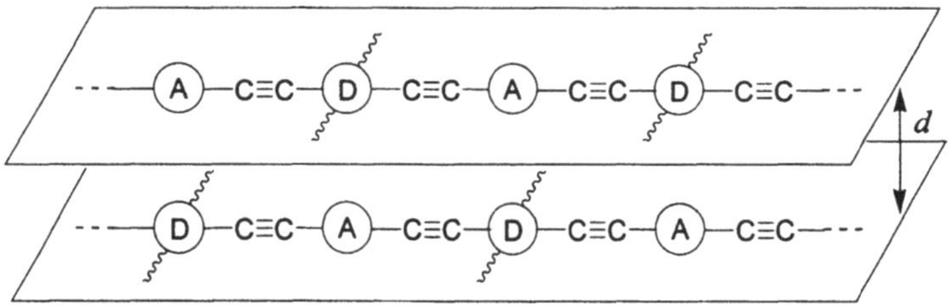

Scheme 4 Proposed stacking structure. The plane-to-plane distance *d* for the polymer shown in No. 19 is about 3.8 Å [93–95], which is somewhat shorter than the *d* value (about 3.85 Å) reported for head-to-tail-type poly(3-alkylthiophene) HT-P3RTh [128, 129] and somewhat longer than the *d* values (3.6–3.65 Å) observed with head-to-head-type poly(4-alkylthiazole-2,5-diyl) [124, 129, 130] and the charge-transfer-type copolymer of thiophene and 4-alkylthiazole [131]. Poly(ethylenedioxythiophene) gives a similar *d* value (3.8 Å) [132]. The existence of S in the aromatic ring seems to make the distance *d* longer

When a π-conjugated polymer has long alkyl or alkoxy side chains and forms the stacked structure, it tends to be aligned on the surface of substrates (e.g., Pt plate) with the alkyl or alkoxy chains oriented toward the surface of the substrate [95, 128, 134, 135], presumably because of a strong tendency for the alkyl chain to stand upright on the surface of substrates [95, 136]. Actually, the PAEs shown in No. 19 in Table 2 are aligned on a Pt plate with the OR chains oriented toward the surface of the Pt plate [95]. A model for the alignment is depicted in Fig. 1.

Control of alignment of π-conjugated polymers on the substrate is important for excellent performance of the polymer in electronic devices (e.g., higher mobility of carrier in field-effect transistors [134, 136]). Details of the molecular structure and molecular assembly of PAEs will be discussed in other chapters.

π-Conjugated polymers with metal complexes are the subject of recent interest [111, 113, 137–148]. PAEs with chelating ligand units such as 2,2′-bipyridyl or metal complex units are shown in Nos. 5–9, 21–23, 28, and 37–41 in Table 2. Recently, PAEs with triphenylamine blocks have also been reported [149].

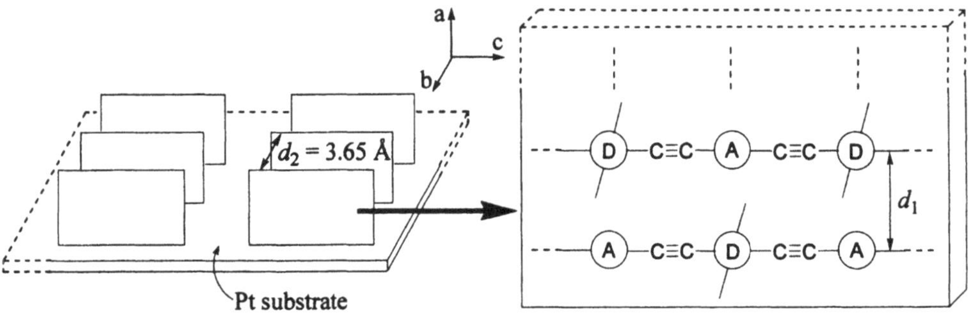

Fig. 1 Model for alignment of PAE No. 19 in Table 2 on the surface of a Pt plate. The powder XRD pattern indicates both d_1 and d_2 peaks, whereas the XRD pattern of a cast film, measured with a reflection mode, gives only the d_1 peak: d_1=(1.18 n+10) Å (n=number of carbon atoms in the OR group) [93, 95]

4
PAEs Containing Other Elements (Si, Fe, and B)

Table 3 lists PAEs containing Si, Fe, and B. Silole containing Si in the five-membered ring [157] has a low π–π* transition energy because of its low LUMO level [158], and its homopolymers have been prepared [159, 160]. 2,5-Dilithiosiloles, which can be prepared by the ring-closing reaction of diethynylsilane, serve as starting materials for various 2,5-disubstituted siloles such as 2,5-dibromosiloles and 2,5-di(trialkylstannyl)siloles [161, 162]. PAEs with the silole ring have been reported as shown in Nos. 1–4 in Table 3. PAE-type polymers with Si atoms in the main chain have also been prepared (Nos. 12–16), and their optical properties including photoconductivity have been revealed [155, 156].

PAEs containing ferrocene units (Nos. 5–9) are electrochemically active due to the redox reaction of the ferrocene unit, and cyclic voltammetry (CV) of the polymers suggests the presence of electronic interaction between the ferrocene units through the π-conjugation system [152]. The CV current (i) obtained with solutions of the ferrocene-containing PAEs is proportional to the square root of the scanning rate ($\nu^{1/2}$), indicating that diffusion of the polymer molecule in the solution is a determining factor for i, similar to redox-active compounds with low molecular weights [152, 153]. Chujo reported that a coupling reaction between B–OMe and –C≡CLi bonds could be applied to the synthesis of PAE-type polymers consisting of a –C≡C–B(Ar)– (Ar=arylene) unit [154]. UV–vis data indicated expansion of the electron system across the –B(Ar)– unit, and the polymer showed stability toward air and moisture. Chujo has been concerned with the synthesis of boron-containing polymers by hydroboration of dienes [163].

Table 3 PAEs containing other elements (Si, Fe, and B)

no.	monomer		conditions		molecular weight[a]	absorption / nm		emission / nm		ref.
	diyne	dihalide	catalyst	solvent		solution	film	solution	film	
1	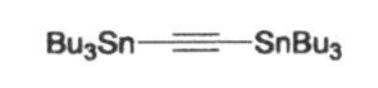	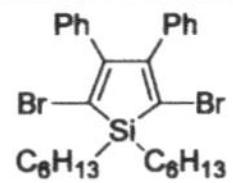	$Pd_2(dba)_3$[b] $P(furyl)_3$	THF	insoluble	not measured	not measured	not measured	not measured	150, 151
2	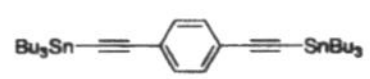	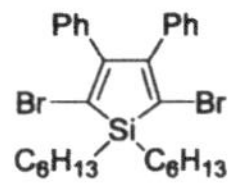	$Pd_2(dba)_3$[b] $P(furyl)_3$	THF	$M_n = 9000$ $M_w = 64000$	505, 527	not measured	not measured	not measured	150, 151
3	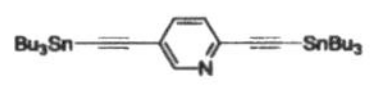	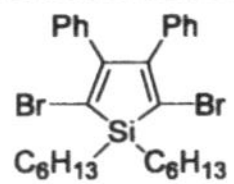	$Pd_2(dba)_3$[b] $P(furyl)_3$	THF	not measured	not measured	not measured	not measured	not measured	150
4	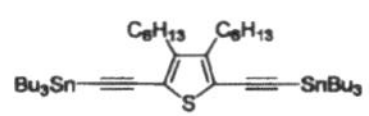	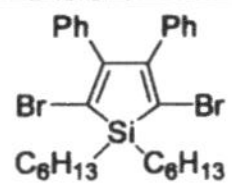	$Pd_2(dba)_3$[b] $P(furyl)_3$	THF	$M_n = 13000$ $M_w = 63000$	576, (605)	not measured	not measured	not measured	150
5	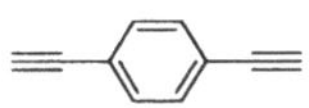	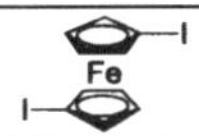	$Pd(PPh_3)_4$ CuI	$^i Pr_2NH$	$M_w = 1500$[c]	325, 415	not measured	not measured	not measured	152, 153
6	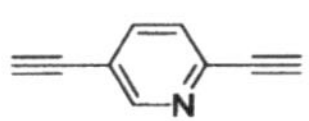	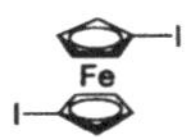	$Pd(PPh_3)_4$ CuI	$^i Pr_2NH$	$M_w = 21000$[c]	332, 460	not measured	not measured	not measured	152, 153

[a] Determined by GPC; [b] Determined by light scattering method; [c] DMF-soluble part; [d] Determined by elemental analysis.

Table 3 (continued)

no.	monomer		conditions		molecular weight[a]	absorption / nm		emission / nm		ref.
	diyne	dihalide	catalyst	solvent		solution	film	solution	film	
7	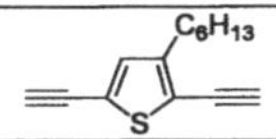		$Pd(PPh_3)_4$ CuI	$^{i}Pr_2NH$	$M_w = 34000$[c]	351, 420	not measured	not measured	not measured	152, 153
8	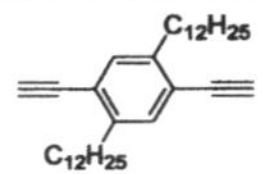	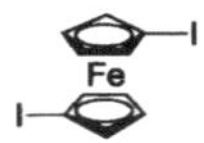	$Pd(PPh_3)_4$ CuI	$^{i}Pr_2NH$	$M_n = 6900$	335, 442	not measured	not measured	not measured	152, 153
9	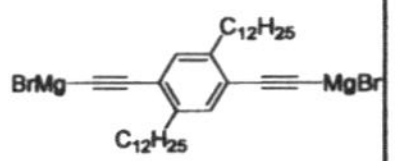	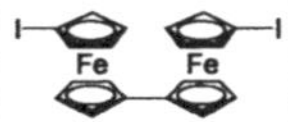	$Pd(OAc)_2$ PPh_3	THF	not measured	313, 465	not measured	not measured	not measured	152
10	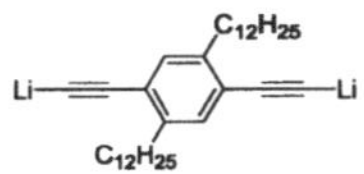	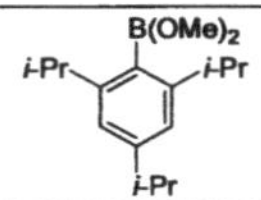	none	THF	$M_n = 2700$ $M_w = 3400$	397	not measured	456	not measured	154
11	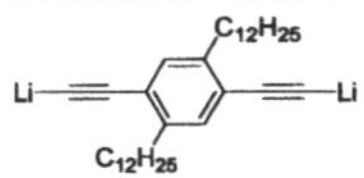	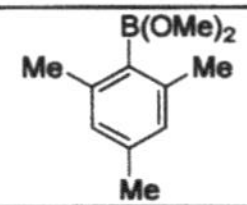	none	THF	$M_n = 1800$ $M_w = 2000$	not measured	not measured	not measured	not measured	154
12	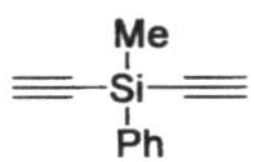	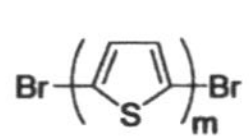 m = 1-3	$Pd(PPh_3)_4$ CuI	toluene Et_3N	m, M_w, PDI 1, 3600, 1.8 2, 2800, 1.6 3, 1200, 1.2	340 400 422	not measured	not measured	not measured	155

Table 3 (continued)

no.	monomer		conditions		molecular weight[a]	absorption / nm		emission / nm		ref.
	diyne	dihalide	catalyst	solvent		solution	film	solution	film	
13	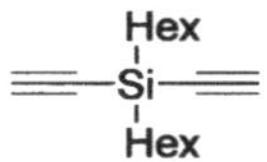	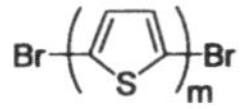 m = 1-3	$Pd(PPh_3)_4$ CuI	toluene Et_3N	m, M_w, PDI 1, 5600, 1.7 2, 4800, 1.6 3, 2000, 1.8	315, 334 378, 400 411, 424	not measured	not measured	not measured	155
14	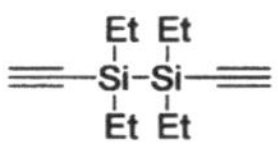	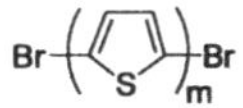 m = 1-4	$Pd(PPh_3)_4$ CuI	toluene Et_3N	m, M_w, PDI 1, 6900, 2.9 2, 6300, 2.9 3, 12000, 2.3 4, 18700, 3.2	340 363, 375 395, 425 425	not measured	not measured	not measured	155
15	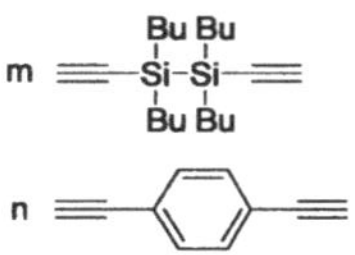	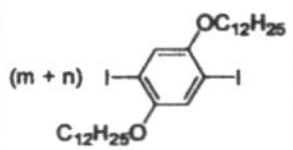	$PdCl_2(PPh_3)_2$ CuI PPh_3	toluene Et_3N	m : n, M_w 1 : 2, 10400 1 : 1, 17000 2 : 1, 12000 1 : 0, 13300	406 398 388 346	424, 445 408 394 -	448, 470 450, 475 445, 470 400	466, 494 446, 488 511 480	156
16	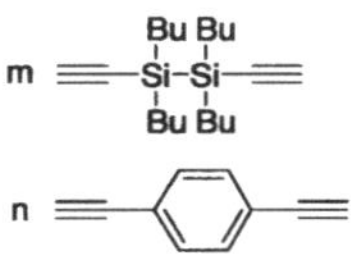	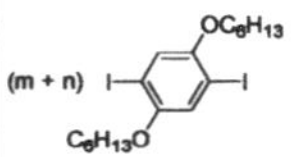	$PdCl_2(PPh_3)_2$ CuI PPh_3	toluene Et_3N	m : n, M_w 1 : 2, 41700 1 : 1, 14200 2 : 1, 5570 1 : 0, 9960	404 398 388 350	not measured	448, 475 445 450 423, 440	not measured	156

5
Summary and Conclusion

Various PAEs can be obtained by choosing the heteroaromatic components in the polymer. They are usually susceptible to electrochemical reduction (n-doping) because of the electron-withdrawing properties of the –C≡C– group. The –C≡C– group undergoes an addition reaction with HBr and receives the hydrogenation reaction. The heteroaromatic PAEs exhibit interesting optical properties such as light emitting properties, similar to those of aromatic hydrocarbon PAEs. Because of their important electronic and optical functionalities, they are expected to support future electronic and optical industries.

References

1. Sonogashira K, Tohda Y, Hagihara N (1975) Tetrahedron Lett 4467
2. Dieck HA, Heck FR (1975) J Organomet Chem 93:259
3. Cassar L (1975) J Organomet Chem 93:253
4. Brase S, Kirchhoff JH, Kobberling J (2003) Tetrahedron 59:885
5. Blaser H, Indolese A, Schnyder A (2000) Curr Sci 78:1336
6. Negishi E, Dumond Y (2002) Handbook of organopalladium chemistry for organic synthesis 1. Wiley, Hoboken
7. Remmele H, Koellhofer A, Plenio H (2003) Organometallics 22:4098
8. Osakada K, Sakata R, Yamamoto T (1997) Organometallics 16:5354
9. Osakada K, Hamada M, Yamamoto T (2000) Organometallics 19:458
10. Osakada K, Yamamoto T (2000) Coord Chem Rev 198:379
11. Osakada K, Yamamoto T (1999) Rev Heteroatom Chem 21:163
12. Yamamoto T, Yamamoto A, Ikeda S (1971) J Am Chem Soc 93:3350
13. Yamamoto T, Yamamoto A, Ikeda S (1971) J Am Chem Soc 93:3360
14. Yamamoto T, Nakamura Y, Yamamoto A (1976) Bull Chem Soc Jpn 49:191
15. Yamamoto T, Abla M, Murakami Y (2002) Bull Chem Soc Jpn 75:1997
16. Yamamoto T (2003) Synlett 425;
    Yamamoto T, Kokubo H (2005) Electrochim Acta 50:1453
17. Exner O (1972) In: Chapman NB, Shorter J (eds) Advances in linear free energy relationships. Plenum, London
18. Corriu RJP, Masse JP (1972) J Chem Soc Chem Commun 144
19. Tamao K, Sumitani K, Kumada M (1972) J Am Chem Soc 94:4374
20. Kumada M, Tamao K, Sumitani K (1978) Org Synth 58:127
21. Kiso Y, Tamao K, Miyake N, Yamamoto K, Kumada M (1974) Tetrahedron Lett 3
22. Yamamoto T, Yamamoto A (1973) J Organomet Chem 57:127
23. Sanechika K, Yamamoto T, Yamamoto A (1984) Bull Chem Soc Jpn 57:752
24. Sanechika K, Yamamoto T, Yamamoto A (1981) Polym Prepr Jpn 30:160
25. Trubo DL, Marvel CS (1986) J Polym Sci A Polym Chem 24:2311
26. Bunz UHF (1999) Top Curr Chem 201:131
27. Bunz UHF (2000) Chem Rev 100:1605
28. Bunz UHF (2001) Acc Chem Res 34:998
29. Giesa R (1996) Macromol Sci Rev Macromol Chem Phys 36:631
30. Yamamoto T (1992) Prog Polym Sci 17:1153
31. Yamamoto T (1999) Bull Chem Soc Jpn 72:621

32. Yamamoto T, Hayashida N (1998) React Funct Polym 37:1
33. Yamamoto T (2002) Macromol Rapid Commun 23:583
34. Bunz UHF, Kloppenburg L (1999) Angew Chem Int Ed 38:478
35. Nishihara Y, Ando J, Kato T, Mon A, Hiyama T (2000) Macromolecules 33:2779
36. Mori A, Kondo T, Kato T, Nishihara Y (2001) Chem Lett 286
37. Giardina G, Rosi P, Ricci A, Lo Sterzo C (2000) J Polym Sci A Polym Chem 38:2603
38. Pizzoferrato R, Berliocchi M, Di Carlo A, Lugli P, Venanzi M, Micozzi A, Ricci A, Lo Sterzo C (2003) Macromolecules 36:2215
39. Takagi M, Kizu K, Miyazaki Y, Maruyama T, Kubota K, Yamamoto T (1993) Chem Lett 913
40. Yamamoto T, Takagi M, Kizu K, Maruyama T, Kubota K, Kanbara H, Kurihara T, Kaino T (1993) J Chem Soc Chem Commun 797
41. Yamamoto T, Honda K, Ooba N, Tomaru S (1998) Macromolecules 31:7
42. Yamamoto T, Muramatsu Y, Shimuzu T, Yamada W (1998) Macromol Rapid Commun 19:263
43. Pang Y, Li J, Barton TJ (1998) J Mater Chem 8:1687
44. Yamamoto T, Shimuzu T (1997) J Mater Chem 7:1967
45. Watanabe Y, Mihara T, Koide N (1998) Macromol Chem Phys 199:977
46. Babudri F, Colangiuli D, Di Lorenzo PA, Farinola GM, Omar OH, Naso F (2003) Chem Commun 130
47. Zhan X, Liu Y, Yu G, Wu X, Zhu D, Sun R, Wang D, Epstein AJ (2001) J Mater Chem 11:1606
48. Yamamoto T, Honda K (1998) J Polym Sci A Polym Chem 36:2201
49. Kang BS, Kim DH, Lim SM, Kim J, Seo ML, Bark KM, Shin SC, Nahm K (1997) Macromolecules 30:7196
50. Li J, Pang Y (1997) Macromolecules 30:7487
51. Hayashi H, Yamamoto T (1997) Macromolecules 30:330
52. Altmann M, Enkelmann V, Lieser G, Bunz UHF (1995) Adv Mater 7:726
53. Li J, Pang Y (1998) Macromolecules 31:5740
54. Yamamoto T, Yamada W, Takagi M, Kizu K, Maruyama T, Ooba N, Tomaru S, Kurihara T, Kaino T, Kubota K (1994) Macromolecules 27:6620
55. Zhang Y, Murphy CB, Jones WE Jr (2002) Macromolecules 35:630
56. Moroni M, Le Moigne J, Luzzati S (1994) Macromolecules 27:562
57. Williams VE, Swager TM (2000) Macromolecules 33:4069
58. Mangel T, Eberhardt A, Scherf U, Bunz UHF, Müllen K (1995) Macromol Rapid Commun 16:571
59. Wautelet P, Moroni M, Moigne JL, Pham A, Bigot JY (1996) Macromolecules 29:446
60. Zuideveld MA, Swennenhuis BH, Boele MDK, Guari Y, van Strijodouck, Peek JNH, Kamer PCJ, Goubitz K, Fraanje J, Lutz M, Spek AL, van Leeuwen PMW (2002) Dalton Trans 2308
61. Yamamoto T, Abe M, Takahashi Y, Kawata K, Kubota K (2003) Polym J 35:603
62. Yamamoto T (1996) J Polym Sci A Polym Chem 34:997
63. Meerhold K, Gregorius H, Müllen K, Heinze K (1994) Adv Mater 6:671
64. Yamamoto T, Xu Y, Inoue T, Yamaguchi I (2000) J Polym Sci A Polym Chem 38:1493
65. Yamamoto T, Takeuchi M, Kubota K (2000) J Polym Sci B Polym Phys 38:1348
66. Yamamoto T, Maruyama T, Zhou ZH, Ito T, Fukuda T, Yoneda Y, Begum F, Ikeda T, Sasaki S, Takezoe H, Fukuda A, Kubota K (1994) J Am Chem Soc 116:4832
67. Pearson DL, Schumm JS, Tour JM (1994) Macromolecules 27:2348
68. Tour JM (1996) Chem Rev 96:537
69. Tour JM (1994) Trends Polym Sci 2:332
70. Matsuda K, Stone MT, Moore JS (2002) J Am Chem Soc 124:11836

71. Zhao D, Moore JS (2002) J Am Chem Soc 124:9996
72. Kuebel C, Mio M, Moore JS, Martin DC (2002) J Am Chem Soc 124:8605
73. Nishigata T, Tanatani A, Oh K, Moore JS (2002) J Am Chem Soc 124:5934
74. Hill D, Mio M, Prince RB, Hughes TS, Moore JS (2001) Chem Rev 101:3893
75. Tanatani A, Mio M, Moore JS (2001) J Am Chem Soc 123:1792
76. Prest PJ, Prince RB, Moore JS (1999) J Am Chem Soc 121:5933
77. Yamamoto T, Fujiwara Y, Fukumoto H, Nakamura Y, Koshihara S, Ishikawa T (2003) Polymer 44:4487
78. Montali A, Bastiaansen C, Smith P, Weder C (1998) Nature 392:261
79. Devadoss C, Bharathi P, Moore JS (1996) J Am Chem Soc 118:9635
80. Yamamoto T, Yoneda Y, Kizu K (1995) Macromol Rapid Commun 16:549
81. Watkins DM, Fox MA (1994) J Am Chem Soc 116:6441
82. Yang JS, Swager TM (1998) J Am Chem Soc 120:5321
83. Bradley DDC, Mon H (1989) In: Kuzmany H, Mehring M, Roth S (eds) Electronic properties of conducting polymers. Springer, Berlin Heidelberg New York
84. Yamamoto T, Lee BL, Kokubo H, Kishida H, Hirota K, Wakabayashi T, Okamoto H (2003) Macromol Rapid Commun 24:440
85. Ooba N, Asobe M, Tomaru S, Kaino T, Yamada W, Takagi M, Yamamoto T (1996) Nonlinear Optics 15:481
86. Zhan X, Liu Y, Zhu D, Jiang X, Jen AKY (2001) Synth Met 124:323
87. Cunningham GB, Li Y, Liu S, Schanze KS (2003) J Phys Chem B 107:12569
88. Egbe DAM, Klemm E (1998) Macromol Chem Phys 199:2683
89. Beginn C, Grazulevicius JV, Strohriegl P, Simmerer J, Haarer D (1994) Macromol Chem Phys 195:2353
90. Al-Higari M, Birckner E, Heise B, Klemm E (1999) J Polym Sci A Polym Chem 37:4442
91. Hayashi H, Yamamoto T (1998) Macromolecules 31:6063
92. Bangcuyo CG, Ellsworth JM, Evans U, Myrick ML, Burn UHF (2003) Macromolecules 36:546
93. Yamamoto T, Fang Q, Morikita T (2003) Macromolecules 36:4262
94. Morikita T, Yamaguchi I, Yamamoto T (2001) Adv Mater 13:1862
95. Yamamoto T, Kokubo H, Morikita T (2001) J Polym Sci B Polym Phys 39:1713
96. Bangcuyo CG, Evans U, Myrick ML, Bunz UHF (2001) Macromolecules 34:7592
97. Wilson JN, Bangcuyo CG, Erdogan B, Myrick ML, Burn UHF (2003) Macromolecules 36:1426
98. Leung ACW, Chong JH, Patrick BO, MacLachlan MJ (2003) Macromolecules 36:5051
99. Bouchard J, Belletête M, Durocher G, Leclerc M (2003) Macromolecules 36:4624
100. Arias-Marin E, Le Moigne J, Maillou T, Guillon D, Moggio I, Geffroy B (2003) Macromolecules 36:3570
101. Liu B, Yu WL, Pei J, Liu SY, Lai YH, Huang W (2001) Macromolecules 34:7932
102. Bangcuyo CG, Rampey-Vaughn ME, Quan LT, Angel SM, Smith MD, Bunz UHF (2002) Macromolecules 35:1563
103. Jégou G, Jenekhe SA (2001) Macromolecules 34:7926
104. Yamamoto T, Fukushima N, Nakajima H, Maruyama T, Yamaguchi I (2000) Macromolecules 33:5988
105. Morikita T, Hayashi H, Yamamoto T (1999) Inorg Chim Acta 296:254
106. Grummt UW, Birckner E, Klemm E, Egbe DAM, Heise B (2000) J Phys Org Chem 13:112
107. Ley KD, Whittle CE, Bartberger MD, Schanze KS (1997) J Am Chem Soc 119:3423
108. Ley KD, Schanze KS (1998) Coord Chem Rev 171:287
109. Politis JK, Curtis MD, Gonzalez L, Martin DC, He Y, Kanicki J (1998) Chem Mater 10:1713

110. Kanbara T, Kushida T, Saito N, Kuwajima I, Kubota K, Yamamoto T (1992) Chem Lett 583
111. Hayashida N, Yamamoto T (1999) Bull Chem Soc Jpn 72:1153
112. Yamamoto T, Ito T, Sanechika K, Hishinuma M (1988) Synth Met 25:103
113. Yamamoto T, Maruyama T, Zhou ZH, Ito T, Fukuda T, Yoneda Y, Begum F, Ikeda T, Sasaki S, Takezoe H, Fukuda A, Kubota K (1994) J Am Chem Soc 116:4832
114. Schiavon C, Zotti G, Bontempelli G (1984) J Electroanal Chem 161:323
115. Yamamoto T, Sugiyama K, Kushida T, Inoue T, Kanbara T (1996) J Am Chem Soc 118:3930
116. Kanbara T, Saito N, Yamamoto T, Kubota K (1991) Macromolecules 24:5883
117. Saito N, Kanbara T, Nakamura Y, Yamamoto T (1994) Macromolecules 27:756
118. Tomat R, Zecchin S, Schiavon G, Zotti G (1988) J Electroanal Chem 252:215
119. Yamamoto T, Suganuma H, Maruyama T, Inoue T, Muramatsu Y, Arai M, Komarudin D, Ooba N, Tomaru S, Sasaki S, Kubota K (1997) Chem Mater 9:1217
120. Yamamoto T, Suganuma H, Maruyama T, Kubota K (1995) J Chem Soc Chem Commun 1613
121. Zotti G, Schiavon G (1984) J Electroanal Chem 163:385
122. Yamamoto T, Kanbara T, Mori C, Wakayama H, Fukuda T, Inoue T, Sasaki S (1996) J Phys Chem 30:12631
123. Kanbara T, Yamamoto T (1993) Chem Lett 419
124. Yamamoto T, Sugiyama K, Kanbara T, Hayashi H, Etori H (1998) Macromol Chem Phys 199:1807
125. Choi BK, Yamamoto T (2003) Electrochem Commun 5:566
126. Yamamoto T, Saito Y, Anzai K, Fukumoto H, Yasuda T, Fujiwara Y, Choi BK, Kubota K, Miyamae T (2003) Macromolecules 36:6722
127. Saitoh Y, Yamamoto T (1995) Chem Lett 785
128. McCullough RD, Tristam-Nagle S, Williams SP, Lowe RD, Jayaman M (1993) J Am Chem Soc 115:4910
129. Yamamoto T, Komarudin D, Arai M, Lee BL, Suganuma H, Asakawa N, Inoue Y, Kubota K, Sasaki S, Fukuda T, Matsuda H (1998) J Am Chem Soc 120:2047
130. Yamamoto T, Lee BL, Suganuma H, Sasaki S (1998) Polym J 30:853
131. Yamamoto T, Arai M, Kokubo H, Sasaki S (2003) Macromolecules 36:7986
132. Yamamoto T, Shiraishi K, Abla M, Yamaguchi I, Groenendaal LB (2002) 43:711
133. Yamamoto T, Kimura T, Shiraishi K (1999) Macromolecules 32:8886
134. Sirringhaus H, Brown PJ, Friend RH, Nielson MM, Bechgaard K, Langeveld-Voss BMW, Spiering JH, Janssen RAJ, Meijer EW, Hartwig P, de Leeuw DM (1999) Nature 401:685
135. Yamamoto T, Kokubo H (2002) Mol Cryst Liq Cryst 381:113
136. Grell M, Knoll W, Lupo D, Meisel A, Miteva T, Neher D, Nothofer HG, Scherf U, Yasuda A (1999) Adv Mater 11:671
137. Yamamoto T, Zhou ZH, Maruyama T, Kanbara T (1990) Chem Lett
138. Yasuda T, Yamamoto T (2003) Macromolecules 36:7513
139. Sato Y, Kagotani M, Yamamoto T, Souma Y (1999) Appl Catal A Gen 185:219
140. Wolf MO, Wrighton MS (1994) Chem Mater 6:1526
141. Zhu SS, Swager TM (1997) J Am Chem Soc 119:12568
142. Kingsborough RP, Swager TM (1999) Prog Inorg Chem 48:123
143. Nishihara H (1997) J Org Synth Jpn 55:410
144. Matsuda J, Aramaki K, Nishihara H (1995) J Chem Soc Faraday Trans 91:477
145. Iijima T, Yamamoto T (2004) Macromol Rapid Commun 25:669
146. Yamamoto T, Yoneda Y, Maruyama T (1992) J Chem Soc Chem Commun 1652
147. Maruyama T, Yamamoto T (1995) Inorg Chim Acta 238:9
148. Yamamoto T, Saitoh Y, Anzai K, Fukumoto H, Yasuda T, Fujiwara Y, Choi BK, Kubota K, Miyamae T (2003) Macromolecules 36:6722

149. Kim SW, Shim SC, Kim DY, Kim CY (2001) Synth Met 122:363
150. Yamaguchi S, Iimura K, Tamao K (1998) Chem Lett 89
151. Chen W, Ijadi-Maghsoodi S, Barton TJ (1997) Polym Prepr 38:189
152. Yamamoto T, Morikita T, Maruyama T, Kubota K, Katada M (1997) Macromolecules 30:5390
153. Morikita T, Yamamoto T (2001) J Organomet Chem 637–639:809
154. Matsumi N, Umeyama T, Chujo Y (2000) Polym Bull 44:431
155. Kakimoto M, Kashihara H, Yamaguchi Y, Takiguchi T (2000) Macromolecules 33:760
156. Li H, West R (1998) Macromolecules 3 1:2866
157. Dubac J, Laporterie A, Manuel G (1990) Chem Rev 90:2 15
158. Yamaguchi S, Tamao K (1996) Bull Chem Soc Jpn 69:2327
159. Yamaguchi S, Jin RZ, Itami Y, Goto T, Tamao K (1999) J Am Chem Soc 121:10420
160. Shimar S, Ijadi-Maglsoodi S, Ni QX, Pang Y, Barton TJ (1989) Synth Met 28:593
161. Tamao K, Yamaguchi S, Shiro M (1994) J Am Chem Soc 116:11715
162. Yamaguchi S, Tamao K (1998) J Synth Org Chem Jpn 56:500
163. Chujo Y (1997) Macromol Symp 118:111

Adv Polym Sci (2005) 177: 209–248
DOI 10.1007/b101379

Electronic Properties of PAEs

Gabriela Voskerician · Christoph Weder (✉)

Department of Macromolecular Science and Engineering, Case Western Reserve University, Cleveland, OH 44106-7202, USA
christoph.weder@case.edu

Abstract Poly(arylene ethynylene)s (PAEs) represent an important family of conjugated polymers with interesting optical and electronic properties. A great deal of attention has been devoted to the synthesis, physicochemical characteristics, and optical properties of these materials, but their electrically (semi)conducting nature has received comparably little attention. Only during the last decade have significant research efforts been devoted to this subject in research laboratories around the world, and PAEs have eventually established themselves as a versatile class of polymeric semiconductors. Focusing on the subjects of electrical conductivity, charge transport, and electroluminescence, this review article attempts to provide a comprehensive summary of the electronic properties of PAEs and their potential applications in "plastic electronic" devices.

Keywords Charge transport · Electrical conductivity · Electroluminescence · Light-emitting diode · Semiconductor

Abbreviations

E_g	Band gap
k	Boltzman constant
μ	carrier mobility
t_{tr}	carrier transit time
CB	Conduction band
CuPc	Copper phthalocyanine
I	Current
CV	Cyclic voltammogram

$\hat{\sigma}$	Diagonal (energetic) disorder parameter
E	Electric field
σ	Electrical conductivity
EL	Electroluminescence
I_{EL}	Electroluminescence intensity
EA	Electron affinity
ET	Electron-transporting
FET	Field-effect transistor
HOMO	Highest occupied molecular orbital
HT	Hole-transporting
ITO	Indium tin oxide
IP	Ionization potential
LED	Light-emitting diode
LEC	Light-emitting electrochemical cell
LCD	Liquid crystal display
LUMO	Lowest unoccupied molecular orbital
M_n	Number-average molecular weight
Σ	Off-diagonal (positional) disorder parameter
PL	Photoluminescence
PA	Poly(arylene)
PAE	Poly(arylene ethynylene)
PAV	Poly(arylene vinylene)
PDA	Poly(diacetylene)
EHO-OPPE	Poly[2,5-dioctyloxy-1,4-diethynyl-phenylene-*alt*-2,5-bis(2′-ethylhexyloxy)-1,4-phenylene]
PEDOT	Poly(3,4-ethylenedioxythiophene)/poly(styrenesulfonic acid)
PPE	Poly(*p*-phenylene ethynylene)
PPV	Poly(*p*-phenylene vinylene)
PLED	Polymer light-emitting diode
μ_0	Prefactor mobility
L	Sample thickness
SCE	Saturated calomel electrode
T	Temperature
TOF	Time-of-flight
UV–Vis	Ultraviolet–visible
EV	Vacuum level
VB	Valence band
V	Voltage
λ_{max}	Wavelength of maximum absorption or emission

1 Introduction

In the past three decades, π-conjugated (semi)conducting polymers have attracted significant interest, since these materials may combine the ease of processibility and outstanding mechanical properties of polymers with the exceptional, readily tailored electronic and optical properties of functional organic molecules [1, 2]. Especially the potential use of these "synthetic metals" as electrical conductors [3] in light-emitting diodes (LEDs) [4–6], field-effect transistors (FETs) [7], photorefractive devices [8], and photovoltaic cells [9] has motivated the development of synthesis and processing methods of conjugated polymer materials with unique field-responsive properties [3, 10, 11]. Among a plethora of materials platforms, poly(arylene ethynylene) (PAE) derivatives (Fig. 1) have attracted the attention of numerous research groups. As the name implies, PAEs feature aromatic rings and ethynylene groups in the polymer backbone. The connectivity of these moieties results in an alternating sequence of single and multiple bonds and gives rise to π-conjugation along the macromolecules. PAEs are closely related to poly(arylene)s (PAs) [12], poly(arylene vinylene)s (PAVs) [12], and poly(diacetylene)s (PDAs) [13], which all represent important classes of conjugated polymers (Fig. 1). The chemical structure of PAEs can be readily and significantly manipulated, for example via the choice of the aromatic moiety, the connectivity of the latter (e.g., *meta* vs *para* substitution), the introduction of heteroatoms or metals, the nature of solubilizing side chains, and noncovalent interactions with metals. The possibility of integrating conjugated moieties other than arylenes and ethynylenes (for example vinylene groups, nonconjugated aliphatic spacers, etc.) represents another synthetic tool, which leads to PAE copolymers. These structural changes allow one to tailor the property matrix of these polymers over a wide range, as discussed in more detail in other chapters of this volume.

Hundreds of different PAEs and PAE copolymers have been reported to date, and during the last 10 years this family of materials has established itself as an important class of conjugated polymers with interesting optical and electronic properties. A number of excellent texts have reviewed the synthesis, physicochemical characteristics, and optical properties of these materials. The early work on PAEs was summarized in 1996 by Giesa [14]. The most comprehensive review on the synthesis, properties, structures, and application of PAEs was published by Bunz in 2000 [15], and several complements and updates to this outstanding compilation have appeared since [16–18]. Yamamoto has reviewed the important subject of heteroaromatic PAEs [19, 20]. PAE electrolytes are part of an outstanding article by Pinto and Schanze [21]. Swager and coworkers have published a number of excellent texts that discuss the application of PAEs and other polymers in sensors [22–25]. The intriguing supramolecular architectures that can be created with PAEs have been addressed by Moore et al. [26, 27]. Linear arylene ethynylene oligomers of well-defined architecture represent a

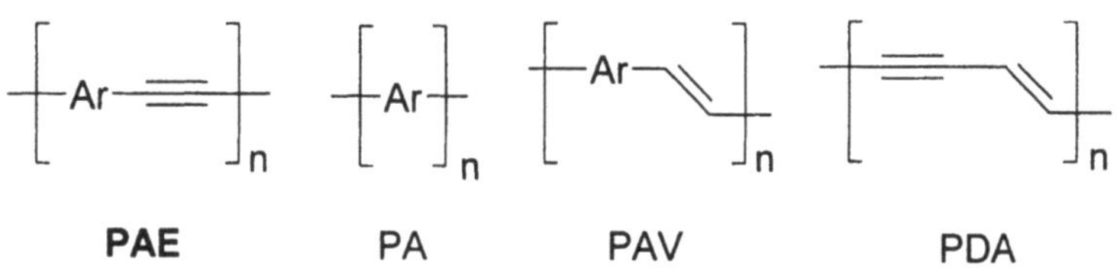

Fig. 1 General schematic structures of poly(arylene ethynylene)s (PAEs), poly(arylene)s (PAs), poly(arylene vinylene)s (PAVs), and poly(diacetylene)s (PDAs)

class of materials that are closely related to the polymers discussed here, although their low-molecular nature positions them beyond the scope of this volume. Their synthesis and optical and electronic properties have been summarized in scholarly contributions by Tour [28], while the work on cyclic analogues was compiled by Haley et al. [29].

Interestingly, the electrically (semi)conducting nature of PAEs has traditionally received comparably little attention. Only during the last decade have increasing research efforts been devoted to this subject, and PAEs have eventually been recognized as a versatile class of polymeric semiconductors with interesting property profiles. Bunz et al. pointed out that one of the most appealing attributes of PAEs originates from the efficient electronic communication along their conjugated polymer backbone [30]. This propensity can be attributed, at least in part, to the cylindrical symmetry of the acetylene bond (–C≡C–), which is able to maintain conjugation between adjacent phenyl groups regardless of the relative orientations of their aromatic planes [31]. The electron-withdrawing nature of the acetylene group is another key feature that exerts a significant influence on the electronic properties of PAEs and makes electron injection comparably easy. Thus, focusing on the subjects of electrical conductivity, charge transport, and electroluminescence (EL), this review article attempts to provide a comprehensive summary of the electronic properties of PAEs and their potential applications in "plastic electronic" devices. In view of the wealth of data, the subject is discussed by using selected, illustrative examples. The chapter has also deliberately been limited to PAEs and copolymers or hybrids comprising repeat units other than arylene ethynylenes; low-molecular arylene ethynylene, and arylene ethynylene oligomers have not been included, except for a few important and illustrative examples.

2 Electrical Conductivity

2.1 General Considerations

As is the case for all classes of conjugated polymers, the electrical conductivity of PAEs is a direct consequence of delocalized chemical bonding [32]. The absence of sp^3 hybridized carbon atoms leads to a situation where overlap of p orbitals on successive carbon atoms enables the delocalization of π electrons along the polymer backbones. The large number of atomic orbitals in the macromolecules translates into a large number of molecular orbitals, which form a band of energies. In the case of a metal this energy band is a continuum; due to the high density of electronic states with electrons of relatively low binding energy, "free electrons" can easily redistribute and under an applied electric field move easily from atom to atom. The material is therefore electrically conducting. Hückel's theory predicts that in conjugated macromolecules π electrons are delocalized in a similar fashion over the entire chain, so one would expect that the electronic properties of a polymeric material composed of sufficiently long conjugated chains is also described well by a continuous energy band. However, as a result of the Peierls instability, the density of π electrons in conjugated organic molecules is not the same between all atoms; there is a distinct alternation between single and multiple bond character, as schematically shown in Fig. 2 for the example of poly(*p*-phenylene ethynylene) (PPE). Thus, the electronic properties of conjugated polymers in their neutral oxidation state are usually better described by a filled valence band (π band, bonding) formed by the highest occupied molecular orbitals (HOMOs) and an empty conduction band (π* band, antibonding) formed by the lowest unoccupied molecular orbitals (LUMOs). Because the energy difference between the highest occupied and the lowest unoccupied band, referred to as band gap (E_g), is usually not near zero, and because there are no partially filled bands, conjugated polymers are typically *semi*conductors in their neutral, *undoped* state. E_g depends on the molecular structure of the polymer's repeat unit and can be controlled via modification of the latter (vide infra).

In their pioneering work on polyacetylene, MacDiarmid, Shirakawa, Heeger et al. demonstrated that *doping* allows one to increase the electrical conductivity of conjugated polymers by many orders of magnitude so that metallic properties are achieved [33]. Doping refers to either removing (oxidation, p-doping) or adding electrons (reduction, n-doping) to the polymer, as shown schematically in Fig. 2 for the example of PPE. This can be accomplished by either conventional chemical or electrochemical means. In principle, two major defect types can be formed upon doping: radical cations or anions (referred to as polarons) and dications or dianions (referred to as bipolarons). As a result of the doping process, the electrochemical potential (Fermi level) of the polymer is moved into an energy regime with a high density of electronic states. Charge neutrality is maintained through counterions, and the doped polymer effectively becomes a salt. The extra electrons or vacancies (holes) introduced through doping act as charge carriers. The delocalization of these carriers can be quite limited, partly because of Coulomb attraction to their counterions, and

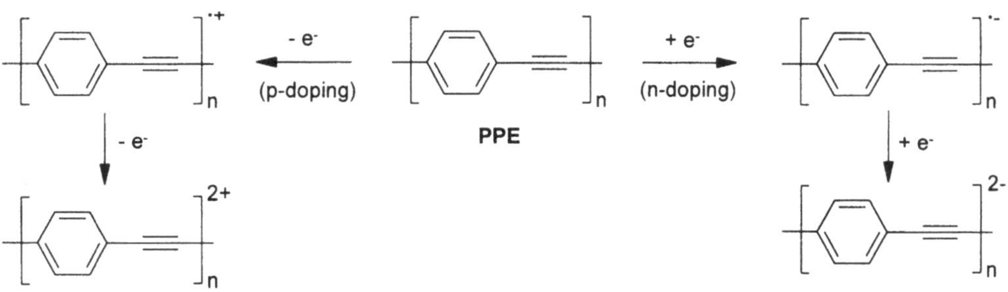

Fig. 2 Schematic representation of the doping of poly(*p*-phenylene ethynylene) (PPE) under formation of polarons (radical cations, radical anions) and bipolarons (dications, dianions). Note that multiple carriers can coexist on each macromolecule

partly because of a local change in the equilibrium geometry of the doped relative to the neutral molecule. However, since every repeat unit is a potential redox site, the doping level of conjugated polymers can usually be rather well controlled. For many materials platforms high levels of electrical conductivity can be achieved at high doping levels, which are concomitant with a high density of charge carriers.

2.2 Electrical Conductivity of PAEs

The most widely investigated representatives of PAEs are poly(*p*-phenylene ethynylene)s, PPEs, (Fig. 2) and their derivatives, in particular PPEs that carry solubilizing side chains such as alkyl or alkyloxy chains [14, 15]. Early accounts of the electrical conductivity of chemically oxidized PPEs have not revealed high conductivity for the unsubstituted polymer (Table 1, entry 1). McDiarmid et al. reported a conductivity σ of 10^{-3} Scm^{-1} for PPE p-doped with AsF_5 [34]. In another study conducted by Yamamoto et al. a conductivity of $6.6 \cdot 10^{-8}$ Scm^{-1} was measured for the same system [35]. The significant difference may be related to the structural purity of the polymer (vide infra) or the fact that the samples were characterized by different doping levels. The study conducted by Yamamoto et al. further quoted a conductivity of $4 \cdot 10^{-5}$ Scm^{-1} for PPE that was p-doped with I_2 vapors. In yet another study, Aramaki et al. reported conductivities of $3 \cdot 10^{-7}$ and 70 Scm^{-1} for PPE doped with I_2 and SO_3 vapors, respectively [36]; however, the PPE employed for these experiments was prepared by a somewhat unusual method (the electrochemical reduction of hexachloro-*p*-xylene) and its chemical structure has not been confirmed. Cava et al. investigated poly(2,5-thienylene ethynylene) (PTE, Table 1, entry 2) but, somewhat surprisingly, no appreciable conductivity could be measured when this material was doped with I_2 [37]. It should be noted that underivatized PAEs such as PPE or PTE are essentially insoluble, and therefore a comprehensive structural characterization, which includes the determination of molecular weights and an assessment of structural purity, is difficult, if not impossible. This situation, together with the uncertainty regarding the doping level, probably explains why the conductivity data reported in these early studies display a significant variation.

A pioneering study of the electrical conductivity of poly(2,5-dibutoxy-*p*-phenylene ethynylene), a fully soluble PPE derivative, was conducted by the groups of Shinar and Barton (Table 1, entry 3) [38]. The polymer employed displayed a band gap, E_g, of ~2.5 eV, which is typical for 2,5-dialkyloxy-PPEs. A conductivity of ~10^{-9} Scm^{-1} was reported for the undoped polymer. This value was later confirmed in an independent study by Lo Sterzo et al., who reported a conductivity of $4 \cdot 10^{-9}$ Scm^{-1} for the same polymer [39]. It was reported that upon exposure to I_2 vapor at a pressure of ~1 Torr, the room-temperature conductivity of poly(2,5-dibutoxy-*p*-phenylene ethynylene) increased by about three orders of magnitude to ~10^{-6} Scm^{-1} [38]. It increased further and imme-

Table 1 Doping conditions and electrical conductivity of various PAEs

Entry	Polymer	Dopant	Comments	Conductivity [Scm^{-1}]	Reference
1	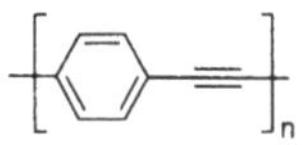	AsF_5	-	10^{-3}	[34]
		AsF_5	-	$6.6 \cdot 10^{-8}$	[35]
		I_2	-	$4 \cdot 10^{-5}$	[35]
		I_2	chemical structure not confirmed	10^{-7}	[36]
		SO_3	chemical structure not confirmed	70	[36]
2	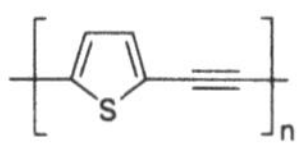	I_2	-	No appreciable conductivity	[37]
3	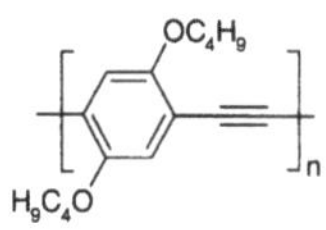	I_2	room temperature	10^{-6}	[38]
		I_2	80°C	$5 \cdot 10^{-3}$	[38]
		undoped	room temperature	$\sim 10^{-9}$	[38]
		undoped	room temperature	$4 \cdot 10^{-9}$	[39]
4	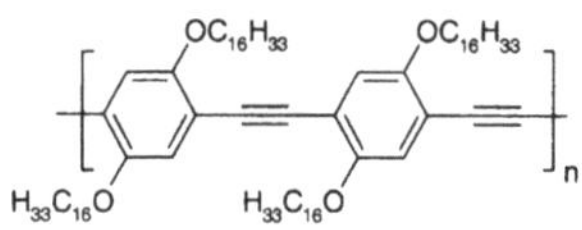	electrochemically in SO_2 at -70°C	n_{ave}=12	0.18±0.04	[40]
		electrochemically in SO_2 at -70°C	n_{ave}=20	0.29±0.17	[40]
5	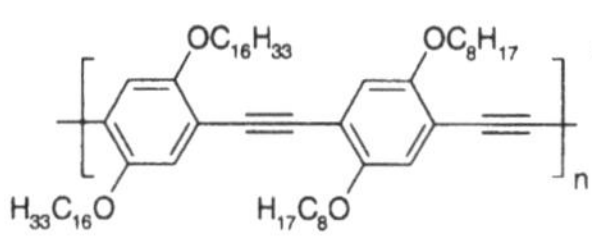	electrochemically in SO_2 at -70°C	n_{ave}=20	1.6±0.8	[40]
6	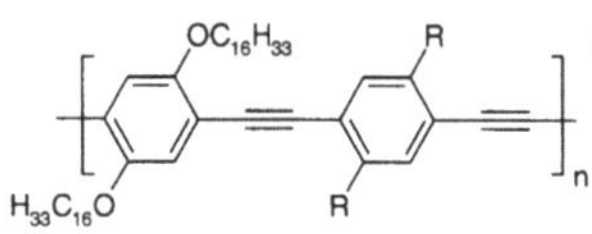 R = 50 % C_8H_{17}; 50 % CH_3	electrochemically in SO_2 at -70°C	n_{ave}=100	4.5±0.8	[40]
		undoped in SO_2 at -70°C	n_{ave}=100	$\leq 5 \cdot 10^{-6}$	[40]
7	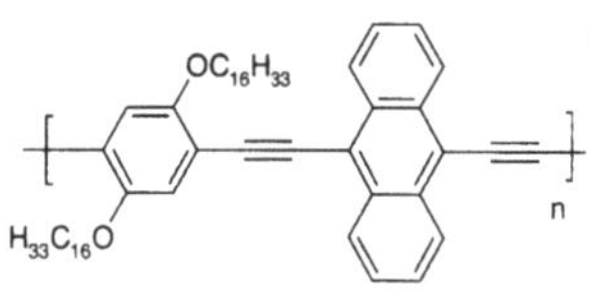	electrochemically in SO_2 at -70°C	n_{ave}=11	0.12±0.03	[40]

diately to ~10^{-3} Scm^{-1} when the sample was heated to ~80 °C. However, when the doped polymer was cooled back to ambient temperature, the I_2 rapidly desorbed and σ decreased below a measurable limit [38]. Thus, the study unequivocally demonstrated that I_2 could be employed to p-dope dialkyloxy-PPEs, presumably via the formation of I_3^- and PPE radical cations (polarons). However, the affinity of the polymer toward iodine was believed to be relatively weak, since the doping was found to be reversible, and under ambient conditions rapid de-doping of the polymer was observed.

Wrighton et al. conducted a detailed study on the conductivity of various poly(2,5-dialkyloxy-*p*-phenylene ethynylene)s (Table 1, entries 4–7) that were electrochemically doped at –70 °C in liquid SO_2 containing $[(n\text{-Bu})_4N]AsF_6$ as electrolyte [40]. This first electrochemical study of PPE thin films paid special attention to the influence of the degree of polymerization and long-range order (Table 1, entries 4–6). Cyclic voltammetry (CV) experiments revealed that the onset of the oxidation of alkyloxy-PPEs is independent of the nature of the alkyloxy group and occurs at ~1.05 V vs a saturated calomel electrode (SCE). This puts the upper edge of the valence band (or in molecular terms, the HOMO) to about 5.9 eV below the vacuum level. The experiments further revealed that the oxidation process was reversible, if the potential was kept below ~2 V vs SCE. The conductivity of the neutral polymer was reported to be $\leq 5 \cdot 10^{-6}$ Scm^{-1} [40]; this value is significantly higher than the one reported previously for another alkyloxy-PPE (~10^{-9} Scm^{-1} [38, 39], vide supra) but it is unclear whether it indeed represents the actual conductivity or merely indicates the lower conductivity limit that could be properly measured with the setup employed. Upon oxidation, all polymers displayed finite potential windows of width ~0.55 V in which a substantial increase of the electrical conductivity was observed as a result of electrochemical doping. Conductivities of between ~0.18 (Table 1, entry 4) and ~4.5 Scm^{-1} (Table 1, entry 6) were observed at a potential of ~1.6 V vs SCE. At this potential the oxidation level was around 0.3 electrons per arylene ethynylene unit. A further increase of the potential reduced the conductivity to a level of ~10^{-4} Scm^{-1} at ~2 V vs SCE. At such a positive potential the oxidation level was around 0.7 electrons per ethynyl arene unit, but the polymers were observed to slowly degrade. A PAE featuring phenylene ethynylene and ethynyl anthracene units in alternating fashion (Table 1, entry 7) had a slightly lower onset of oxidation (~0.8 V vs SCE) and a lower potential of maximum conductivity (~1.5 V vs SCE) than the alkyloxy-PPEs. It also featured a broader potential window of conductivity (0.85 V), but displayed the lowest conductivity (~0.12 Scm^{-1}) of the series under investigation.

The structural characterization of the PPEs studied by Wrighton et al. revealed that the polymers exhibited different degrees of order and crystallinity, depending on the nature of the side chains. In line with other studies [41, 42], it was shown that the polymers with less regular structure (Table 1, entries 6 and 7) were less ordered than the derivatives with regularly spaced linear side chains, which adopted lamellar architectures (Table 1, entries 4 and 5). Wrighton et al. concluded that for the materials studied, higher conductivity was associated with lower long-range order and that the number-average degree of polymerization had little influence on the electrical conductivity. However, this interpretation contrasts with numerous studies [43–45] that have unequivocally shown that a high degree of long-range order usually leads to high carrier mobility and high conductivity. This apparent contradiction may be explained with the authors' finding that greater crystallinity accounted for sluggish electrochemical response. Indeed, the electrochemical activity depended on the penetration by solvent and electrolyte into the polymer films. Thus, the trend of the conductivities may ultimately be associated with kinetic effects, and reflect that doping is slower in the case of the more crystalline samples. The interpretation of the data is further complicated by the fact that polymers with different molecular weights were investigated. The highest conductivity was observed for the sample with the highest degree of polymerization (Table 1, entry 6), and the data seem to suggest that there might be a correlation between the degree of polymerization and conductivity. In comparison with other studies, the conductivities reported by Wrighton et al. are by far the highest reported to date for any alkyloxy-PPE. However, the study also unequivocally demonstrates that alkyloxy-PPEs display low ionization potentials and are very difficult to oxidize. This feature is of interest for their potential application in light-emitting diodes (vide infra), but it represents an obstacle for their exploitation as p-dopable electrical conductors.

Another electrochemical study was recently undertaken by Myrick et al. who investigated PPEs that were derivatized with alkyl side chains [46]. No conductivity experiments were reported, but the authors conducted a detailed spectroelectrochemical investigation that has provided interesting insights regarding the nature of the species produced upon oxidation. The cyclic voltammograms (CVs) of the investigated alkyl-PPEs (Fig. 3) display anodic peaks at 0.59 and 1.19 V vs Fc/Fc^+ (~1.0 and ~1.6 V vs SCE). These features are very similar to those found in the CVs shown by Wrighton et al. for alkyloxy-PPEs [40]. Interestingly, the first anodic peak is only reversible (reduction at 0.54 V) if the upper voltage is limited to 0.85 V vs Fc/Fc^+. When the upper voltage limit is increased to 1.45 V vs Fc/Fc^+ only one reduction is observed at –0.26 V vs Fc/Fc^+. With the help of in situ ultraviolet–visible (UV–Vis) absorption spectra and deconvolution of the spectra by means of iterative target transformation factor analysis, Myrick et al. assigned the first anodic peak to the oxidation of the neutral PPE (absorption maxima, λ_{max}=404 nm/3.07 eV and 440 nm/2.82 eV) to a polaron (radical cation, λ_{max}=645 nm/1.92 eV), which is reversible if the potential is kept below 0.85 V vs Fc/Fc^+. The second oxidation induces conversion of polarons to bipolarons (dication, λ_{max}=545 nm/2.27 eV). This process is irreversible in that the bipolarons cannot be reduced back to polarons; however, reduction to the neutral PPE is possible, albeit at a much lower potential of –0.26 V vs Fc/Fc^+. At potentials above 0.9 V vs Fc/Fc^+, UV–Vis absorption spectra revealed another species with a λ_{max} of 380 nm, which was thought to be formed by further reaction of bipolarons.

Thus, a comparison of the data generated by Wrighton et al. [40] and Myrick et al. [46] suggests that high electrical conductivity in PPEs is related to

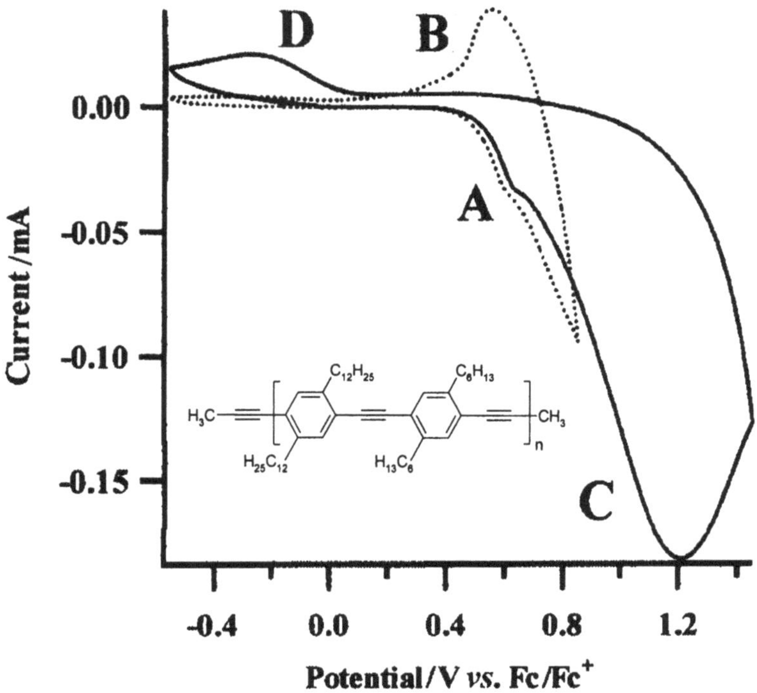

Fig. 3 Cyclic voltammogram for the oxidation of an alkyl-PPE (see *inset* for chemical structure) in CH_2Cl_2/0.1 M $[(n\text{-}Bu)_4N]ClO_4$ showing the effect of upper potential limit on the reversibility. Peak A: oxidation from neutral PPE to polaron. Peak B: reduction of polaron to neutral PPE. Peak C: oxidation from neutral PPE to bipolaron. Peak D: reduction of bipolaron to neutral PPE. Reproduced with permission from [46]

polarons, which are formed at intermediate doping levels. Higher doping levels lead to the formation of bipolarons, concomitant with a significant reduction of the electrical conductivity, compared to the window of maximum conductivity. The data also indicate that dialkyloxy-PPEs and dialkyl-PPEs display similarly large barriers for hole injection, which appear to be, at least in part, related to the electron-withdrawing nature of the ethynylene moieties [47].

Bunz et al. pointed out that it would be of interest to develop materials that combine the stability, electron affinity, and high emissive quantum yield of PPEs with the excellent hole injection capabilities of poly(*p*-phenylene vinylene)s (PPVs) [48]. In line with this notion, recent synthetic activities have focused on the engineering of the band gap, conduction band, and valence band of PAEs with the objective to render these materials more useful for practical applications that exploit their electrically (semi)conducting nature. Examples of materials that emerged from these efforts are discussed in detail in other portions of this volume (in particular the chapters by Bunz, Klemm, and Yamamoto). They include, among others, poly(heteroarylene ethynylenes) such

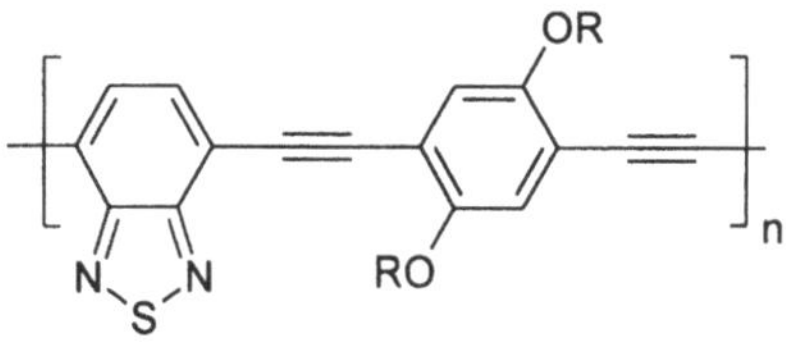

Fig. 4 Chemical structure of a family of PAEs developed independently by the groups of Bunz [57] and Yamamoto [52]. The polymers feature electron-poor 2,1,3-benzothiadiazole and electron-rich 2,5-dialkoxybenzene moieties in an alternating fashion

as the polymers investigated by the groups of Yamamoto [47, 49–52], Schanze [53], Pang [54], Klemm [55, 56], and Bunz [57, 58], cross-conjugated PPE–PPV hybrids investigated by the Bunz group [48, 59], and the linear PAE–PPV hybrids studied by Klemm and Bunz [60]. While no electrical conductivities have as yet been reported for any of these materials, some candidates appear to exhibit intriguing electronic features. For example, Bunz and Yamamoto have independently developed a family of PAEs that comprise electron-poor 2,1,3-benzothiadiazole units and electron-rich 2,5-dialkyloxybenzene moieties in an alternating fashion (Fig. 4; note that polymers with different alkyloxy side chains were investigated) [52, 57]. The polymers display an intriguing metallic luster in the solid state and display a band gap (optical band gap: ~2.0 eV) that is significantly reduced when compared to alkyloxy-PPEs (optical band gap: ~2.5 eV). The HOMO of these polymers appears to be essentially the same as in case of alkyloxy-PPEs, but electrochemical studies show that electrochemical oxidation is an irreversible process. Thus, p-doping does not appear to be a viable option [52, 57]. The LUMOs are significantly reduced when compared to those of alkyloxy-PPEs; CV spectra display a reversible reduction wave at ~−1.42 to +1.49 V vs Ag^+/Ag, which might correspond to the formation of polarons [52, 57]. Yamamoto further reports a second reversible reduction at −1.90 V vs Ag^+/Ag that is interpreted by the formation of bipolarons [52], while Bunz found a second irreversible reduction at −2.45 V vs Ag^+/Ag) [57]. While this apparent discrepancy needs to be resolved, these polymers clearly combine a low band gap with a low-lying LUMO, which makes them very interesting as n-type semiconductors, and possibly n-dopable electrical conductors. Unfortunately, their solid-state luminescence appears to be rather inefficient, so that their usefulness in LED applications is somewhat questionable.

In summary, it has been established that undoped PPEs are semiconductors, which exhibit an electrical conductivity of the order of $\sim 10^{-9}$ Scm^{-1}. Various derivatives have been p-doped by chemical or electrochemical means, and conductivities of up to ~5 Scm^{-1} were demonstrated. The charge carriers in these polymers were identified as polarons. Unfortunately, PAEs are notoriously difficult to oxidize, and no PAE derivative has yet been identified as sufficiently stable in its p-doped form to allow for technological exploitation under ambient conditions. PAEs which contain electron-poor groups such as 2,1,3-benzothiadiazole units have recently been shown to exhibit low-lying LUMOs. These polymers appear to combine good stability toward (photo)oxidation with the possibility for electrochemical n-doping. While electrical conductivities have yet to be reported for these materials, this materials platform promises to be a most interesting class of n-type (semi)conductors.

3 Charge-Transport Characteristics of PAEs

A detailed study of the charge-transport properties of poly[2,5-dioctyloxy-1,4-diethynylphenylene-*alt*-2,5-bis(2′-ethylhexyloxy)-1,4-phenylene] (EHO-OPPE, Fig. 5) was recently conducted by Weder and Singer et al. [61, 62]. The alkyloxy-PPE employed in this study, which appears to be the only carrier-mobility study as far as PAEs are concerned, is representative of this family of conjugated polymers [42]. The charge transport in EHO-OPPE of a number-average molecular weight, M_n, of about 10,000 gmol^{-1} was investigated using time-of-flight (TOF) measurements on indium tin oxide (ITO)/polymer/gold sandwich structures [63]. In this technique, a short light pulse incident on the polymer through a semitransparent electrode creates a thin sheet of charge carriers and, depending on the polarity of the electric field E applied between the electrodes, electrons or holes are driven across the sample. The absorption depth of the optical excitation is small compared to the sample thickness L, and the duration of the optical pulse is short compared to the transit time t_{tr} of the charge carriers. Thus, using $\mu = L/t_{tr} \cdot E$ the carrier mobility μ can be determined from the displacement photocurrent transients. ITO/EHO-OPPE/gold sandwich structures were produced by solution-casting the polymer from toluene, leading to samples in which the EHO-OPPE had a thickness of ~6.5–12 μm.

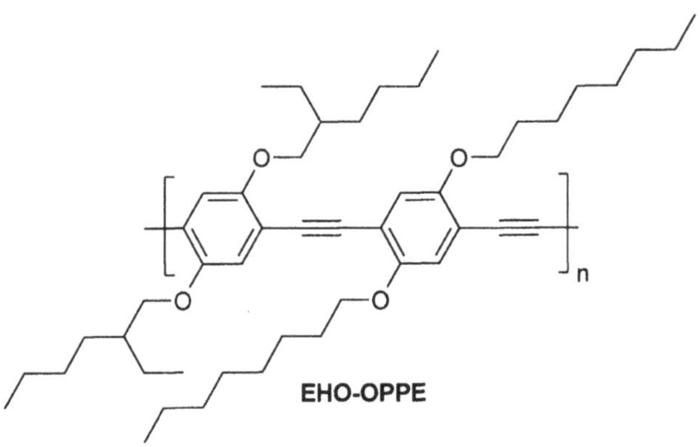

Fig. 5 Chemical structure of poly[2,5-dioctyloxy-1,4-diethynylphenylene-*alt*-2,5-bis(2′-ethylhexyloxy)-1,4-phenylene] (EHO-OPPE), the alkyloxy-PPE employed in the charge-transport studies by Weder and Singer et al. [61, 62]

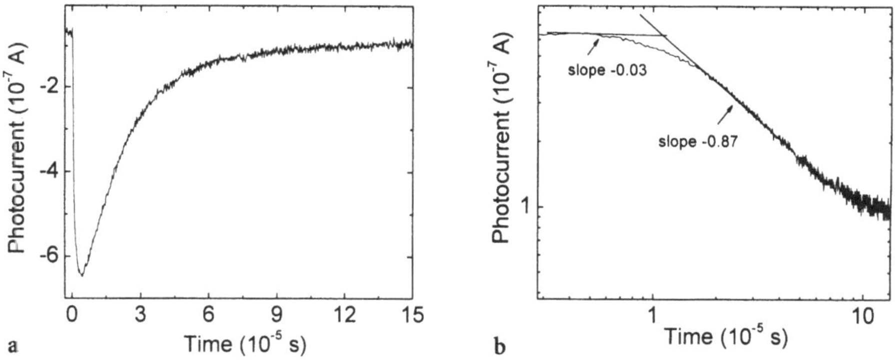

Fig. 6 Electron time-of-flight photocurrent transients of solution-cast film of EHO-OPPE (L=8 µm), measured at 295 K and an electric field of $2.5\cdot10^5$ V cm^{-1} in (**a**) linear and (**b**) double logarithmic plots. Reproduced with permission from [61]

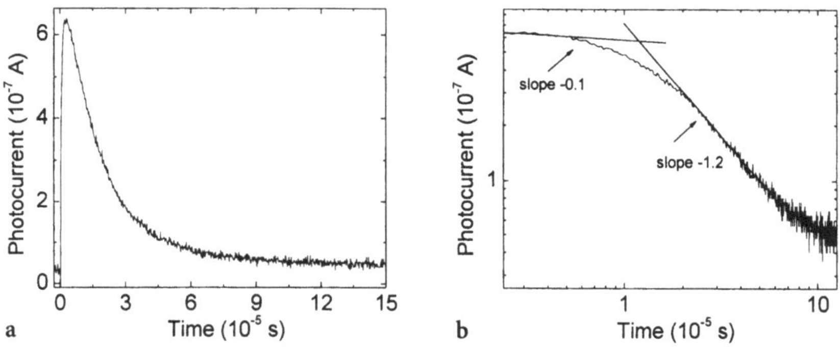

Fig. 7 Hole time-of-flight photocurrent transients of a solution-cast film of EHO-OPPE (L=8 µm), measured at 295 K and an electric field of $2.5\cdot10^5$ V cm^{-1} in (**a**) linear and (**b**) double logarithmic plots. Reproduced with permission from [61]

Typical photocurrent transients are shown in Fig. 6 for electrons and in Fig. 7 for holes. The shape of these curves is representative for all transients observed in the study and is characteristic of dispersive transport [64–68]. The carrier mobility μ was determined from the inflection point in the double logarithmic plots (cf. Fig. 6b and Fig. 7b) [74]. TOF measurements were performed as a function of carrier type, applied field, and film thickness (Fig. 8). As can be seen from Fig. 8, the drift mobility is independent of L, demonstrating that the photocurrents are not range-limited but indeed reflect the drift of the carrier sheet across the entire sample. Both the independence of the mobility from L, and the fact that the slopes of the tangents used to determine the mobility (Fig. 6 and Fig. 7) do not add to −2 as predicted by the Scher–Montroll theory, indicate that the Scher–Montroll picture of dispersive transients does not adequately describe the transport in amorphous EHO-OPPE [69]. The dispersive nature of the transient is due to the high degree of disorder in the sample and its impact on car-

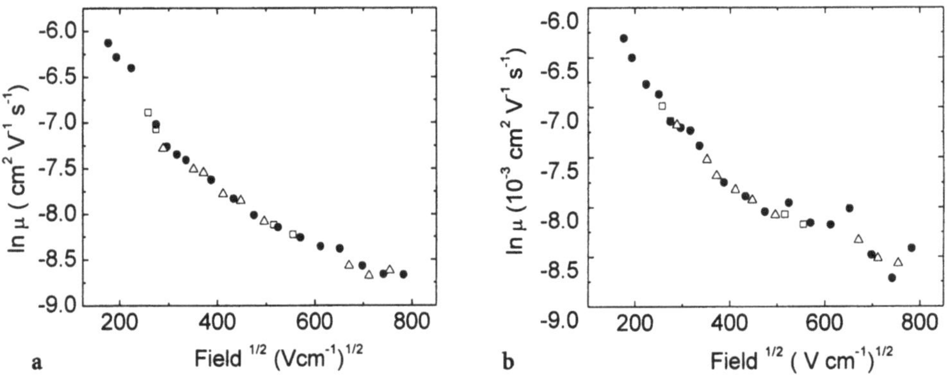

Fig. 8 Electron (**a**) and hole (**b**) mobilities of EHO-OPPE films as a function of electric field at various film thicknesses (*open triangles*: L=6.5 μm, *filled circles*: L=8 μm, *open squares*: L=12 μm). Reproduced with permission from [61]

rier transport [74]. Although the energetic disorder described by a Gaussian density of states certainly contributes to the dispersive shape of the transients [70], the independence of μ from L indicates that other types of disorder will also contribute [71]. The observed disorder likely arises from the large degree of positional disorder as described below, and perhaps polaron or intrachain mechanisms in these conjugated polymers.

High electron ($2.2 \cdot 10^{-3}$ cm^{2}V^{-1}s^{-1}) and hole ($1.8 \cdot 10^{-3}$ cm^{2}V^{-1}s^{-1}) mobilities were found at low field ($3.1 \cdot 10^4$ V cm^{-1}). This finding is most intriguing, since these values compare favorably with the values observed for other ambipolar conjugated polymers [72, 73]. The data plots and fits in Fig. 9 indicate that the temperature dependence is well-described by the Gaussian disorder model summarized below. As is evident from Fig. 8a,b, the mobility strongly depends on the field strength, and decreases with increasing bias. This somewhat exceptional

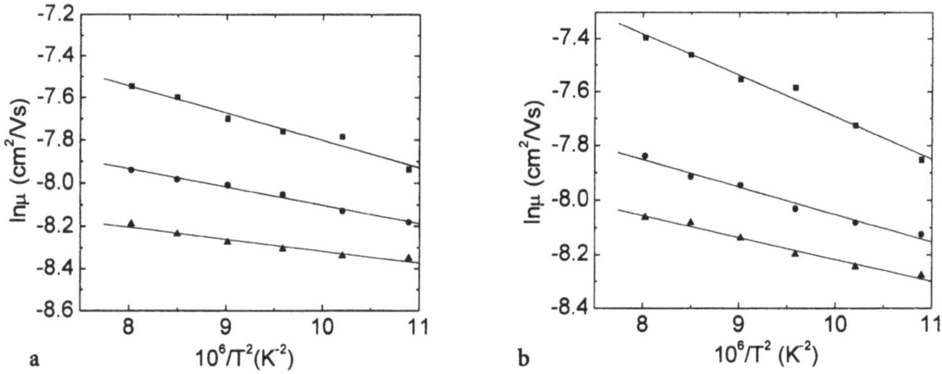

Fig. 9 Temperature dependence of (**a**) electron and (**b**) hole mobilities of an EHO-OPPE film (L=8 μm) measured at E=$2.0 \cdot 10^5$ (*squares*), $3.0 \cdot 10^5$ (*circles*), and $4.0 \cdot 10^5$ V cm^{-1} (*triangles*). Reproduced with permission from [61]

behavior has been observed before [74, 66], particularly at low electric fields, and is consistent with a hopping transport model that accounts for off-diagonal (positional) disorder caused by variations of intersite distances in addition to diagonal (energetic) disorder in the transport manifold [75, 76]. The disorder transport model considers charge transport in organic solids as time-independent random walks within a distribution of localized hopping states broadened by disorder effects. The large off-diagonal disorder may result in a negative field dependence of the mobility at low fields, because a stronger applied electric field favors forward hopping and inhibits more facile routes for carriers involving hops transverse to the applied electric field. The negative field dependence was also predicted for quasi one-dimensional transport in the presence of defects and barriers [77].

To analyze the negative field dependence of the mobility in EHO-OPPE within the Gaussian disorder transport formalism and to determine the diagonal (energetic) disorder parameter $\hat{\sigma}$ and the off-diagonal (positional) disorder parameter $\hat{\sigma}$, the following relation between the charge mobility μ and the disorder parameters was employed [75]:

$$\mu(\hat{\sigma}, \Sigma, E) = \mu_0 \exp\left[-\left(\frac{2}{3}\hat{\sigma}\right)^2\right] \exp C(\hat{\sigma}^2 - \Sigma^2)\, E^{1/2} \tag{1}$$

where μ_0 is the prefactor mobility, E is the electric field, and C is an empirical constant. The diagonal disorder parameter $\hat{\sigma}$ is related to the width of the Gaussian density of states σ, as $\hat{\sigma}=\sigma/kT$, where k is the Boltzman constant and T is the temperature. Equation 1 can be represented in the form:

$$\ln \mu = \alpha(E)\frac{1}{T^2} + g(E) \tag{2}$$

where

$$\alpha(E) = \frac{\sigma^2}{k^2}\left(C\sqrt{E} - \frac{4}{9}\right) \tag{3}$$

and

$$g(E) = \ln \mu_0 - C\sqrt{E}\,\Sigma^2 \tag{4}$$

The experimental temperature dependence of the mobility parametric in the applied electric field (Fig. 9) allows for calculating α and g for different values of the electric field. Because $\alpha(E)$ scales linearly with $\sqrt{E}$ the value $\alpha_{E=0}$ can be determined by plotting α vs $\sqrt{E}$ and then determining the intersection of the linear fit of the data with E=0. Knowing $\alpha_{E=0}$ allows one to determine the width of density of states σ from the relation

$$\sigma = \frac{3}{2}k^2\sqrt{\alpha_{E=0}} \tag{5}$$

Finally, the implicit function $g(\alpha)$ should be linear, because the derivative $\frac{\partial g}{\partial \alpha}$ is a constant:

$$\frac{\partial g}{\partial \alpha} = \frac{\dfrac{\partial g}{\partial \sqrt{E}}}{\dfrac{\partial \alpha}{\partial \sqrt{E}}} = \frac{-C\Sigma^2}{C\dfrac{\sigma^2}{k^2}} = -\frac{\Sigma^2 k^2}{\sigma^2} = \varphi \tag{6}$$

Therefore, calculating the slope $\varphi=\frac{\partial g}{\partial \alpha}$ of the plot $g(\alpha)$ allows for the determination of the value of the off-diagonal disorder parameter Σ:

$$\Sigma^2 = -\varphi\frac{\sigma^2}{k^2} \tag{7}$$

Table 2 shows the calculated values for the density of states at zero field $\sigma_{E=0}$, the disorder parameters $\hat{\sigma}$ and Σ, and the constant C for holes and electrons in EHO-OPPE. Figure 10 displays the temperature dependence of the diagonal disorder parameter $\hat{\sigma}$. It is apparent from the data presented in Table 2 and Fig. 10 that the off-diagonal disorder parameter Σ is larger than the diagonal disorder parameter over the entire temperature range experimentally investigated. The temperature-independent off-diagonal disorder parameter Σ was 3.4 and 3.6 for electrons and holes, respectively, while the range of the temperature-dependent diagonal disorder parameter $\hat{\sigma}$ was between 2.8 and 2.5 and 3.0 and 2.7 in the temperature range between 20 and 50 °C for electrons and holes, respectively. Monte Carlo simulations of carrier hopping in the transport manifold with energy levels with a Gaussian distribution [74, 76] predicted that for large Σ, $\partial \ln \alpha/\partial \sqrt{E}$ becomes negative at low electric fields ($E \leq 5\cdot 10^5$ V cm^{-1}) and does not obey conventional Poole–Frenkel law $\ln \mu \sim \sqrt{E}$. At higher fields, $\partial \ln \alpha/\partial \sqrt{E}$ becomes positive and gives rise to Poole–Frenkel-like behavior. Unfortunately, it was impossible to measure the mobility at higher fields

Table 2 Results of data fits to the Gaussian disorder model of charge transport in EHO-OPPE[a]

Charge carrier	$\sigma_{E=0}$ (eV)[b]	Σ[c]	$\hat{\sigma}_{20\,°C}$[d]	C (cm/V)$^{1/2}$
Holes	0.072	3.6	3.0	5.4×10^{-4}
Electrons	0.068	3.4	2.8	5.7×10^{-4}

[a] See text and Eq. 1 for a detailed description.
[b] Width of the Gaussian density of states.
[c] Off-diagonal (positional) disorder parameter.
[d] Diagonal (energetic) disorder parameter at 20 °C.

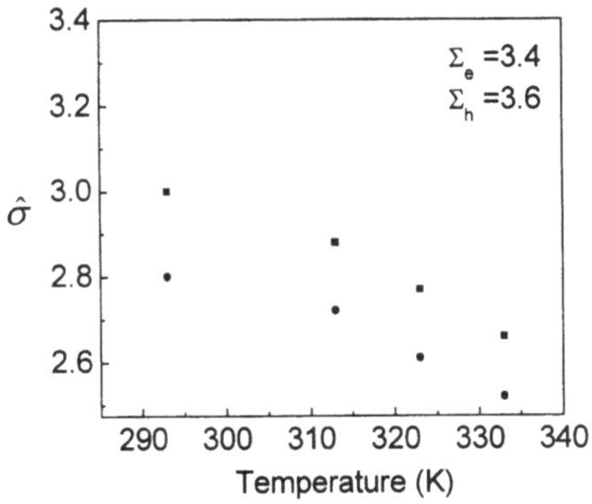

Fig. 10 Temperature dependence of the diagonal disorder parameter $\hat{\sigma}$ for holes (*squares*) and electrons (*circles*). Parameters have been obtained by fitting the experimental data (Fig. 9) to Eq. 1 and Eq. 5. Reproduced with permission from [61]

because of the dielectric breakdown of the polymer. However, the observed negative field dependence of the mobility in EHO-OPPE with high off-diagonal disorder is consistent with the disorder transport formalism applied to a system with both positional and energetic disorders.

Thus, EHO-OPPE exhibits a comparably high ambipolar carrier mobility, but the charge transport in this material is – as in virtually all amorphous conjugated polymers [43–45, 78] – unequivocally limited by disorder effects, which prevent efficient interchain coupling and lead to one-dimensional electronic properties. Rapid charge transport, however, is important for the exploitation of these materials in electronic devices [2]. Enhanced interchain interactions, which concur with improved charge-carrier mobility, are observed in highly ordered structures [43–45], but their fabrication is usually intricate. The Weder group has embarked on an orthogonal approach, and shown recently that rapid charge transport can also be achieved by introducing conjugated cross-links between conjugated macromolecules [62]. Interestingly, this structural motif has received little attention in the past, at least with regard to systematic studies and well-defined systems. This situation may be a direct consequence of the challenge to introduce conjugated cross-links into conjugated polymers and retain adequate processibility. Hence, examples of conjugated polymers with conjugated cross-links based on either covalent bonds [79, 80]or metal complexes between chains [81, 82] are rare, and in many cases have been obtained serendipitously or lack unambiguous characterization. Weder et al. have shown that this problem can be overcome by synthesizing such polymers in the form of spherical particles, which can be processed from (aqueous) dispersions [83, 84]. Applying concepts employed before for the preparation of dispersions of linear conjugated polymers [85–87], exploiting the fact that the metal-catalyzed cross-coupling reaction employed for the synthesis of PPEs is tolerant to the presence of water [88], and employing here 1,2,4-tribromobenzene as cross-linker in the Pd-catalyzed cross-coupling of 2,5-diiodo-4-[(2-ethylhexyl)-

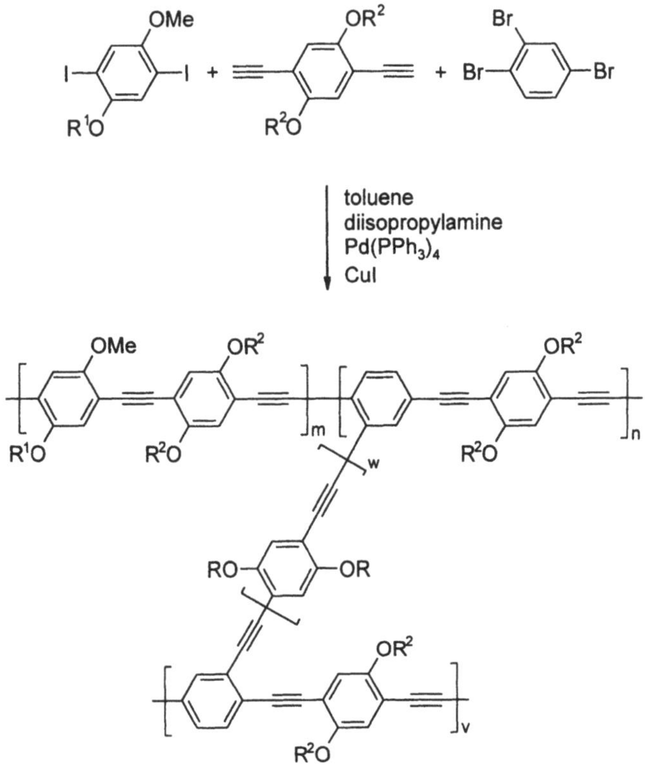

Scheme 1 Synthesis of cross-linked PPEs by the palladium-catalyzed cross-coupling reaction of 2,5-diiodo-4-[(2-ethylhexyl)oxy]methoxybenzene, 1,4-diethynyl-2,5-bis(octyloxy)benzene, and the trifunctional cross-linker 1,2,4-tribromobenzene. R^1=2-ethylhexyl, R^2=*n*-octyl

oxy]methoxybenzene and 1,4-diethynyl-2,5-bis(octyloxy)benzene (Scheme 1), it was demonstrated that cross-linked conjugated PPE particles can be conveniently produced by polymerization in aqueous macro-, micro-, and miniemulsions (Fig. 11) [83, 84].

In another approach, Weder et al. have shown that organometallic networks can readily be prepared by ligand-exchange reactions between conjugated macromolecules and low-molecular metal complexes [62, 89–91]. The conjugated polymer employed in this study was EHO-OPPE (Fig. 5) and the ethynylene groups were employed as "built-in" ligand sites (Scheme 2). After an initial study that focused on (nonconjugated) dinuclear Pt^{II} cross-links [89], Pt^0 was chosen as cross-linker, since it forms stable bis(ethynylene) complexes [90, 92], which due to π-backbonding should allow for electronic conjugation [93]. A styrene solution of $Pt(styrene)_3$ served as the Pt^0 source [94]. An in situ NMR investigation of model reactions of $Pt(styrene)_3$ with diphenylacetylene (DPA) confirmed that the ethynylene moieties in the PPE readily and selectively coordinate to Pt^0 centers with the release of the relatively weakly bound styrene ligands and formation of bis(diphenylacetylene)platinum [90].

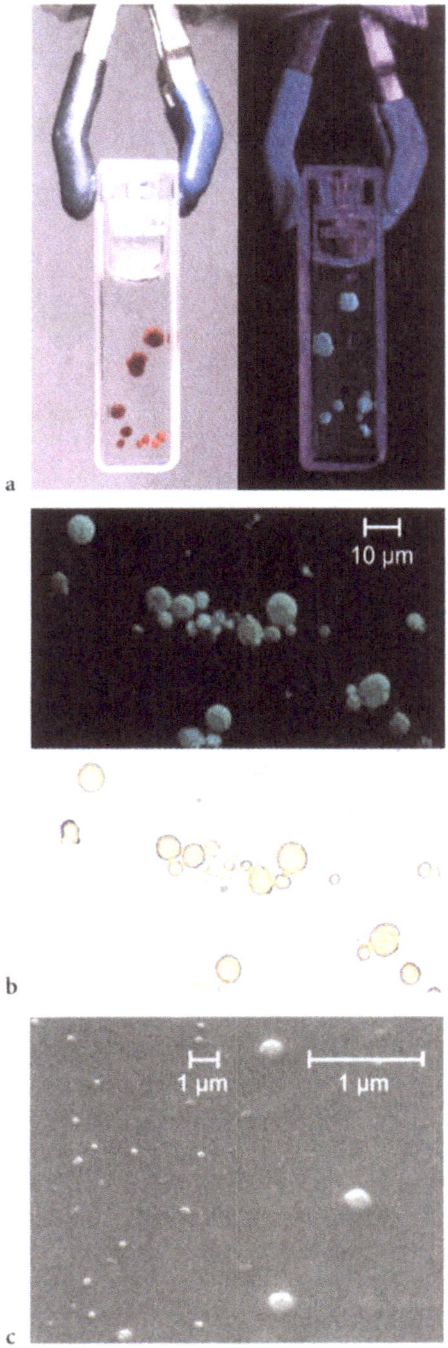

Fig. 11 Photograph (**a**), optical micrograph (**b**), and scanning electron micrograph (**c**) of cross-linked PPE milli- (**a**), micro- (**b**), and nanoparticles (**c**) prepared by palladium-catalyzed cross-coupling reactions in aqueous emulsions. Reproduced with permission from [83]

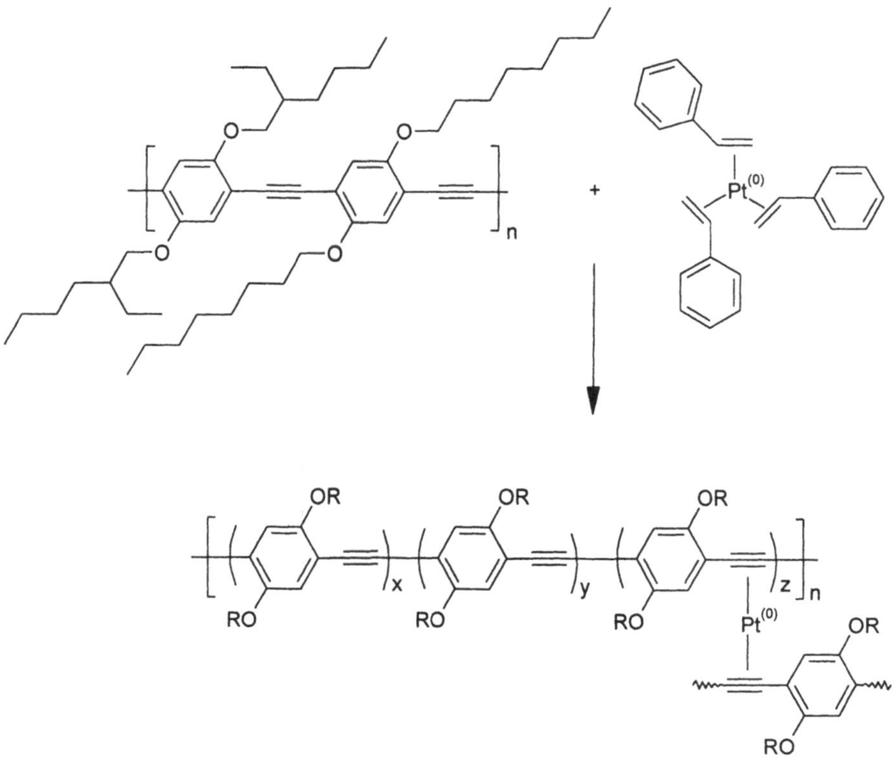

Scheme 2 Ligand-exchange reaction between EHO-OPPE and $[Pt(PhCH{=}CH_2)_3]$ leading to EHO-OPPE-Pt^0 networks

The same framework was subsequently employed for the formation of EHO-OPPE-Pt^0 networks [91]. The ratio of the molar concentrations of Pt^0 and phenylene ethynylene (PE) moieties, $[Pt^0]/[PE]$, was varied between 0.016 and 0.34. The reaction mixtures gelled within seconds, consistent with the formation of cross-linked structures (Scheme 2). Homogeneous films could be produced by spin and solution casting before gelation. On the other hand, gels with high solvent content (>95% w/w) could readily be produced. Consistent with their cross-linked structure, the films were insoluble in solvents for EHO-OPPE. By contrast, styrene was found to readily dissolve EHO-OPPE-Pt^0, demonstrating that the ligand exchange is reversible.

As expected, the coordination of Pt^0 markedly influences the photophysical characteristics of the PPE. The photoluminescence is efficiently quenched, and the absorption maximum in the visible regime experiences a hypsochromic shift. The charge-carrier mobility of different EHO-OPPE-Pt^0 samples was determined by TOF measurements as described above for the neat EHO-OPPE. The shape of the photocurrent transients of all EHO-OPPE-Pt^0 samples was similar to those shown in Figs. 6 and 7 for the neat EHO-OPPE. This indicates that these organometallic conjugated polymers networks are also characterized

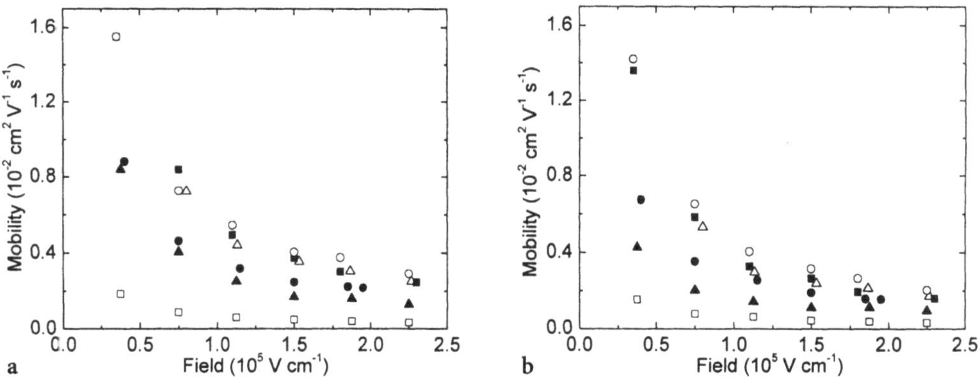

Fig. 12 Electron (**a**) and hole (**b**) mobility of EHO-OPPE-Pt^0 as a function of $[Pt^0]/[PE]$ and electric field F ($[Pt^0]/[PE]$: □=0, ▲=0.016, ●=0.086, △=0.17, ■=0.25, ○=0.34). Reproduced with permission from [62]

by dispersive transport characteristics [62]. The Pt^0 does not act as a dopant or as a trap for charges. Importantly, as can be seen from the data compiled in Fig. 12, the carrier mobility strongly increases upon introduction of Pt^0. While a distinct enhancement of the mobility is observed for EHO-OPPE-Pt^0 with small $[Pt^0]/[PE]$, the effect levels off at a $[Pt^0]/[PE]$ of ~0.17, where a charge carrier mobility of $1.6 \cdot 10^{-2}$ $cm^2V^{-1}s^{-1}$ (electrons) and $1.4 \cdot 10^{-2}$ $cm^2V^{-1}s^{-1}$ (holes) is reached. These values are an order of magnitude higher than those of the neat PPE. Higher Pt^0 contents did not lead to a further increase of μ. Thus, the cross-link density may have reached a saturation limit, or – as indicated by the optical data – the excessive complexation may limit the effective conjugation length of the PPE. The EHO-OPPE-Pt^0 networks display the same field dependence as the parent PPE, which suggests that introduction of Pt^0 preserves the large off-diagonal disorder in the network [61, 62].

In summary, it has been established that EHO-OPPE exhibits high ambipolar mobility. The positional (off-diagonal) and energetic (diagonal) disorder parameters have been calculated from the experimental temperature and field dependences of the hole and electron mobilities. The observed negative field dependence of the mobility was explained within the Gaussian disorder formalism to originate from high off-diagonal disorder. The experimental temperature- and field-dependent mobilities are consistent with a Gaussian disorder transport formalism applied to systems exhibiting both positional and energetic disorders. The introduction of conjugated cross-links between the conjugated macromolecules leads to a substantial increase of the charge-carrier mobility. The charge-carrier mobilites of the EHO-OPPE-Pt^0 networks discussed represent the highest mobilities yet observed in disordered conjugated polymers and also compare well to the hole mobilities of ordered materials (~10^{-1} $cm^2V^{-1}s^{-1}$) [45]. The ease of processing and the ambipolar characteristics of the organometallic PAE networks are particularly intriguing, and may allow a new generation of higher-performance semiconducting devices.

4 Light-Emitting Diodes

4.1 General Considerations

Light-emitting diodes (LEDs) are rectifying semiconductor devices that convert electrical energy into electromagnetic radiation and emit light in the ultraviolet, visible, or infrared regime of the electromagnetic spectrum. Conventional LEDs, commercially introduced in the 1960s, are based on inorganic semiconductors such as GaAs, GaP, AlGaAs, GaAsP, InGaP, etc. The role of polymers in LEDs has traditionally been associated with their use as a structural material. Epoxy resins are routinely applied as the transparent "packaging material" in conventional LEDs, in which light is generated at the junction of two *inorganic* semiconductors. The polymer provides the structural integrity of the device, and serves as the optical medium that enables the extraction of light [95]. During the last 15 years, however, LEDs have been developed in which the semiconducting, electroluminescent (EL) material itself is a polymer. These devices are usually referred to as polymer LEDs or PLEDs, and have attracted significant attention in both academic and industrial environments [4–6, 96]. The principal interest in the use of EL polymers is based on their promise to combine the ease of processibility and mechanical flexibility of macromolecular materials with the exceptional, readily tailored properties of organic semiconductors. PLEDs offer the prospect of low production costs, mechanical flexibility, light weight, large area, fast switching times, low power consumption, high brightness, large viewing angle, and crisp colors. Thus, the new technology may feature distinct advantages over existing flat-panel display devices such as conventional emissive displays, inorganic LEDs, and liquid crystal displays (LCDs). As a result, the list of potential applications of PLEDs ranges from small, low-information-content displays for simple electronic devices such as pagers and portable phones to high-resolution video-rate displays for use in desktop monitors and ultrathin television sets. In addition, PLEDs may be employed as flat lighting devices, for example in automotive rear lights or as backlights in LCDs.

Electroluminescence in organic materials was originally discovered by Pope et al. in the early 1960s [97], but useful devices were only demonstrated some 20 years later by Tang and VanSlyke [98] and Adachi and coworkers [99]. Electric-field-induced light emission from a fluorescent, semiconducting polymer was first observed in 1990, when Burroughes et al. applied an electrical field to a thin film of poly(*p*-phenylene vinylene) (PPV, Fig. 13) [100]. The LED reported in this initial paper is an embodiment of the most simple device configuration of a polymer LED, which, in accordance with the number of organic layers used, is usually referred to as a single-layer device (Fig. 14). The semiconducting polymer film (typical thickness 50–500 nm) is sandwiched between two electrodes, of which at least one has to be (semi)transparent in order to allow the generated light to escape from the device. Adopting a standard Schottky configuration, a

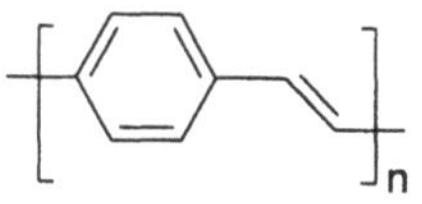

Fig. 13 Chemical structure of poly(*p*-phenylene vinylene) (PPV)

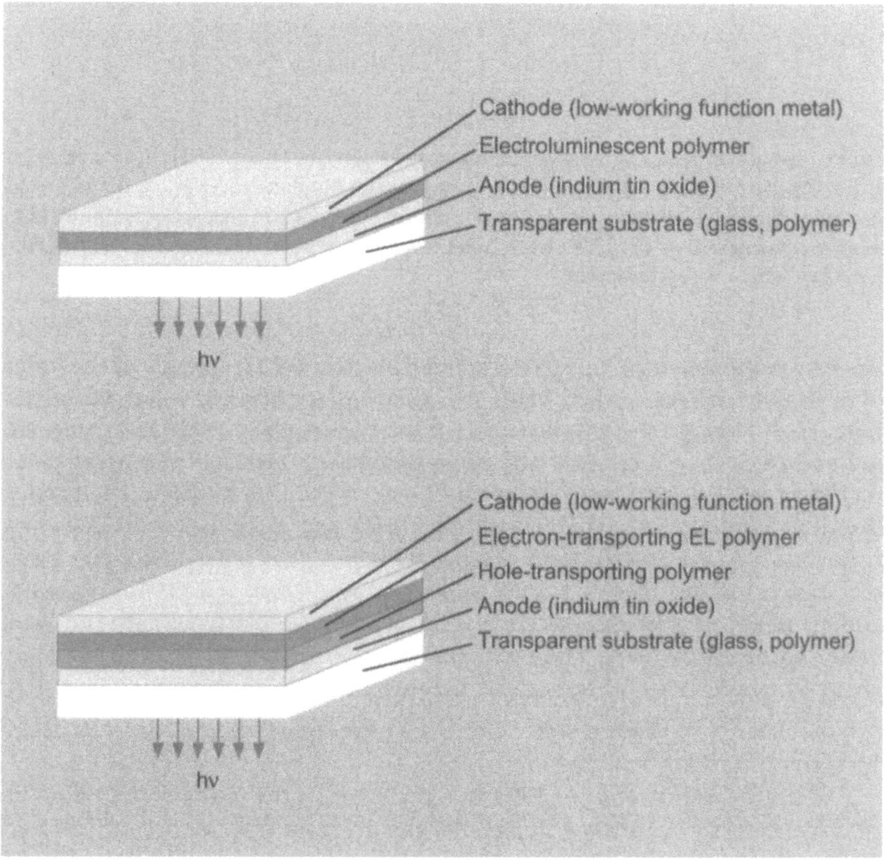

Fig. 14 Schematic drawings of the configurations of a single-layer PLED (*top*) and a two-layer PLED (*bottom*)

high-work-function anode such as semitransparent indium tin oxide, and a low-work-function cathode such as Al, Mg, or Ca are employed, leading to non-linear rectification and, thus, rendering the device into a diode. If an external voltage, or bias, is applied, oppositely charged carriers – holes and electrons – are injected above a threshold voltage from the electrodes into the valence and conduction bands, respectively, of the semiconducting polymer (Fig. 15). These electronic levels correspond to the ionization potential (IP) and electron affin-

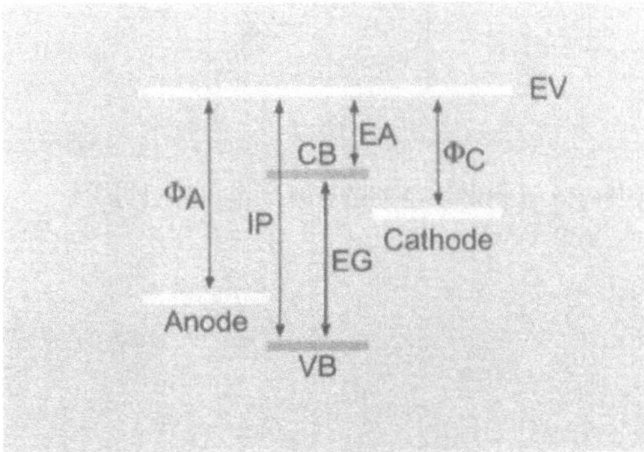

Fig. 15 Simplified schematic representation of the electronic energy levels in a single-layer PLED. *CB* and *VB* are the conduction band and valence band, respectively, of the semiconducting polymer, which correspond to the ionization potential (*IP*) and electron affinity (*EA*) relative to vacuum level (*EV*). The work functions for anode (Φ_A) and cathode (Φ_C) and the band gap (*EG*) are also indicated

ity (EA) of the polymer. Due to the applied electric field (typically of the order of 10^6 V cm^{-1}, corresponding to an operating voltage of a few volts), these carriers travel through the polymer toward the oppositely charged electrode. On their way, the charge carriers will either recombine within the emitting polymer layer with an opposite charge and form singlet and triplet excited states (excitons), or they reach the opposite electrode and discharge without having contributed to the EL effect. The singlet excited states created through charge recombination are identical with those generated through photoexcitation, and, as in the case of fluorescence, they relax to the electronic ground state under the emission of light. The color of the latter is governed by the band gap of the EL polymer, i.e., the energy difference between the singlet excited state and the electronic ground state. The triplet excited states, by contrast, decay through nonradiative processes.

The quantum efficiency of a PLED is governed by the (solid-state) photoluminescence (PL) quantum efficiency of the polymer, the fraction of injected charge carriers that form excited states, and the fraction of singlet excited states created (as opposed to triplets). Besides maximizing the PL quantum yield, the optimization of charge injection and transport (both of which directly influence the fraction of injected charge carriers that form excited states) are important aspects for the design of efficient PLEDs. In order to allow for efficient charge injection and, concomitantly, low operating voltages, the work functions of cathode and anode should match the electron affinity and ionization potential (or in semiconductor terminology, conduction and valence band) of the semiconducting emitter, respectively, as closely as possible. Unequal barrier heights at the two electrode–polymer junctions are undesirable, because one carrier will be injected preferentially, leading to an increased driving voltage and a reduced luminous efficiency. On the other hand, most organic semiconductors are *p*-type conductors, allowing holes to be transported more efficiently than electrons. Consequently, substantial research efforts have been made regarding the identification of optimal electrode–polymer combinations. A general possibility, of course, is the chemical modification of the emitting polymer, with the intent to tailor – besides the color-determining band gap – its electron affinity, ionization potential, and/or charge-transport characteristics. Another strategy is the fabrication of two-layer or three-layer devices, in which a hole-transporting (HT) polymer and an electron-transporting (ET) polymer are combined (Fig. 14). The resulting heterojunction between the two semiconducting polymers allows control of the charge-injection rates and helps to confine the charge recombination to the emitting layer (in a two-layer configuration either HT or ET material also serves as the emitting layer, while in a three-layer device the emitting layer is sandwiched between HT and ET layers). In addition, this architecture effectively reduces problems associated with the migration of oxygen and metal ions from the electrodes into the organic layer and the related creation of quenching sites in the emission zone. Consequently, the two-layer architecture has become the preferred PLED embodiment.

4.2 Light-Emitting Diodes Based on PAEs

Many PAEs are strongly photoluminescent and display high quantum efficiency in solution as well as in the solid state [14, 15]. This particular feature, in combination with their semiconducting nature and good thermal and optical stability [39, 101, 102], should make PAEs excellent candidates for application as the emitting layer in LEDs. Triggered by the initial success and widespread use of PPVs in LED applications, a number of studies have focused on PPEs, i.e., dehydrogenated congeners of the latter. The first study on PPE-based LEDs was conducted by the groups of Shinar and Barton [103–106]. The LEDs were fabricated under ambient conditions by spin-coating films of poly(2,5-dihexyloxy-*p*-phenylene ethynylene) or poly(2,5-dibutyloxy-*p*-phenylene ethynylene) (thickness 50–500 nm) onto ITO-coated glass substrates and applying Ca:Al or Al as cathode (Table 3, entries 1 and 2) [103]. When operated under ambient conditions and with no heat sink applied, devices with a Ca:Al cathode degraded within minutes. These devices displayed a somewhat higher stability (lifetime of the order of hours) when operated under He atmosphere at low temperatures. Under these conditions an ITO/poly(2,5-dihexyloxy-*p*-phenylene ethynylene)/Ca:Al device displayed highly non-ohmic current–voltage (*I*-*V*) curves, which reflect the rectifying character of the devices. The reported onset voltage for EL of 40 V is relatively high, but it appears to be related to the significant thickness (500 nm) of the emitting polymer layer. The EL spectrum showed a maximum around 600 nm, which was redshifted by about 20 nm with respect to the PL spectrum. In contrast to the PL spectrum, the phonon side bands of the EL spectrum were broad and poorly resolved. A broad, structureless band was observed around 800 nm, which was absent in the PL spectrum, and has been explained

Table 3 Device architecture and performance of light-emitting diodes with PAE emitting layer

Entry	Polymer	Device Structure[1]	Onset Voltage [V]	Luminance [cd/m^2]	Efficiency [cd/A] / [%]	λ_{max} [nm]	Comments	Reference
1	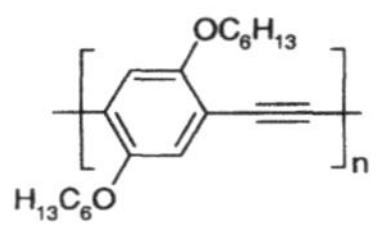	ITO/PPE(~500 nm)/Ca:Al	40	$5\cdot10^{-4}$ mW	0.1-0.5 %	~600, 800		[103]
		ITO/PPE(~500 nm)/Al	n.a.	n.a.	0.002 %	~600, 800	performance is increased upon annealing at 150°C	
2	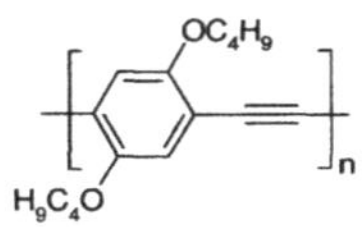	ITO/PPE(~500 nm)/Al	n.a.	n.a.	n.a.	~575	performance is increased upon annealing at 150°C	[103]
3	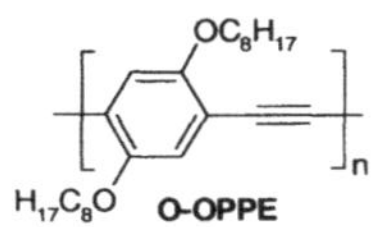 O-OPPE	ITO/PPE(100 nm)/Ca	14.2	35	0.02 %	535		[107]
		ITO/PPE(100 nm)/Al	11	80	0.032 %	535		
4	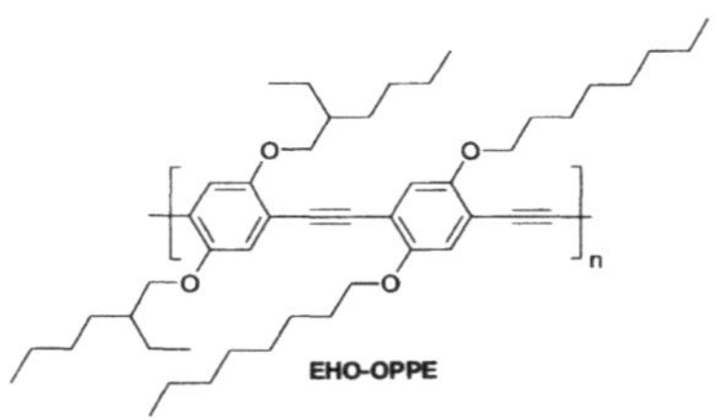 EHO-OPPE	ITO/PPE(100 nm)/Ca	14.5	38	0.023 %	535		[107]
		ITO/PPE(100 nm)/Al	10.8	80	0.035 %	535		
		ITO/PPE(100 nm)/Cr	19.7	33	0.015 %	535		

[1] See text for more detailed descriptions. Where available, the thickness of the EL layer was included.

Table 3 (continued)

Entry	Polymer	Device Structure	Onset Voltage [V]	Luminance [cd/m^2]	Efficiency [cd/A] / [%]	λ_{max} [nm]	Comments	Reference
5	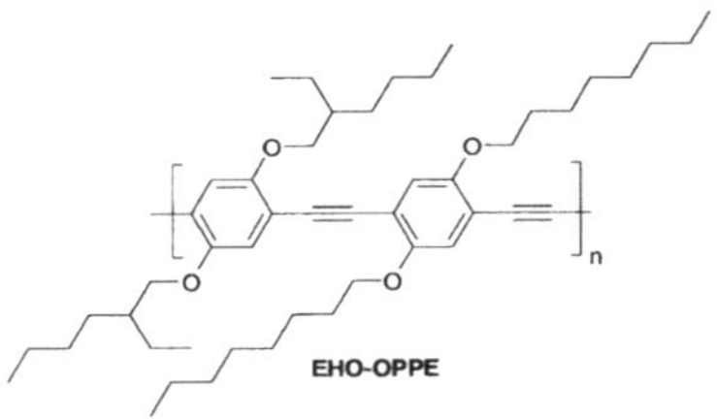 EHO-OPPE	ITO/PPE(100 nm)/Poly-TPD(60nm)/Al	10	19	0.016 cd/A	n.a.		[111]
		ITO/75%PPE:25%Poly-TPD (110nm)/Al	9	33	$4 \cdot 10^{-4}$ cd/A	n.a.		
		ITO/50%PPE:50%Poly-TPD (110nm)/Al	8.5	31	0.002 cd/A	n.a.		
		ITO/25%PPE:75%Poly-TPD (110nm)/Al	9.5	146	0.02 cd/A	n.a.		
	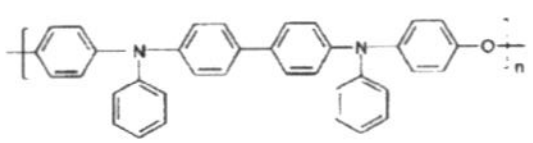 poly-TPD	ITO/10%PPE:90%Poly-TPD (110nm)/Al	9.5	4	0.007 cd/A	n.a.		
		ITO/25%PPE:75%Poly-TPD (110nm)/spiro-Qux (48nm)/Al	4.5	257	0.15 cd/A	n.a.		
	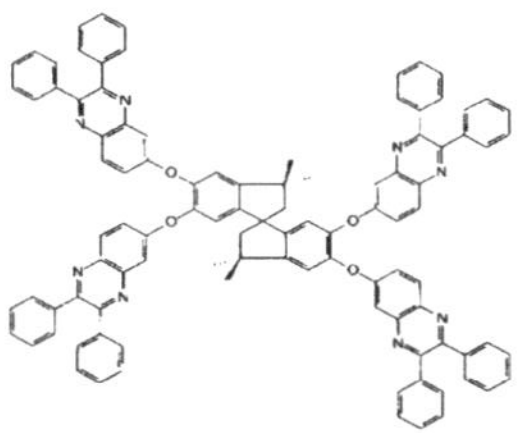 spiro-Qux							

Table 3 (continued)

Entry	Polymer	Device Structure	Onset Voltage [V]	Luminance [cd/m²]	Efficiency [cd/A] / [%]	λ_{max} [nm]	Comments	Reference
6	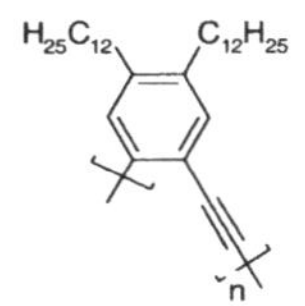	ITO/PAE/Al	n.a.	n.a.	n.a.	~400 (3.1 eV)		[112]
7	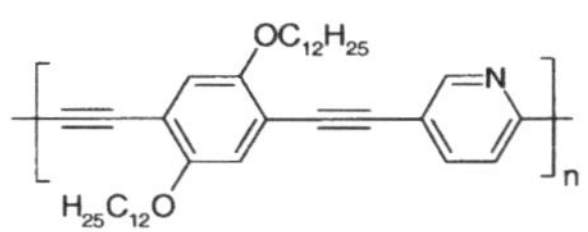	ITO/PPE/Al	n.a.	n.a.	n.a.	~480 (2.6 eV)		[114]
8	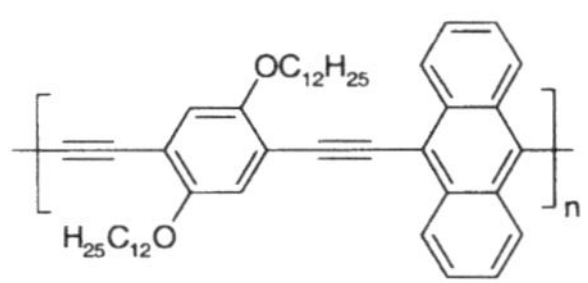	ITO/PAE/Al	n.a.	n.a.	n.a.	~590 (2.1 eV)		[114]
9	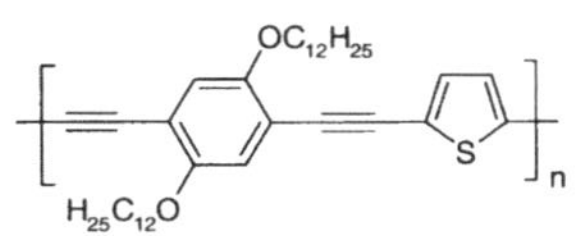	ITO/PAE/Al	n.a.	n.a.	n.a.	~530 (2.3 eV)		[114]

Table 3 (continued)

Entry	Polymer	Device Structure	Onset Voltage [V]	Luminance [cd/m^2]	Efficiency [cd/A] / [%]	λ_{max} [nm]	Comments	Reference
10	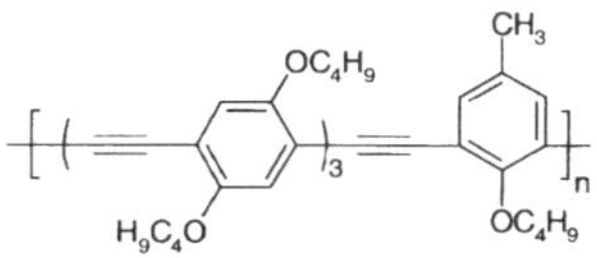	ITO/PEDOT/PAE/Ca:Al	8	21	0.013 %	~547-577	EL spectra display voltage-dependence	[116]
11	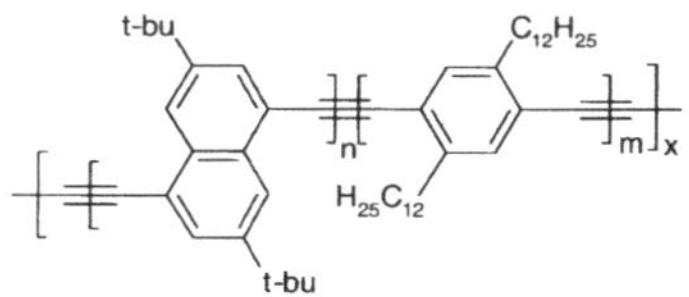	ITO/PEDOT/PAE (n:m=1:2, 80 nm)/LiF/Al	5.5	100 (10V)	0.1 Cd/A	472		[118]
		ITO/PEDOT/PAE (n:m=1:1, 80 nm)/LiF/Al	~7	20 (16 V)	0.02 Cd/A	428		
12	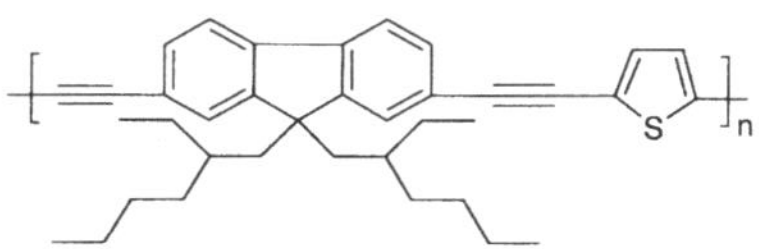	ITO/CuPc/PAE/Ca:Ag	11	n.a.	0.01 %	541, 585	broad, almost-white emission	[119]

Table 3 (continued)

Entry	Polymer	Device Structure	Onset Voltage [V]	Luminance [cd/m^2]	Efficiency [cd/A] / [%]	λ_{max} [nm]	Comments	Reference
13	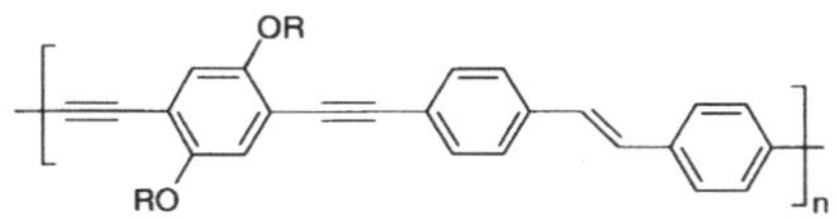	ITO/PAE(100-200 nm)/Al	12	n.a.	~0.05 %	~540(2.3 eV)	I-V curves under forward and reverse bias are symmetric	[120]
14	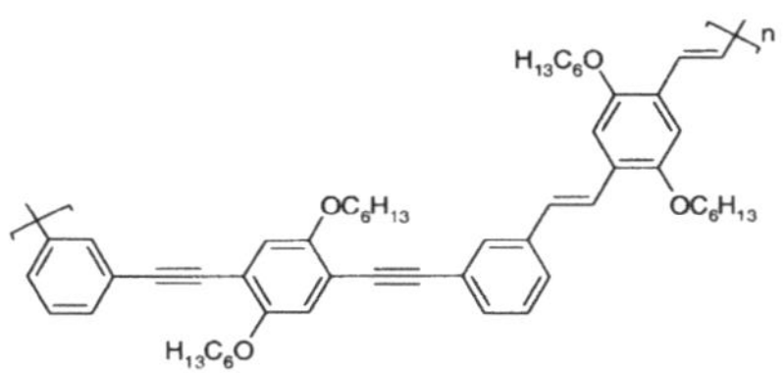	ITO/PPV/PAE/Ca	8	300	0.15 cd/A	513,529	EL spectra display voltage-dependence	[121]
		ITO/PEDOT/PAE/Ca	11	130	0.32 cd/A	498,523		
15	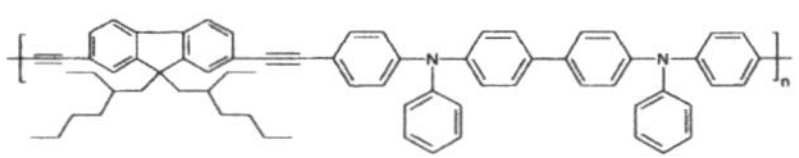	ITO/PAE:PEO:LiTf/Al	2.9		0.47 cd/A		electrochemical cell	[124]

by structural defects at the polymer/metal interface. A maximum brightness of ~$5 \cdot 10^{-4}$ mW was reported for high current densities, but apparently the lifetime under these conditions was limited to seconds. The reported (presumably external) photon/electron yield was in the 0.1–0.5% range, which is similar to early results reported for ITO/PPV/Ca devices.

An LED based on poly(2,5-dihexyloxy-*p*-phenylene ethynylene) but with an Al instead of a Ca:Al cathode was reported to be stable for a few hours at ambient temperature. Its performance was found to improve significantly upon annealing at 150 °C for 2–6 h; however, the device efficiency was reduced to about 0.002%, which the authors explained by the higher work function of the metal. Interestingly, the emission spectrum remained similar to that of the device with a Ca:Al cathode, suggesting that the near-IR band centered around 800 nm does not originate from specific polymer–metal defects. The characteristics of ITO/poly(2,5-dibutyloxy-*p*-phenylene ethynylene)/Al devices were similar to those of ITO/poly(2,5-dihexyloxy-*p*-phenylene ethynylene)/Al diodes, but unfortunately no specific data sets were reported. Interestingly, however, the EL spectra of these devices were void of any near-IR bands, but exclusively displayed a broad emission band centered around 575 nm, suggesting that structural defects or impurities in the poly(2,5-dihexyloxy-*p*-phenylene ethynylene) employed may have caused the near-IR emission. The comparably poor performances of these early devices may have triggered the view [4] that PPEs are not very promising candidates for LED applications.

A detailed study on LEDs based on dialkyloxy-PPEs was subsequently conducted by Weder et al. [107]. Investigated were single-layer devices based on two different PPE derivatives featuring only *n*-octyloxy side chains (O-OPPE) and *n*-octyloxy and 2-ethylhexyloxy side chains in an alternating pattern (EHO-OPPE), cf. Table 3 entries 3 and 4. Essentially identical characteristics were observed for LEDs based on these two polymers, in agreement with previous studies, which revealed similar photophysical properties for *amorphous* films of these materials [42]. The *I*–*V* and EL intensity–voltage (I_{EL}–*V*) curves of all devices investigated show typical rectifying behavior. All devices were found to emit yellow-green light (Fig. 16) with an emission maximum at

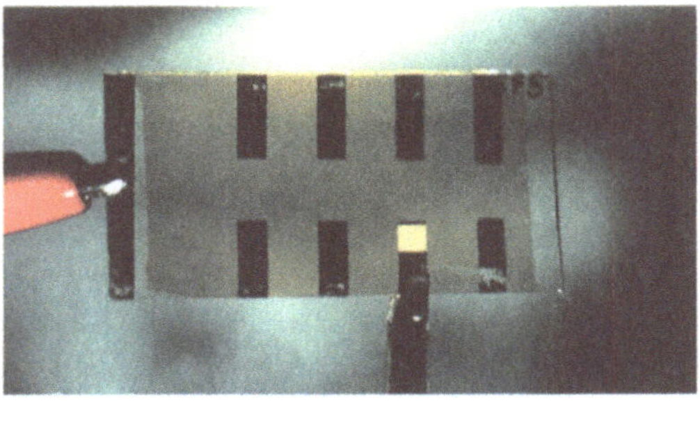

Fig. 16 ITO/EHO-OPPE/Al LED in operation. The size of the device was 9 mm^2; the picture was taken under ambient illumination. Reproduced with permission from [107]

535 nm and a brightness of up to 80 cd/m^2. The EL emission spectra essentially match the PL spectra, but show a shoulder that reaches out into the near-IR regime.

In contrast to the above-discussed earlier studies [103–106], the nature of the cathode material, when comparing Al and Ca, was found to influence the characteristics of single-layer devices with an ITO anode only marginally. Interestingly, LEDs fabricated with Al cathodes were found to have a slightly higher external quantum efficiency (0.035%) and lower onset voltage (11 V) for EL than those with a Ca cathode (14 V, 0.020%). Devices comprising a Cr cathode exhibit a clearly higher EL threshold voltage (19.7 V) and also a lower quantum efficiency (0.015%). The fact that devices with an Al cathode exhibit a better performance than those with a Ca cathode contrasts with the situation observed for many single-layer devices based on PPEs [103] and other emitters [4–6]. Typically a change from an Al to a Ca cathode leads to a lower EL threshold voltage and improved quantum efficiency, due to the enhanced electron injection from Ca, which has a significantly different work function than Al (–2.9 and –4.2 eV, respectively [108]). However, the behavior observed for the devices studied by Weder et al. [107] is consistent with the potentials of the conduction (~–3.6/–3.9 eV) and valence band (~–5.8/–6.3 eV) edges of the dialkyloxy-PPEs employed (Fig. 17), which were determined from the optical band gap and ultraviolet photoelectron spectroscopy and cyclic voltammetry, respectively. A comparison of these values with the work functions of ITO, and the cathode materials employed (Fig. 17) reveals that a significantly lower energy barrier has to be overcome by the electrons being injected from the cathode into the conduction band of the PPEs than by the holes injected from the ITO anode. Thus, the increased efficiency of LEDs with an Al cathode compared to those with a Ca cathode appears to result from a more balanced charge injection; but ultimately, the device performance is limited by a significant barrier for hole injection.

In order to overcome this problem, a subsequent study focused on devices in which EHO-OPPE was used in combination with a hole-conducting poly-

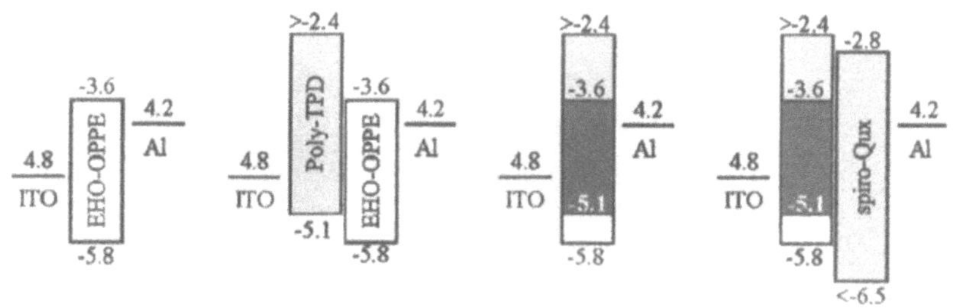

Fig. 17 Schematic representation of the device structures described in Refs. 107 and 111: **a** single-layer EHO-OPPE, **b** two-layer EHO-OPPE/poly-TPD, **c** single-layer EHO-OPPE:poly-TPD blend, and **d** two-layer EHO-OPPE:poly-TPDblend with additional spiro-Qux hole-blocking layer, and their corresponding energy-level diagrams. The working functions of Ca (2.9 eV) and Cr (4.5 eV) were omitted. Reproduced with permission from [111]

(triphenyldiamine) (poly-TPD, [109]) and an electron-transporting/hole-blocking tetrameric spiroquinoxaline ether (spiro-Qux, [110]) (Table 3, entry 5) [111]. Poly-TPD and spiro-Qux were selected because these compounds display adequate electronic potentials and charge-transport properties to facilitate a more balanced electron and hole injection into the EHO-OPPE. Different device configurations were investigated, which all relied on ITO and Al as electrodes. The diodes investigated included EHO-OPPE/poly-TPD bilayer structures, single-layer structures based on phase-separated EHO-OPPE:poly-TPD blends, and bilayer structures based on EHO-OPPE:poly-TPD blends and spiro-Qux. A combinatorial approach was employed to determine optimal layer thicknesses and blend compositions. The best performance was observed for a bilayer device based on a 25:75% EHO-OPPE:poly-TPD blend and a rather thick (48 nm) spiro-Qux layer; a brightness of 257 cd/m^2, an onset voltage of 4.5 V, and a photometric efficiency of 0.15 cd/m^2 were reported as key parameters. This device appears to be one of the brightest and most efficient PPE LEDs reported to date. The maximum brightness and photometric efficiency of this device are significantly lower than the figures of merit achieved by the highly optimized systems that are beginning to enter the market [4–6], but it should be noted that comparably little effort was devoted to materials and device optimization.

A number of studies have also focused on PAE LEDs with arylene segments other than *p*-phenylene ethynylene. Yoshino and Onoda (Table 3, entry 6) studied devices based on poly(3,4-dialkyl-1,6-phenylene ethynylene). The introduction of *o*-phenylene segments causes a rather large band gap (3.1 eV) and leads to the emission of blue light [112]. The same groups also investigated a number of alternating copolymers comprising 2,5-didodecyloxy-1,4-phenylene diethynylene and either 2,5-pyridinylene (Table 3, entry 7), 9,10-anthracenylene (Table 3, entry 8), or 2,5-thienylene (Table 3, entry 9) moieties [113, 114]. The EL emission spectra of LEDs based on these polymers ranged from green (2,5-pyridinylene) to orange (9,10-anthracenylene) and, except for some broadening, matched the PL spectra of the respective polymers reasonably well. Unfortunately, no other details regarding the device performance of these LEDs have been given. LEDs based on copolymers comprising 2,5-dialkyloxy-1,4-phenylene diethynylene and 1,3-phenylene or 2,5-dialkyloxy-1,3-phenylene moieties (Table 3, entry 10) were reported by the groups of Pang and Karasz [115, 116]. The EL spectrum reported for single-layer LEDs based on the latter polymer showed a remarkable blueshift (30 nm) when the driving voltage was increased from 12 to 17 V, which the authors explained on the basis of Joule heating and thermochromism. LEDs based on thermally cross-linked copolymers comprising 1,4-phenylene diethynylene or 1,3-phenylene diethynylene and 9,9-di-*n*-hexylfluorenylene moieties were reported by Kim et al. but, again, little information regarding the performance of the LEDs has been disclosed [117].

Bunz at al. have investigated a series of random PAE copolymers featuring 2,5-didodecylphenylene ethynylene and 3,7-di-*tert*-butylnaphthalene units

(Table 3, entry 11) that were prepared by alkyne metathesis with simple catalyst systems that were pioneered by Bunz [118]. These polymers display efficient solid-state emission in the blue regime, with PL quantum yields between 0.16 and 0.76, and emission maxima between 422 and 443 nm. Thus, the emission characteristics, which can be fine-tuned via the content of the 3,7-di-*tert*-butylnaphthalene units, differ significantly from those of standard dialkyl-PPEs, which emit at lower energies and generally display lower PL quantum efficiencies. It was suggested that the effective and blue-shifted solid-state emission is the consequence of disrupted solid-state alignment, diminished π–π interactions among the macromolecules, and a reduced planarization of the latter, all resulting from the introduction of the 3,7-di-*tert*-butylnaphthalene units. The HOMO values of copolymers containing between 17 and 66 mol% of the 3,7-di-*tert*-butylnaphthalene units are between –5.6 and –5.8 eV vs vacuum level. These values are similar to the electrochemical HOMO values reported by Montali et al. for dialkyloxy-PPEs (–5.8 eV, vide supra) [107]. The large band gap puts the LUMO of these polymers to around –2.3 eV.

Multilayer LEDs were fabricated with polymers comprising 33 and 50 mol% of 3,7-di-*tert*-butylnaphthalene units. Devices were based on an ITO anode, a PEDOT hole-injecting layer, a LiF hole-blocking layer, and either Ca or Al as the cathode. Interestingly, it was observed that devices based on an Al cathode had larger efficiencies than those with a Ca cathode. While a similar trend was observed by Weder et al. for devices based on dialkyloxy-PPEs [107], this particular situation is somewhat surprising, since the position of the LUMO (–2.3 eV) gives rise to a significant barrier for electron injection if Al (work function –4.3 eV) is employed as the cathode as opposed to Ca (work function –2.9 eV). The device based on the copolymer comprising 33% 3,7-di-*tert*-butylnaphthalene units displayed bluish-green electroluminescence (λ_{max}=472 nm) and was characterized by a low turn-on voltage (5.5 V), an appreciable peak brightness of 100 cd/m^2, and an efficiency (0.1 cd/A/0.1–0.15% external quantum efficiency) that is among the best for any PAE LED. Upon increasing the concentration of the 3,7-di-*tert*-butylnaphthalene units in the copolymer to 50%, the emission maximum shifted to higher energy (λ_{max}=428 nm), but maximum brightness (20 cd/m^2) and efficiency (0.02 cd/A) were considerably reduced. In view of the rather subtle differences of the PL quantum efficiencies, HOMOs, and LUMOs of the two polymers employed, it appears that the limited device performance observed in the case of the copolymer comprising 50% 3,7-di-*tert*-butylnaphthalene is related to inefficient charge transport through the light-emitting polymer.

Jen et al. have investigated a copolymer PAE comprising electron-accepting 9,9-bis(2-ethylhexyl)-2,7-diethynyl fluorene and hole-transporting thiophene moieties in an alternating fashion (Table 3, entry 12) [119]. The design was based on the assumption that the combination of fluorenyl and thiophene moieties would lead to a balanced transport of electrons and holes. The sterically hindered substituents were introduced with the hope that they would prevent intermolecular interactions and, thus, keep the PL quantum efficiency high.

However, while the PL spectrum in solution shows a strong blue emission around 444 nm, the solid-state emission is green, with an emission maximum around 520 nm and a tail that stretches out to well above 700 nm. On the basis of this significant redshift and the broadening of the emission band, the authors explained the emission characteristics by the formation of excimers, indicating considerable aggregation processes. Using cyclic voltammetry, the HOMO and LUMO energy levels and band gap of the polymer were determined to be −5.31, −2.48, and 2.83 eV, respectively, so that Ca suggests itself as a cathode material. Double-layer devices were fabricated with ITO as the anode, Ca:Al as the cathode, copper phthalocyanine (CuPc) as a hole-transport layer, and the fluorene–thiophene copolymer as the emissive layer. The devices were characterized by typical nonlinear *I*–*V* characteristics, a turn-on voltage of 11 V, and an external quantum efficiency of 0.01%. A very broad, nearly white emission was observed, ranging from ~440 nm into the near IR and displaying maxima at 541 and 585 nm; unfortunately no brightness was reported.

Shinar et al. reported on the performance of LEDs based on a 2,5-didodecyloxy derivative of a PPV/PPE hybrid polymer (Table 3, entry 13) [120]. The single-layer devices comprised ITO and Al as electrodes and show typical rectifying behavior. The reported onset voltage was 12 V and the quantum efficiency was estimated to be of the order of 0.05%. Interestingly, in contrast to comparable devices comprising only PPV or only alkyloxy-PPE, the *I*–*V* curves and EL intensity–voltage (I_{EL}–V) curves under forward and reverse bias are symmetric. The spectra of the EL light generated under both biases are identical (broad, poorly structured peak with maximum around 2.3 eV, 540 nm) and very similar to the PL spectra. This unusual behavior is inconsistent with the conventional picture that the barrier heights for electron and hole injection at the polymer/electrode interfaces are dictated by the work functions of the electrodes, since the work functions of the electrodes used are dissimilar (4.3 and 5.0 eV for Al and ITO, respectively). Thus, the authors suggest that Fermi energy pinning by a high density of deep defect states at the polymer/electrode interfaces is at play. These defects could capture charge carriers from the bulk, thereby creating barriers at the interfaces.

PPV/PPE hybrid polymer LEDs were also reported by Karasz et al. [121]. The polymers investigated comprise short phenylene vinylene and phenylene ethynylene segments that are bridged via sharing a common *m*-phenylene ring (Table 3, entry 14). A number of derivatives with different side chains were investigated; their electronic properties in solution (quantum efficiency=0.55, solution λ_{max}=446 nm) are essentially identical, but their solid-state emission spectra are somewhat dependent on the nature of the side chains. LEDs were based on one derivative featuring hexyloxy side chains (Table 3, entry 14). The single layer devices comprised ITO and Ca as electrodes, and PPV or PEDOT were used as hole-injection layer. In particular, the devices comprising PEDOT display interesting device characteristics. The diodes show typical rectifying behavior, the EL spectra match the PL spectra reasonably well (although the emission characteristics are slightly voltage-dependent, which the authors explain on the basis of Joule heating and thermochromism), and the external device efficiency of 0.32 cd/A is quite high.

Light-emitting electrochemical cells (LECs) offer an alternative approach for EL from conjugated polymers [122, 123]. By blending a polymer electrolyte into the emissive layer, the electrochemical doping property of luminescent conjugated polymers can be exploited in such EL devices. When a sufficient bias is applied to LECs, the polymer is simultaneously p-doped and n-doped at the anode and cathode side of the device, respectively. As a result of the electrochemical doping, ohmic contacts are formed at the polymer/electrode interfaces. Thus, facile and balanced electron and hole injection is readily achieved from inert electrodes at low turn-on voltages, and the charge recombination mechanism is essentially identical to the one observed in "conventional" diodes. As a result of the ohmic contacts, however, the devices can be operated in both forward and reverse bias modes.

Using a copolymer comprising electron-accepting 9,9-bis(2-ethylhexyl)-2,7-diethynyl fluorene and electron-donating tetraphenyldiaminobiphenyl units, Sun et al. recently adopted this approach for a PAE [124]. The copolymer was blended with a poly(ethylene oxide)/lithium triflate ion conductor and devices were fabricated by sandwiching this blend between an ITO and an Al electrode (Table 3, entry 15). The PL spectrum of the blend displays a broad peak between ~400 and ~700 nm with a maximum at ~500 nm, which has been explained by the formation of aggregates and pronounced intermolecular electronic interactions. The EL spectrum was slightly broadened when compared to the PL spectrum, but otherwise matched the latter well. The LECs displayed exceptionally low onset voltages of 2.9 V (forward bias) and –3.2 V (reverse bias), which, of course, are directly related to the operating principle of the device. The LEC displayed a maximum EL efficiency of 0.47 cd/A, which is the highest yet reported for any light-emitting devices based on a PAE. However, the maximum light intensity was very low (<3 cd/m^2) and due to the operating mechanism the switching times are exceedingly slow (of the order of 10 s).

In summary, it has been demonstrated that PAEs can effectively be used as the emitting layer in polymer LEDs. The highest external quantum efficiencies reached are of the order of 0.1–0.3 cd/A, with a brightness of up to 300 cd/m^2. In comparison to the performance of highly optimized PLEDs that have been trimmed to the point of commercial exploitation [4–6] these values may appear low, but it should be recognized that comparably little effort has been devoted to the optimization of PAE LEDs. However, the data available now have probably corrected the earlier misconceptions regarding the unsuitability of PAEs as the emitting layer in PLEDs. The high oxidative stability of PAEs represents a tremendously important advantage over alternative systems. Future work focused on achieving a more balanced charge injection should be a very rewarding activity that may allow one to capitalize on this asset.

5 Summary and Conclusions

The last decade has witnessed significant research activities focused on the exploration and exploitation of the (semi)conducting nature of PAEs. Focusing on the subjects of electrical conductivity, charge transport, and electroluminescence, we have here attempted to summarize and highlight the most relevant aspects regarding the electronic properties of PAEs and their potential applications in "plastic electronic" devices. In view of the rather unique property matrix of many representatives of this class of polymers – in particular the combination of high oxidative stability and high charge-carrier mobility – it appears that research in this arena has merely begun. Indeed, the spectacular progress made on many frontiers has propelled PAEs from a few research labs into the scientific mainstream and many applications, including PAE transistors, solar cells, and more, have been spurred, but are still waiting to be realized.

Acknowledgements The authors thank Dr. P. Iyer and A. Kokil for helpful suggestions and comments and for proof-reading this manuscript. Recent and current PAE work in the Weder group was/is made possible through generous financial support from the Case Presidential Research Initiative, the Case School of Engineering, DuPont (Aid To Education Award, Young Professor Grant), the Goodyear Tire and Rubber Company, the Hayes Investment Fund, the National Science Foundation (NSF DMR-0215342), and the Petroleum Research Fund (ACS-PRF 38525-AC). C.W. acknowledges fruitful and stimulating collaborations in the arena of PAEs with F. Bangerter, PD Dr. C. Bosshard, PD Dr. W. Caseri, Prof. Dr. F.N. Castellano, S. Dellsperger, Dr. F. Dötz, Prof. Dr. P. Günter, Dr. G.S. He, Prof. Dr. B. Hecht, E. Hittinger, C. Huber, A. Kokil, Dr. A. Montali, Dr. A.R.A. Palmans, Prof. Dr. P.N. Prasad, Dr. A. Renn, Prof. Dr. S.J. Rowan, Prof. Dr. H.-W. Schmidt, Dr. C. Schmitz, Dr. I. Shiyanovskaya, Prof. Dr. K.D. Singer, Prof. Dr. P. Smith, Dr. D. Steiger, PD Dr. M. Thelakkat, Dr. W. Trabesinger, Prof. Dr. U.P. Wild, Prof. Dr. M.S. Wrighton, and P. Yao.

References

1. Friend RH (1993) Conjugated polymers and related materials, the interconnection of chemical and electronic structure. In: Salaneck WR, Lundström I, Ranby B (eds) Proceedings of the 81st Nobel symposium. Oxford University Press, New York, p 285
2. Heeger AJ (2001) J Phys Chem 105:8475
3. Skotheim TJ, Elsenbaumer RL, Reynolds JR (eds) (1998) Handbook of conducting polymers, 2nd edn. Dekker, New York
4. Kraft A, Grimsdale AC, Holmes AB (1998) Angew Chem Int Ed 37:403
5. Mitschke U, Bäuerle P (2000) J Mater Chem 10:1471
6. Greiner A, Weder C (2003) Light-emitting diodes. In: Kroschwitz JI (ed) Encyclopedia of polymer science and technology, 3rd edn, vol. 3. Wiley-Interscience, New York, p 87
7. Horowitz G (1998) Adv Mater 10:365
8. Moerner WE, Silence SM (1994) Chem Rev 94:127
9. Brabec CJ, Sariciftci NS, Hummelen JC (2001) Adv Funct Mater 11:15
10. Nalwa HS (ed) (1996) Handbook of organic conductive molecules and polymers. Wiley, New York

11. Mark JA (ed) (1996) Physical properties of polymers handbook. American Institute of Physics, New York
12. Scherf U (1999) Top Curr Chem 201:163
13. Cantow HJ (ed) (1998) Advances in polymer science, vol 63. Springer, Berlin Heidelberg New York
14. Giesa R (1996) J M S Rev Macromol Chem Phys C36:631
15. Bunz UHF (2000) Chem Rev 100:1605
16. Bunz UHF (2001) Acc Chem Res 34:998
17. Bunz UHF (2002) The ADIMET reaction: synthesis and properties of poly(dialkylpara-phenyleneethynylene)s. In: Astruc D (ed) Modern arene chemistry. Wiley-VCH, p 217
18. Bunz UHF (2003) Acyclic diyne metathesis utilizing in situ transition metal catalysts: an efficient access to alkyne-bridged polymers. In: Grubbs RH (ed) Handbook of metathesis, vol 3. Wiley-VCH, p 345
19. Yamamoto T (1999) Bull Chem Soc Jpn 72:621
20. Yamamoto T (2003) Synlett 425
21. Pinto MR, Schanze KS (2002) Synthesis 1293
22. Swager TM (1998) Acc Chem Res 31:201
23. McQuade DT, Pullen AE, Swager TM (2000) Chem Rev 100:2537
24. Swager TM (2002) Chem Res Toxicol 15:125
25. Wosnick JH, Swager TM (2002) Curr Opin Chem Biol 4:715
26. Moore JS (1997) Acc Chem Res 30:402
27. Hill DJ, Mio MJ, Prince RB, Hughes TS, Moore JS (2001) Chem Rev 101:3893
28. (a) Tour JM (1996) Chem Rev 96:537; (b) Tour JM (2000) Acc Chem Res 33:791
29. (a) Haley MM, Pak JJ, Brand SC (1999) Top Curr Chem 101:81; (b) Haley MM, Marsden JA, Palmer GJ (2003) Eur J Org Chem:2355
30. Levitus M, Schmieder K, Ricks H, Shimizu KD, Bunz UHF, Garcia-Garibay MA (2001) J Am Chem Soc 12:4259
31. (a) Halkyard CE, Rampey ME, Kloppenburg L, Studer-Martinez SL, Bunz UHF (1998) Macromolecules 31:8655; (b) Miteva T, Palmer L, Kloppenburg L, Neher D, Bunz UHF (2000) Macromolecules 33:652
32. Roth S, Caroll D (2004) One-dimensional metals, 2nd edn. Wiley-VCH, Weinheim
33. (a) Chiang CK, Fincher CR Jr, Park YW, Heeger AJ, Shirakawa H, Louis EJ, Gau SC, MacDiarmid AG (1977) Phys Lett 60A:375; (b) Shirakawa H, Louis EJ, MacDiarmid AG, Chiang CK, Heeger AJ (1977) Chem Commun 578
34. Lakshmikantham MV, Vartikar J, Jen KY, Cava MP, Huang WS, MacDiarmid AG (1983) Polym Prepr 24:75
35. Sanechika K, Yamamoto T, Yamamoto A (1984) Bull Chem Soc Jpn 57:752
36. Tateishi M, Nishihara H, Aramaki K (1987) Chem Lett 1727
37. Tormos GV, Nugara PN, Lakshikantham MV, Cava MP (1993) Synth Met 53:271
38. Ni QX, Swanson LS, Lane PA, Shinar J, Ding YW, Ijadi-Maghsoodi S, Barton TJ (1992) Synth Met 49–50:453
39. Pizzoferrato R, Berliocchi M, Di Carlo A, Lugli P, Venanzi M, Micozzi A, Ricci A, Lo Sterzo C (2003) Macromolecules 36:2215
40. Ofer D, Swager TM, Wrighton MS (1995) Chem Mater 7:418
41. Rodriguez-Prada JM, Duran R, Wegner G (1989) Macromolecules 22:2507
42. Weder C, Wrighton MS (1996) Macromolecules 29:5157
43. Wang ZH, Li C, Scherr EM, MacDiarmid AG, Epstein AG (1991) Phys Rev Lett 66:1745
44. Bao Z, Dodabalapur A, Lovinger AJ (1996) Appl Phys Lett 69:4108
45. Sirringhaus H, Brown PJ, Friend RH, Nielsen MM, Bechgaard K, Langeveld-Voss BMW, Spiering AJH, Janssen RAJ, Meijer EW, Herwig P, de Leeuw DM (1999) Nature 401:685

46. Evans U, Soyemi O, Doescher MS, Bunz UHF, Kloppenburg L, Myrick ML (2001) Analyst 126:508
47. Yamamoto T (1999) Bull Chem Soc Jpn 72:621
48. Wilson JN, Windschaeif PM, Evans U, Myrick ML, Bunz UHF (2002) Macromolecules 35:8681
49. Hayashi H, Yamamoto T (1998) Macromolecules 31:6063
50. Yamamoto T, Kokubo H, Morikita T (2001) J Polym Sci B Polym Phys 39:1713
51. Morikita T, Yamaguchi I, Yamamoto T (2001) Adv Mater 13:1862
52. Yamamoto T, Fang Q, Morikita T (2003) Macromolecules 36:4262
53. Li YT, Whittle CE, Walters KA, Ley KD, Schanze KS (2001) Pure Appl Chem 73:497
54. Li J, Pang Y (1998) Macromolecules 31:5740
55. Egbe DAM, Klemm E (1998) Macromol Chem Phys 199:2683
56. Grummt UW, Birckner E, Klemm E, Egbe DAM, Heise B (2000) J Phys Org Chem 13:112
57. Bangcuyo CG, Evans U, Myrick ML, Bunz UHF (2001) Macromolecules 34:7592
58. Bangcuyo CG, Ellsworth JM, Evans U, Myrick ML, Bunz UHF (2003) Macromolecules 36:546
59. Wilson JN, Josowicz M, Wang Y, Bunz UHF (2003) Chem Commun 24:2962
60. (a) Egbe DAM, Bader C, Nowotn J, Günther W, Klemm E (2003) Macromolecules 36:5459; (b) Brizius G, Pschirer NG, Steffen W, Stitzer K, zur Loye HC, Bunz UHF (2000) J Am Chem Soc 122:12435
61. Kokil A, Shiyanovskaya I, Singer KD, Weder C (2003) Synth Met 138:513
62. Kokil A, Shiyanovskaya I, Singer KD, Weder C (2002) J Am Chem Soc 124:9978
63. Shiyanovskaya I, Singer KD, Twieg RJ, Sukhomlinova L, Gettwert V (2002) Phys Rev E 65:41715
64. Borsenberger PM, Weiss DS (1998) Organic photoreceptors for xerography. Marcel Dekker, New York
65. Lebedev E, Dittrich T, Petrova-Koch V, Karg S, Brütting W (1997) Appl Phys Lett 71:2686
66. Hertel D, Bässler H, Scherf U, Hörhold HH (1999) J Chem Phys 110:9214
67. Campbell IH, Smith DL, Neef CJ, Ferraris JP (1999) Appl Phys Lett 74:2809
68. Inigo AR, Tan CH, Fann W, Huang YS, Perng GY, Chen SA (2001) Adv Mater 13:504
69. Scher H, Montroll EW (1975) Phys Rev B 12:2455
70. Borsenberger PM, Pautmeier L, Bässler H (1992) Phys Rev B 46:12145
71. Scott JC, Pautmeier L, Schein LB (1992) Phys Rev B 46:8603
72. Babel AM, Jenekhe SA (2002) Adv Mater 14:371
73. Babel AM, Jenekhe SA (2003) J Am Chem Soc 125:13656
74. Borsenberger PM, Pautmeier L, Bässler H (1991) J Chem Phys 94:5447
75. Bässler H (1993) Phys Stat Sol B 175:15
76. Pautmeier L, Richert R, Bässler H (1990) Synth Met 37:271
77. Movaghar B, Murray DW, Donovan KJ, Wilson EG (1984) J Phys C Solid State Phys 17:1247
78. Aleshin A, Kiebooms R, Menon R, Wudl F, Heeger AJ (1997) Phys Rev B 56:3659
79. Kumar A, Reynolds JR (1996) Macromolecules 29:7629
80. Joo J, Lee JK, Lee SY, Jang KS, Oh EJ, Epstein AJ (2000) Macromolecules 33:5131
81. Wright ME (1989) Macromolecules 22:3256
82. Hirao T, Yamaguchi S, Fukuhara S (1999) Tetrahedron Lett 40:3009
83. Hittinger E, Kokil A, Weder C (2004) Angew Chem 116:1844
84. Hittinger E, Kokil A, Weder C (2004) Macromol Rapid Commun 25:710
85. Landfester K, Montenegro R, Scherf U, Güntner R, Aswapirom U, Patil S, Neher D, Kietzke T (2002) Adv Mater 14:651
86. Marie E, Rothe R, Antonietti M, Landfester K (2003) Macromolecules 36:3967

87. Groenendaal BL, Jonas F, Freitag D, Pielartzik H, Reynolds JR (2000) Adv Mater 12:481
88. Genet JP, Savinac M (1999) J Organomet Chem 576:305
89. Huber C, Bangerter F, Caseri W, Weder C (2001) J Am Chem Soc 123:3857
90. Huber C, Kokil A, Caseri WR, Weder C (2002) Organometallics 21:3817
91. Kokil A, Huber C, Caseri WR, Weder C (2003) Macromol Chem Phys 204:40
92. Green M, Grove DM, Howard JAK, Spencer J L, Stone FGA (1976) J Chem Soc Chem Commun 759
93. Malatesta L, Cenini S (1974) Zerovalent compounds of metals. Academic, London
94. Caseri W, Pregosin PS (1988) Organometallics 7:1373
95. Craford MG, Steranka FM (1994) In: Trigg GL (ed) Encyclopedia of applied physics, vol 8. VCH, Weinheim, p 485
96. Bradley D (1996) Curr Opin Solid State Mater Sci 1:789
97. Pope M, Kallmann H, Magnante P (1963) J Chem Phys 38:2042
98. Tang CW, VanSlyke SA (1987) Appl Phys Lett 51:913
99. Adachi C, Tsutsui T, Saito S (1990) Appl Phys Lett 56:799
100. Burroughes JH, Bradley DDC, Brown AR, Marks RN, Mackay K, Friend RH, Burn PL, Holmes AB (1990) Nature 347:539
101. Kocher C, Montali A, Smith P, Weder C (2001) Adv Funct Mater 11:31
102. Cumpston BH, Jensen KF (1998) J Appl Polym Sci 69:2451
103. Swanson LS, Lu F, Shinar J, Ding YW, Barton TJ (1993) Proc SPIE 1910:101
104. Shinar J, Swanson LS (1993) Proc SPIE 1910:147
105. Swanson LS, Shinar J, Ding YW, Barton TJ (1993) Synth Met 55–57:1
106. Shinar J, Swanson LS, Lu F, Ding YW (1994) US Patent 5,334,539
107. Montali A, Smith P, Weder C (1998) Synth Met 97:123
108. Lide DR (ed) (1995) CRC handbook of chemistry and physics, 76th edn. CRC, Boca Raton
109. Thelakkat M, Fink R, Haubner F, Schmidt HW (1997) Macromol Symp 125:157
110. Schmitz C, Pösch P, Thelakkat M, Schmidt HW (1999) Phys Chem Chem Phys 1:1777
111. Schmitz C, Pösch P, Thelakkat M, Schmidt HW, Montali A, Feldmann K, Smith P, Weder C (2001) Adv Funct Mater 11:41
112. Yoshino K, Tada K, Onoda M (1994) Jpn J Appl Phys 2 33:L1785
113. Tada K, Onoda M, Hirohata M, Kawai T, Yoshino K (1996) Jpn J Appl Phys 2 35:L251
114. Hirohata M, Tada K, Kawai T, Onoda M, Yoshino K (1997) Synth Met 85:1997
115. Pang Y, Li J, Hu B, Karasz FE (1998) Macromolecules 31:6730
116. Chu Q, Pang Y, Ding L, Karasz FE (2002) Macromolecules 35:7569
117. Hong JM, Cho HN, Kim DY, Kim CY (1999) Synth Met 102:933
118. Pschirer NG, Miteva T, Evans U, Roberts RS, Marshall AR, Neher D, Myrick ML, Bunz UHF (2001) Chem Mater 13:2691
119. Zhan X, Liu Y, Zhu D, Jiang X, Jen AK-Y (2001) Synth Met 124:323
120. Jeglinski SA, Amir O, Wi X, Vardeny ZV, Shinar J, Cerkvenik T, Chen W, Barton TJ (1995) Appl Phys Lett 67:3960
121. Chu Q, Pang Y, Ding L, Karasz FE (2003) Macromolecules 36:3848
122. Pei QB, Yu G, Zhang C, Yang Y, Heeger AJ (1995) Science 269:1086
123. Pei QB, Yang Y, Yu G, Zhang C, Heeger AJ (1996) J Am Chem Soc 118:3922
124. Sun Q, Zhan X, Zhang B, Yang C, Liu Y, Li Y, Zhu D (2004) Polym Adv Technol 15:70

Author Index Volumes 101–177

Author Index Volumes 1–100 see Volume 100

Subject Index

springeronline.com
The interactive website for all Springer books and journals

FSC
www.fsc.org
MIX
Papier aus verantwortungsvollen Quellen
Paper from responsible sources
FSC® C105338

If you have any concerns about our products,
you can contact us on
ProductSafety@springernature.com

In case Publisher is established outside the EU,
the EU authorized representative is:
Springer Nature Customer Service Center GmbH
Europaplatz 3, 69115 Heidelberg, Germany

Printed by Libri Plureos GmbH
in Hamburg, Germany